Vietnam.
Part 2

A Research-based Cross-cultural Socio-political Analysis of

The War

To Quang
Please tell your
family & children about
the war.

Thank you for your support:
HÀ H. TUONG, Ed. D.
Vietnam: Peace or Freedom

4/17/22

Written by Dr. Ha H. Tuong

Former ARVN Officer
Retired Minneapolis Public Schools Administrator

Vietnam: Peace or Freedom
Part 2

A Research-based Cross-cultural Socio-political Analysis of The War

Copyright © 2020 by Dr. Ha H Tuong

All rights reserved. No part of this book may be reproduced, transmitted, excerpted or translated in any form or by any means without written permission from the author.

ISBN: 978-0-578-63659-7

Publisher: Vietnam: Peace or Freedom, Minnesota U.S.A

45 Years of Remembrance - Limited Edition

Printed in the United States of America

Note and disclaimer: This book is a non-partisan research-based analysis of the Vietnam War, based on observation of events, verifiable documentaries, interviews, and personal accounts of political connections that are bound together by cross-cultural psychological rational reasoning. Information is gathered from various sources, many of which might be controversial or contradicting each other; nevertheless, the main purpose is for balanced triangulation. By all means, its focus is not ethno-centric (meaning "focus is not on a specific ethnicity such as South Vietnamese…"). Even though it is an account of socio-political and military events in Vietnam, as a partially scientific research, it does discuss those in Laos and Cambodia as well, because the War encompasses all three countries. All personal stories are true, and historical events are accounted for as referenced. Most names have not been changed, but some are to protect the individuals' privacy. Names of locations, organizations and groups or tribes are the same ones used during the actual époque when the events take place before April 30th, 1975. Conspiracy theories as presented based on the most plausible information retrieved from internet researches, and declassified materials. Most importantly, the triangulation process of information from different sources has been used to suggest a rational logical conclusion (socio-political qualitative research methods). Such conclusions might differ or even contradict readers' belief, but they present another view that is socio-scientifically reasonable, which cannot be omitted.

Foreword

Dr. Ha Tuong's book, Part 2, The War follows his Part 1 - Two Minnows, an account of his personal experiences growing up in Vietnam, fighting in the Vietnam War, and finally escaping from his homeland to the West in 1975, after the fall of Sài-gòn. Compelling and riveting as it was, Two Minnows only begins to tell the story about "Dr. Ha's" journey to freedom and his feelings about the war that tore his homeland apart and turned he and his family's lives upside down.

In this book Part 2 - The War, he writes about the history of his native homeland, thousands of years of foreign oppression and its impact on the historical political and social development of Vietnam which eventually leads to the division of his homeland and finally, the Vietnam War. The War is not only an exploration of the Vietnam War and its causes; it is an intense research-based account about Dr. Tuong and many of his own personal experiences in the war itself, from the beginning to the end. He writes about the historical divisions of his homeland, North, Central, South, and how the people in those regions have participated in its nations past conflicts, concluding with the Vietnam War. Dr. Tuong pulls no punches regarding the war and his feelings about it. Among many of the events written in the book about what occurred in the war, Dr. Tuong examines, through fact and his own opinion: Who started it? How could the conflict have been prevented? What was America's role in the war and once involved in it, did it back its commitments? What about the hidden war in neighboring countries? Were the war goals and how to fight it aligned between the Republic of South Vietnam and all of its allies? Why and how were the Army of South Vietnam and its soldiers so terribly disparaged and misunderstood? Who were the War major players? How effective was propaganda used by all sides? How instrumental was the world and United States media to the war final outcome? How were the veterans on all sides treated in the aftermath of the war? How do the Vietnam War refugees today view it and how they now feel about their lives in their new countries (mostly in America)? Due to the people changing attitudes in today's Vietnam, could there be a potential political change coming to this country? And finally, what "Peace" was really achieved by the Paris Peace Accords?

Dr. Tuong writes about war in a unique way. He utilizes the psychological theories of Abraham Maslow's Hierarchy of Needs and Daniel Goleman's Emotional Intelligence to help the readers make sense why the participants acted the way they did during the war. While battles and accounts of them are neatly discussed through maps and pictures, the reader is introduced to an entirely different way to understand and discuss the Vietnam War: the qualitative research way. Vietnam Veterans and others interested in the study of the war will appreciate this different approach. If nothing else, it provides a rare insight and perspective from a former Vietnamese officer and combatant. It is a griping personalized account from the start to the finish.

Finally, imagine, under dangerous and perilous conditions, he and his wife escaping from the end of the war in 1975 (his escape and journey was detailed in <u>Two Minnows</u>) and to the present day, 2019 which spans 44 years, without him telling about his feelings concerning his own experiences. While the war may have been "over" in 1975, it really had not been completely over for him. In this book and <u>Two Minnows</u>, he finally puts in writing not only his account of the war, but his feelings and opinions about it. He aims to set the "records straight" with well-researched facts and his opinions about the war, and how he feels it has been misreported and misrepresented in past books and media accounts over the years. Perhaps Dr. Tuong's work will serve clear up misconceptions held by people and begin a long overdue healing process.

Interrupted by living his daily working life, this work has taken Dr. Tuong over ten years to complete. The wait for it is worth it!

Dr. Donald Hauck
Retired Educator
USAF Vietnam Veteran 1968-1969, 1970-1971
Menomonie, Wisconsin
November, 2019

Vietnam: Peace or Freedom: Part 2 - The War - A Cross-cultural and Psycho-political Analysis is dedicated:

To The Fallen Heroes of the United States, Allied Forces, and the Army of the Republic of Vietnam
Who have sacrificed all they have to secure Freedom and Democracy. Special Honor is devoted to the ARVN Heroes who would rather be executed than surrender or subjugate selves to the enemy's oppression.

To The Surviving Heroes of the United States, Allied Forces and the Army of the Republic of Vietnam
Who have fulfilled their country Call to Duties
In spite of being betrayed by their own or dearest and entrusted allies. They deserve Respect and Honor as they have done their best for us, the people they barely know.

To The Family of the Heroes above,
As they have faithfully and courageously supported the Heroes, No matter how high or low the tides can be. Special Honor is devoted to the Gold Star families.

To My Parents and my former wife Mười
Who have been on my side through the War and Resettlement in the U.S., Who have never given up on me, for better or for worse, As we survive through the toughest times in our life.

To My Current Wife Carol:
Who has always been besides me, for better or for worse, As I grow in my new life, and as I progress in my emotional journey.

To the 1970-80s Congregation of the University Baptist Church (with First Congregation Church's assistance)
I am truly grateful they have sponsored Mười and me
So that we could start our second life and career in Minneapolis.

And, Finally, to my Friends, Colleagues, and People of Minnesota
Who have been so supportive of me during my silent struggle trying to elevate and rebuild my new life:
Thank you with my whole heart!

To those who used to believe in the Vietnamese Communist Party,
If they have woken up and realized that they have been duped, and That many of them in Vietnam have lost their Freedom or Democracy, Good Luck!
**No matter how difficult obstacles can be,
As long as we keep telling ourselves**

"We can do it and we will do it,"
And, with some help from true friends,
We will soar like an Eagle.

THANK YOU

Amy, Amy, Amy, André, Andrew, Anh Dũng, Anne, Annette, Arica, Art, Audrey, Bailey, Banlang, Bảng, Barbara, Barbara, Bernie, Beth, Bette, Buèche, Britgitt, Carla, Carol, Carol, Chanto, Cheri, Chi, Chín, Choua, Chue, Cliff, Dr. Corky, Dan, Dan, Danh, Daniel, Dr. Dao, Dr. Dave, Dr. David, David, Daud, Debbie, Deidre, Điệu, Dr. Don, Duang, Dung, Edy, Ingrid, Giang, George, Hà, Hán, Hằng, Hằng, Hariette, Hatti, Hattie, Hoeun, Huệ, Hùng, Hùng , Hùng, Huyền Vân, Janet, Janette, Dr. Jermaine, Jean, Jeanine, Jenna, Jeff, Jenna, Jennifer, Jennifer, Jeralyn, Jill, Jim, Jollie, Joyce, Julie, Julie, Karen, Karen, Kasha, Kay, Kay, Khai, Khambay, Khanh, Khao, Kiệm, Kim, Kim, Kim, Lafayette, Lan, Lance, Lang, Larry, Lễ, Len, Leo, Liên, Liên, Ligaya, Lisa, Lộc, Lori, Luis, Lynne, Mary, Mary, MaryLou, Merri Lynn, Michael, Nancy, Neng, Nhạc, Nhất, Như, Ngọc Dung, Ngọc Dung, Ngọc Lan, Ngọc Ngà, Ngọc Thuỷ, Pam, Pamela, Pang, Pat, Pat, Pat, Phước, Quý, Rick, Rick, Roger, Dr. Roger, Roger, Ron, Rosemary, Ruby, Ruby, Scott, Scott, Scott, Shirley, Sonja, Southavilay, Stacy, Steve, Steve, Steve, Sue, Sue, Susan, Tameka, Tân, Teresa, Thanh Hà, Thảo, Thắng, Thas, Thọ, Thu Hương, Thu Hương, Thuý, Thuý, Thuý, Thuý, Thuý Cẩm, Thuý Mai, Toàn, Tony, Trang, Trình, Troy, Truyền, Tươi, Tường Vy, Tuyết, Tuyết, Vickie, Viễn, Vy, Wanda, Wendy, Wendy, William, Winnelle, Xin, Yuping, Yvonne, Zhuojing, and many more…

The War is the second book of the "Vietnam: Peace or Freedom" collection written by the author. The first one is his memoir Two Minnows.

Front cover illustration:
Author's 2016 design based on his original 2010 Memorial banner: the Republic of Vietnam RVN flag with actual insignias of U.S., allies and ARVN units.
Author - Back cover photos:
Left photo: 2nd Lt. "Hal," Regional Forces Infantry, Army of the Republic of Vietnam (1973-1975).
Right photo: "Mr. T," or "uncle T," then "Dr. T," retired school administrator of Minneapolis Public Schools, Minnesota (1992-2006).
Cross reference input and Foreword: Dr. Donald Hauck 2 tours in Nam, retired school administrtor of Wisconsin Public Schools.
Final Format-Editing: Luke Coolbear Both Two Minnows and The War.

About the Author

Having grown up through the most interesting turning points during the time when Việt-nam was devastated by the war and the United States was battling with major social restructuring and economic regression, Dr. Ha Tuong had witnessed drastic socio-political and cultural transformation from the era of French colonization, through both Republics of Vietnam political turmoil, all of which being more or less influenced by a constantly metamorphosing democratic American society. Contradicting values conveyed by different systems of values and knowledge, that originated from sources varying from modern French to modern American, between home and streets, alternating between news and words of mouth, combining ancient Chinese Sun Tzu with modern Western Maslow/ Goleman, floating from student's free-spirited lifestyles to strict military leadership mentality, all had been bombarding his avid mind in order to shape and synthesize his beliefs into a multicultural socio-psychologically integrated resolution that made him the man he had become.

Tuong was a son in a migrant family from North Việt-nam (1940s). He was sent to French total immersion schools by his rich French godfather who was the general director of "la Société des Théâtres d'Indochine." Upon graduation from French Lycée Jean Jacques Rousseau (Sài-gòn), he attended the Sài-gòn University majoring in Math and Science, then switched to the Buddhist University of Vạn Hạnh (Sài-gòn) where he received a Bachelor's Degree in English Education, while serving in the Army of the Republic of Vietnam ARVN (2nd Lieutenant). At first, he was in combat infantry as Regional Forces platoon leader, then, thanks to his Bachelor's Degree in Education, he was re-assigned as the coordinator of the newly created mobile boot-camp training program for Regional Forces "chương-trình huấn-luyện tại chỗ dành cho Địa phương quân" of Gia Định District. Married into a political Southerner family, he was closely exposed to the 2nd Republic of Vietnam policy.

When the war ended in April 1975, he escaped by boat and migrated to Minnesota under the congressionally-approved refugee program for Vietnam War victims, and worked for

This ID photo of author was taken towards the end of 1974. He gave it to his parents so they could use it for his funeral, should he be killed in action.

Minneapolis Public Schools until he retired in 2006. During his career in education, his insatiable pursuit of knowledge and higher education helped him acquire a Master's Degree, and, later on, a Doctorate Degree in Education Leadership, along with a principal license from the Catholic University of St. Thomas. He had successfully evolved from classroom teacher to school principal, touching at least 20,000 lives from all walks of life and cultures, and working closely with some thousands of staff to develop substantial programs for children of all needs. During his active time, he served the immigrant as well as minority communities in the Twin Cities to help them with their socio-political adjustments and advancement. Upon retirement, he participated with Vietnam Veterans and allies in their lobbying efforts to bring out the truth of the War against Communist Aggression.

As he realized there had been so much misleading information available to the people about the Vietnam War, making the bad look good and the good bad, he had decided to commit a decade to researching what was available on the internet and in the media, and to combine the findings with interviews, eyewitness memories, personal experiences, while applying cross-cultural comparative logics to come up with a foundation for "Peace or Freedom." He wrote books and lecture reports, not as a partisan story-teller of or advocate for the RVN, but as a neutral social-scientist and historian, using verified facts and well-founded rational thinking as they pertained to the outcome of the war. As this war was nothing as concrete-sequential and clearly defined as WW2 or the Korean War, which were conventional and, therefore, more predictable than a guerrilla war, it had mislead a great many and ought to be clarified, and rectified. Major cultural differences and historical background components had brewed into a cauldron of conflicts that had totally confused everyone, including the participants, whether they were Western (American, and Western allies) or Eastern (Vietnamese of the North as well as South, and other Asian allies) and, definitely, the American and European "spectators."

With those ideas in mind, Dr. Tuong's first book Two Minnows has presented the fate of the RVN civilians as well as military common soldiers. This second book, The War, will be the companion that unveils the behind the scene backgrounds, leading from one event to the next. Many of which could have been circumvented if something else that has been ignored or hidden has actually been addressed seriously. Like water from upstream, the flow takes different turns whenever it hits an obstacle. As a result, meanders keep appearing and forming. After many times bouncing back and sideway, the force from behind will keep pushing forward and precipitating masses of fish into an unknown future, the inevitable destiny. Such was the life of the minnows!

About the series
"Vietnam: Peace or Freedom"

Peace or Freedom, Part 2: The War - **A Cross-cultural and Psycho-political Analysis**, is the second of the series of two books. It is published in April 2020 in commemoration of the 45 years of Remembrance honoring the South Vietnamese, Royal Lao/ Hmong, and Khmer as well as the allied combatants who have sacrificed for the RVN, the Kingdom of Laos and the Republic of Khmer. It expands the Peace or Freedom Series into a substantial analysis of the war from the nearly four millennium back ground history of Việt-nam to the current events through the lenses of diverse socio-political psychologies that have shaped the mind and guts of all combating parties. Astoundingly, it is a war that has to be viewed, not just militarily or politically, and war researchers, from one political turning point to another, many of which could have helped divert the RVN and the Democratic Republic of Vietnam (DRV) from war. **The War** also exposes many cultural and strategic conflicts between allies as well as many political maneuverings, some rational, others tricky and even dirty, that the governments of all sides have used to outplay and outwit each other.

A few pieces of psychology that are introduced in Two Minnows will be elaborated in this book as they gravely affect the mind of the combatants and other people of interest during and even after the war. The War will explore answers to many questions that are still puzzling people's mind regarding the war outcomes. Who have deliberately or inadvertently done their best to hurt the innocent victims in genocidal efforts? Some answers, many of which are based on newly declassified documents, can be shocking!

Many of the true nationalist heroes in this horrendous war, in this case Americans/ Allies and South-East Asians, are taking turns passing away. Before they are all gone, their stories have to be heard and the untold information about them and what has been unfairly done to them have to be unveiled. Their Legacies have to be known and heard out loud, as

Respect and Honor is long overdue. The author hopes both <u>Vietnam: Peace or Freedom -Two Minnows</u> and <u>The War</u> will bring about Social Justice so those who have sacrificed can be honored and remembered. They have done their best doing what their Countries have asked them to do. Most importantly, the world has to learn about the mistakes that have been made so they will never be repeated again. Millions of people otherwise could have lived and many peace-loving generations could have been spared.

Bloodshed has harmed mostly innocent civilians, and yet more bloodshed will occur in the future if we keep repeating what our ancestors have been doing: make mistakes, and repeat the same mistake. As far as we are concerned, a glimpse at what is happening in the present can tell us that History from many past Wars has not been learned, and millions are still suffering and dying.

"Peace or Freedom, Part 1: Two Minnows," published in April 2018 as 44 Years of Remembrance – Special Edition, is a non-fiction memoir that encompasses the mental struggle of the author, a young South-born North Vietnamese son who has to escape Peace so that he can pursue Freedom. If he chooses Peace, he will have to stay behind and subjugate himself to the incoming communist regime, meaning being incarcerated for 7 to 12 years in harsh concentration camps; nevertheless, in returns, if he comes out of camps alive, he will be able to fulfill his Confucian responsibilities as expected of any eldest son: caring for the elder parents and his younger teenage siblings. **Two Minnows** takes the readers from the author's childhood growing up in the remnant of the old French Indochina system, followed by the confusing era of the first Republic of Vietnam, and finally his exposure to the war under American-influenced 2nd Republic, also as confusing as the first one, that leads to his escape from Việt-nam and sets him up for a lingering life in the Malaysian refugee camp. His rebirth in the Free United States of America helps him become more stabilized so he can mentally reconstruct his story: what has happened spiritually.

<u>**The War**</u> is an expanded elaboration of what has been covered in **Part 1: Two Minnows.** Why has the RVN (RVN) government decided to abandon Central Việt-nam in early 1975? How different is the North Vietnam Army (NVA) from the Army of the Republic of Vietnam (ARVN)? Are we aware that it is totally different from the World War 2? What has

influenced Hà-nội's moves? What is so devilish about the Paris Accords Negotiation? Who have really lost or won the war? Who has caused them to lose or helps them win? What is the impact of the television news media on the outcomes of the war, as compared to WW2? This comprehensive analysis will take the readers and, especially, the history and war researchers, from one political turning point to another, many of which could have changed the course of history. The War also exposes many cultural and strategic conflicts between the allies as well as many political maneuverings, some rational, others tricky and even dirty, that the governments of all sides have used to outplay and outwit each other.

Meanwhile, **Peace or Freedom, Part 1: Two Minnows,** published in April 2018 as the 44 Years of Remembrance – Special Edition, is a non-fiction memoir that encompasses the mental struggle of the author, a young Southern-born North Vietnamese son who has to escape Peace so that he can pursue Freedom. If he chooses Peace, he will have to stay behind and subjugate himself to the incoming Communist regime, meaning being incarcerated for 7 to 12 years in harsh concentration camps, if not executed as his commanding officer was; nevertheless, in returns, if he comes out of camps alive, he will be able to fulfill his Confucian responsibilities as expected of any eldest son: caring for the elder parents and his younger teenage siblings. Two Minnows takes the readers from the author's childhood growing up in the remnant of the old French Indochina system, followed by the confusing era of the first Republic of Vietnam, and finally his exposure to the war under American-influenced 2nd Republic, also as confusing as the first one, that leads to his escape from Vietnam and sets him up for a lingering life in the Malaysian refugee camp. Eventually, his rebirth in the Free United States of America helps him become more stabilized so he can mentally reconstruct history: what has happened to the nearly three million lives left behind that could have been spared. Two Minnows also illustrates the tenacity and tough will that are needed for a phoenix to elevate from an empty-handed no-name simple immigrant to a real Confucian whose sole goal is to serve the community, this time Minnesota and the United States of America.

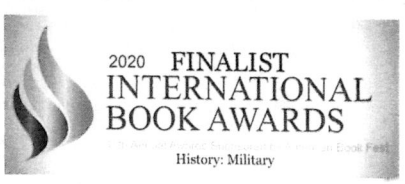

CHRONOLOGY

PEACE OR FREEDOM CHRONOLOGY

Dates	VIETNAM		IMPORTANT EVENTS
	US Presidency and Advisory	Troops	
2879-111BC	A China Protectorate for 2,762 years	NA	**Lạc Việt, Nam Việt, … then later, Giao Chỉ, Giao Châu, Vạn Xuân, Đại Cồ Việt, Đại Việt, Việt Nam, Đại Nam, … An Nam** were the names of the same country under China Protectorate
111BC-938AD	1st, 2nd and 3rd China Domination were opposed by An Nam after 1,049 years	NA	**Bắc Thuộc: During this period, there were three Dominations by China that were all defeated.** The last one was the Han Dynasty army being defeated at **Battle of Bạch Đằng by Giao Châu King Ngô Quyền**, Short Independences.
1258-1288AD	China Invasion	NA	After a short period of independence, 3 Mongol invasions of Đại Việt and Champa (1258AD, 1285AD and 1287AD) ended with Gensis Khan's grandson **Yuan Emperor Ku Bla Khan's army being defeated by Đại Việt Emperor Trần Nhân Tông**
1407-1427AD	4th China Domination: the Last One	NA	The fourth Chinese Domination of An Nam by the Ming Dynasty **lasted only 21 years. It ended with a defeat by An Nam Emperor Lê Lợi** (Legend: the Sword and the Golden Turtle)

1427-1600AD	China Invasion, then An Nam Expansion to the South. Civil War divided An Nam into North and South	NA	Đại Việt (at that time North and part of Central) expanded into Champa (protectorate by An Nam during Mongol invasions) which later became South Vietnam (there are contradicting versions of History regarding this annexing). **Major Civil War between Đại Việt Lords Trịnh (serving An-Nam Emperor Lê Chiêu Thống in North) and Nguyễn (South), while Emperor Lê was reigning in the North.**
1765-1783	American Revolution	NA	
7/4/1776	American Declaration of Independence	NA	
4/30/1789	First Presidency of the USA	NA	Gen. **George Washington** became the first President of the U.S.

1780-1789	China last Invasion of An Nam to rescue Emperor Lê of An Nam. An Nam was divided into Nôth and South	NA	**Lord Nguyễn Huệ defeated both Lords Trinh and Nguyễn,** and became new Emperor Quang Trung. Deposed Emperor Lê went in exile, requesting China to rescue. Emperor Quang Trung postponed Southbound push to counter attack China army in North. China Emperor Qianlong's army, in an attempt to rescue Emperor Lê, was heavily defeated. **This was China last aggression in An Nam.** Imperial Citadel Thăng Long celebrated Independence with a delayed New Year of the Rooster Kỷ Dậu. In spite of victory, Emperor Quang Trung negotiated status quo to maintain Peace with Qing Dynasty.
7/14/1789	French Revolution	NA	**French Revolution** getting rid of the French Nobility
4/25/1882	French Invasion of An-Nam	NA	**French commander Henri Riviere seized the citadel of Hanoi.** Capt. Henri Riviere was later beheaded after he attempted to seize the coal deposits at Ha long Bay. **The outraged French proceeded to colonize Vietnam.**
1914-1918	World War I	NA	4 to 6 Batallions of Indochinese indigenous soldiers, mainly Vietnamese, fought for the French army (1917-1918) in WW1
1887-1954	French Colonization: Indochina	NA	**France colonized** all of Vietnam (Tonkin North VN, An Nam Central VN, Cochinchina South VN), Laos and Cambodia: the French Indochina

Peace or Freedom - Part 2: THE WAR | 15

Date	Event	Stats	Description
1939-1945 AD	**World War II**	NA	Vietnam entangled in **First Indochinese Wars** against France and war against Japan
1933-1945	**Pres. Franklin Roosevelt 1933-45 Democrat - 3 terms**	NA	The **Last Emperor of An Nam Bảo Đại** was puppet Emperor collaborating with the French, then Japanese, then French again until 1945. **5/19/1941 mixed communist and nationalist Viet Minh** (originally 1935 Nanjing) was formed in Pac Bo
6/4/1940	State Sec. Cordell Hull 1933 State Sec. Ed. Stettinius 1944	**French Indochina WW2**	**French and British forces abandoned France at Dunkirk** - German occupation of France with Vichy
1940-1941		Fr 3,000 Jp 36,000	France **Vichy Government** (under Nazi) ceded Vietnam to Japan
1941-42	**Senate% 67D/31R House% 62D/38R**	**French Indochina WW2**	1941: French Indochina under Japanese control - Tonkin, An Nam, Cochinchina , Laos and Cambodia.
Mid 1942	**CODE: Black when Dem were Majority**	US 151,000 Jp 129,000	04/09/1942: USA and Philippines **Bataan Death March**. 05/06/1942: **Japan captured the Philippines.**
6/6/1944	**CODE: Grey when Rep were Majority**		**US and Allies started Normandy Invasion**: Allied Forces 156,000 vs German 50,000.

16 | Peace or Freedom - Part 2: THE WAR

1943-44	Senate% 58D/41R House% 52D/48R		
Beginning 1945	Pres. Harry Truman 1945-53 Democrat - 1st term	Indochina WW2	WW2 Military Casualties: Allies 16,000,000 vs Axis 8,000,000. Total Civilian Death: 49,000,000
Mid 1945	State Sec. Ed. Stettinius 1945 State Sec. James Byrnes 1946		OSS Office of Strategic Services, now CIA, trained Hồ Chí Minh's Guerillas against Japan, and discovered the latter being communist.
			Hồ Chí Minh started preparing for a major insurgence against the French colonial establishment.
7/5/1945			General Douglas MacArthur returned to liberate the Philippines
8/6/1945	State Sec. George Marshall 1947 State Sec. Dean Acheson 1949		US dropped atomic bombs onto Hiroshima and Nagasaki. August 1945: Hồ Chí Minh officially instated Tonkin as the new independent (Communist) Democratic Republic of Vietnam.
8/14/1945	ChiefsStaff John Steelman D	French Indochina WW2 US 151,000 Jp 129,000	Nationalist and Communist Viet Minh launched **August Revolution** against French. Emperor Bao Dai abdicated (8/25/1945)

Date				Notes
9/2/1945				Japan signed unconditional surrender to the US and Allies, then left Vietnam. China and Korea were also liberated from Japan
1945–46	Senate% 56D/43R House% 59D/44R			
1945				The **French came back** and took over control over An Nam, Cochinchina, Laos and Cambodia - with **strong USA support** (1 Billion). French return. Anti-French Insurrection war increased intensity by Việt Minh. **Emperor Bảo Đại abdicated.**
1947–48		Senate% 53R/47D House% 57R/43D		1945-1948 **Remnant of Japanese Imperial Army** hiding in Vietnam helped train Việt Minh fight the French and American.
1949			Pres. Harry Truman 1945-53 Democrat - 2nd term	Author Standley Karnow: **Emperor Bảo Đại abdicating** was a decision that indirectly gave Hồ Chí Minh the Celestial Mandate of taking decisive political action for a take-over
1911-1966			State Sec. Ed. Stettinius 1945 State Sec. James Byrnes 1946	**Stemmed in the Paris Commune in 1871, Communism evolved and spread out into other countries. The ones in China 1911-66, Russia 1917-18, North Vietnam (Tonkin) 1945, Cuba 1959 were all connected. Millions died!**

18 | Peace or Freedom - Part 2: THE WAR

Period	Political	Casualties	Events
	State Sec. George Marshall 1947 State Sec. Dean Acheson 1949		The **Soviet Union** had been Communist since Dec 1922 under Head of State Mikhail Lenin and Sec of State Joseph Stalin. The **COLD WAR started** with many incidents between the US and the Communist block
	Chiefs of Staff John Steelman D		**Independence of Cochinchina**: Anti-communist French-American backed State of Vietnam (South) led by abdicated Bảo Đại. An Nam remained French territory
1949-1950	**Senate% 55D/45R House% 61D/39R**	**Korea WW2** SK 600,000	**Korea War June 1950** started at the end of Pres. Truman's term: US Commander Douglas MacArthur
Mid 1950 End of 1951		US 330,000 UK 14,000	USA signed **Mutual Defense Assistance Program** to provide military aids to France against Communist expansion into Laos, Cambodia and the State of Vietnam (South VN)
Beginning 1952			S Vietnam created her own Army.
1951-52	**Senate% 51R/49D House% 54D/46R**	**N Korea WW2** NK 266,000	**Special Forces Group (SFG)** units were created in Vietnam and, later on, served as core units in Unconventional Warfare
		CN 1.4M Sov 26,000	Korean War ended. Gen. MacArthur's army pushed N. Korean army north of the 38th Parallel, Panmunjorm which became the new De-Militarized Zone DMZ until today

Peace or Freedom - Part 2: THE WAR | 19

Date	Person/Role	Location/Numbers	Event
Mid 1953	Pres. Dwight Eisenhower 1953-61 Republican - 1st term		Besides the US and S. Korea, 16 more countries fought in this war against N. Korea, China and Soviet Union. The two Communist vs Non-Communist sides had officially created their own alliances at the beginning of Korean War. It was the beginning of the **Cold War**
10/22/1953	State Sec. John Dulles 1953 State Sec. Christian Herter 1959	Indochina	**Independence of Laos** from France, followed by **North Vietnam 1st attack** in Laos
11/9/1953	Def. Sec. Ch. Wilson 1953-57		**Independence of Cambodia** from France
Mid 1954	Def. Sec. Neil McElroy 1957-59 Def. Sec. Th. Gates Jr. 1959-61	**Cochinchina** Fr 20,000 VM 50,000 US 17	Upon requests by French Chief of Staff Gen. Paul Paul Ély, the U.S. deployed carriers and aircraft in Hải Phòng, N. Vietnam, but no further action was taken as most of US top brass officers were divided.
	ChiefsStaff Sherman Adams R1953		Congress and British PM **Winston Churchill opposed American** involvement in Điện Biên Phủ.

3/13/1954		Cochinchina Fr 20,000 VM 50,000 US 17	The Việt Minh under the command of **Gen. Võ Nguyên Giáp attacked the French-Vietnamese Allied garrisons at Điện Biên Phủ Battle and ended French rules in Vietnam.** Even though flown by French pilots, US Corsair fighters and DC8 transport planes used in the battle **were still carrying the USA colors.**
1953-54	**Senate% 50R/49D House% 51R/49D**		North Vietnam therefore used this for their **PROPAGANDA THAT THE US WERE INVADING VIETNAM.**
5/7/1954			**France's Defeat at the Battle of Điện Biên Phủ** led to **Independence of Vietnam** from France. French government unconditionally surrendered
7/20/1954			July 20, 1954: **Geneva Conference** on French Indochina
7/20/1954			**Geneva Convention** April-July 1954 divided Vietnam into Communist North and Democratic South at the 17th Parallel (DMZ), and set July 1956 as the **General Election** day for Vietnam unification
7/22/1954			Birth of the **NFL National Front of Liberation (VC)** in the South as Passage to Freedom would allow 1,000,000 refugees fro the North to move to the South

		Passage to Freedom: The Geneva Accords also allowed civilians to seek Freedom and political asylum in the South between Aug 1954 and May 1955 (10 months). Massive exodus of the North Vietnamese to the South (Bắc Kỳ Di Cư 1954, or North Vietnamese Refugees of 1954) as **assisted by the U.S. Navy.**
		Hồ Chí Minh started purging and consolidating Communism in Việt Minh
10/26/1956		PM **Ngô Đình Diệm removed Emperor Bảo Đại** (10/26/1955) in Referendum that made him President of 1st Republic.
	Senate% 51R/49D House% 53D/47R	**U.S. Military Assistance Program** for France was given directly to the newly formed State of VN (South) armed forces. **This added more conspiracy** to Hồ Chí Minh's propaganda
1955-56		
7/1/1956		Ngô Đình Diệm: 1st President of Republic of Vietnam. He **successfully blocked General Referendum Election** (ruled by Geneva Conference in 1954).
		Pres. Eisenhower: His presidency would not be bound by Geneva Accord, and South Vietnam RVN would stay a separate nation from the rest of Vietnam.

4/28/1956				French completely withdrew from Vietnam (4/28/56). USA kept increasing aids to South Vietnam
1957-58	**Senate% 53D/47R** **House% 54D/46R**		**Statistics on troops**	*** Troops source:** http://www.americanwarlibrary.com/vietnam/vwatl.htm
1957	Pres. Dwight Eisenhower 1953-61			**Emperor Bảo Đại was overthrown** and moved to France. US-backed Pres. Ngô Đình Diệm started the first Republic of Vietnam
1957	Pres. Dwight Eisenhower 1953-61			**Emperor Bảo Đại was overthrown** and moved to France. US-backed Pres. Ngô Đình Diệm started the first Republic of Vietnam
1959		State Sec.Christian Herter 1959 Def. Sec. Neil McElroy 1957-59 Def. Sec. T.). Gates Jr. 1959-61 ChiefsStaff Wilton Persons R1958	**VN State** US 760+ adv RVN 243,000 **Laos** Royal 50,000 Merc. 21,000 Miao 23,000	**North Vietnam 2nd attack in Laos in 1959** and expanded support for Pathet Laos in the high land of Truong Son Mountains. **Hồ Chí Minh Trail** was born from Low Land Laos. Hồ Chí Minh committed troops to overthrowing RVN Government. "Tập kết ra Bắc" regroupee South VN pro-communist forces were recruited to the North. Rebel and VietMinh cells were implanting in the South (Viet Cong). Hồ Chí Minh's Trail became active.

Date	Congress / Administration	Forces	Events
1957		Pathet Lao 8,000	**Laos requested US Aids** to fight Pathet Lao. **Formation of Miao CIA-backed army** (changed to Hmong in 1970's) to help rescue downed US pilots in N Laos and fight Pathet Lao. RVN built up modern army which was trained by US Spec. Forces SFG. WW2 equipments (US Aids to French in 1950's) were transferred to ARVN. **OSS CIA also formed RVN commandos** groups to watch HCMTrail
1959-60	Senate% 66D/34R House% 65D/35R		**MACV-SOG Viet SEAL** (later on: RVN Strategic Technical Directorate) commandos started being inserted in Laos and N Vietnam for spy missions (assassination, kidnapping, intel gathering…). It was expanded into Cambodia in late 1960's when HCM Trail became a major nuisance.
1961-62	Senate% 63D/37R House% 60D/40R		
1961	Pres. John F. Kennedy 1961-63 Democrat - Partial Term		
2/2/1962	State Sec. Dean Rusk 1961-63 Def. Sec. R. McNamara 1961-63	RVN & Allies US 3,200adv. RVN 243,000	**Strategic Hamlets "Ấp Chiến Lược"** were organized in the RVN (British Counter-insurgency Advisor: Sir Robert Thompson) RVN campaigned to stabilize agricultural areas "Bình định nông thôn" - Pres. Kennedy committed to protecting RVN

Date	Officials	Senate/House %	Forces	Events
1963-64	ChiefsStaff Kenneth O'Donnell D ChiefsStaff Kenneth O'Donnell D Chair ChiefsStaff Maxwell Taylor	**Senate% 65D/35R** **House% 59D/41R**		**Wall was built separating East from West Berlin**, USA built up all branches of Army. US Marine Corps were reinforcing Spec Forces in RVN. All Spec Forces wore Green Berets. Eventually, creation of SEAL teams 1 & 2
				US-Cuba Missile Crisis (10/16/1962) almost causing a nuclear clash between the US and Cuba (and the Soviet Union)
				ARVN's major attack against NVA base near Cambodia (Áp Bắc and Ấp Tân Thới)
				Buddhist uprising against against Pres. Ngô Đình Diệm and his brothers in Huế (Buddhist Crisis)
5/1/1963	Pres. John F. Kennedy 1961-63 Democrat		ARVN & Allies US 16,300adv. RVN 243,000	**Pres. Kennedy was determined to withdraw all military personnel** (National Security Action Memorandum 263) in spite of Joint Chiefs of Staff's pushing for WWII Korea-type of War.
10/11/1963				Due to his major conflict of policies with the military, JFK resorted to a **back channel** with brother Robert and other friends.

Peace or Freedom - Part 2: THE WAR | 25

Date	Leadership	Forces	Notes
11/1/1963			**RVN Pres. Ngô Đình Diệm, and his brother, powerful Ngô Đình Nhu**, were both assassinated. Both had refused US direct military involvement. The US "closed its eyes" and supported the Coup. CIA had got rid of Pres. Diệm either because of conflict of policies or because of tyranny against his people.
11/22/1963	**Pres. L. B. Johnson 1963-69 Democrat - 1+ Term** State Sec. Dean Rusk 1963-69 Def. Sec. R. McNamara 1963-68	**ARVN & Allies** US 23,300adv RVN 514000 Aus 198 ROK 200 NZ 30 Phil20 Thai 0	**Pres. JF Kennedy was assassinated in Dallas** during his 2nd term semi-campaign visit. Sucessor Pres. LB Johnson was sworn in immediately. The assassin Lee Oswald was also assassinated within a few hours while in custody in Dallas Police headquarters. Case was closed a few days later. **Conspiracy theories:** CIA, VP LB Johnson, Mafia, Fidel Castro, KGB, or a combination.
1/30/1964	Def. Sec. Cl. Clifford 1968-69		**ARVN Gen. Nguyen Khanh took over** government from coup gen. Duong van Minh. Until now, US held **Advisory role** only.
4/27/1964	ChiefsStaff Marvin Watson D1963 ChiefsStaff James R Jones D1968		**ARVN Col. Đỗ Cao Trí** conducting successful large scale division attack on NVA VC tactical base Đỗ Xá in II Corps with US providing transport and artillary support (Quyết Thắng 202). Base Đỗ Xá was destroyed during this 3rd attempt.

Date	Troops	Senate/House	Events
8/2/1964			**Tonkin Gulf incident** between DRV boats and US Navy battleship gave the US a reason to start direct intervention. 12/28/64 Taking advantage of Saigon unrest, NVA launched a large scale attack on ARVN at Bình Giả, Phước Tuy Prov.
3/8/1965			US Army boots on the ground in Da Nang and started engaging VC's and NVA. Operation Rolling Thunder B52 bombing DRV
1965-66	**ARVN & Allies** US 184,300 RVN 642,500	**Senate% 67D/33R House% 68D/32R**	4/6/65 US NSAM 328 increased troops (200,000 to 500,000) followed by Pres. Johnson's speech on Peace without Conquest: Renowned IaDrang Battle (Nov. 1965) between NVA and US, and later ARVN (Lt.Col. Ngo Quang Truong, Maj. N. Schwarzkopf)
End 1965	Aus 1,560 ROK 20,620 NZ 120 Phil 70 Thai 20		**ARVN and US troops started fighting their separate war.** US PressMedia, The U.S. Army was the only one fighting the war in all 4 Army Corps (NBC). RVN internal leadership turmoil
1/1/1967			
1967-68		**Senate% 62D/38R House% 57D/43R**	**US Full Engagement** in Vietnam War: Operation Cedar Falls, Prairie II... Involving most major branches of US Army

Date	Forces		Event
6/27/1967	VC NVA Allies NVA 290,000 VC 120,000 China 170,000 N Korea 200 Cuba ?		**ARVN Battle of Suoi Long.** Death of NVA Gen. Nguyen Chi Thanh
1/13/1968			NVA lauched major attack against Royal Lao Army at Nam Bac, causing the latter to be dismantled. Lima Site 85 Base fell on Mar 10.
1/30/1968	Laos Royal 60,000+ Thai 21,000 Miao 23,000 Pathet 48,000		**Tết Mậu Thân Offensive**: NVA and VC's violated New Year Truce and attacked all major cities in RVN. US stroops and ARVN more of less fought their own battles, to recover invaded parts of cities. US Air and Artillery support for ARVN. This was preempting Paris Negotiations to begin. It was also HCM testing the RVN and the Allied forces.
5/10/1968	RVN & Allies		Preliminary **Paris Accords Negotiatons** started between Averell Harriman (US) and Xuân Thuỷ (DRV).
1969-70	RVN 820,000 US 536,100	**Senate% 57D/43R** **House% 56D/44R**	**List of US Battles in Vietnam 1965-1970** http://hist-sdc.com/images/spotlights/sb_vietnam/vn_list_of_battles.pdf

28 | Peace or Freedom - Part 2: THE WAR

Date			Event
6/5/1968	ROK 50,000 Aus 7,660 NZ 520 Phil 1,580 Thai 6,000		**Democrat Sen. Robert Kennedy**, an icon of modern American Liberalism, was assassinated during his campaign for Presidency. He just won nomination by California and S. Dakota. Running against him were Democrat Hubert Humphrey and Republican Richard Nixon. Assassin Sirhan Sirhan (Palestinian/Jordanian) was apprehended.
	Pres. Richard Nixon 1969-74 Republican	Communists	**Five conspiracy theories**: confusing theories involving a second shooter, the secret mind-controlling "Manchurian Candidate," Palestina Sirhan Sirhan killed Bobby because the latter was supporting the rise of Israel, a third person (lady in polka dot dress), and the CIA. LAPD destroyed evidence!
3/23/1969	State Sec. W. Rogers 1969-73 H. Kissinger 1973-74	VC NVA 220000	**Royal Lao Army launched major counter-attack** in the Plain of Jars and regained lost areas by August 1969. US **Barrol Roll bombing campaign** in Region Militaire 2 (Gen. Vang Pao's Area). By Sept, NVA volunteer regiment pulled back
8/15/1969			**Kissinger started secret meetings with DRV Lê Đức Thọ**
2/1/1970	Def. Sec. M. Laird 1969-72	VC NVA 287,465	**Royal Lao Army took Plain of Jars back. Gen. Vang Pao's army** had to be rescued as his Long Tieng Base was heavily attacked.

1971-72	Senate% 55D/45R House% 59D/41R		
3/18/1970	Def Sec. E. Richardson 1973 J. Schlesinger 1973-74		**Prince Sihanouk was overthrown** by an American-supported coup in Cambodia. US-backed **Pres. Lon Nol took power.**
4/29/1970	ChiefsStaff HR Haldeman R1969 Alexander Haig R1973	**Kampuchea** RepKhm 35,000 KhmRouge 4,000	**Kampuchea incursion by US and ARVN.** Gen. Đỗ Cao Trí lead III Corps ARVN Infantry 18th Div separate from US campaign. Troops and supply bases built up by NVA infiltrated through HCM Trail was destroyed.
2/8/1971 to 3/25/1971			**Lam Son 719 Campaign:** II Corps RVN attacked NVA tactical bases at Tchepone in Southern Laos with US Air support. Poor planning and performance in **Gen. Hoàng Xuân Lãm**'s leadership caused major disaster and withdrawal.

1971-72	**1971 RVN & Allies** RVN 1,046,250 US 156,800 ROK 45,700 Aus 2,000 Thai 6,000 NZ 100 Phil 50	The on-going **Battle of Skyline Ridge in Laos** Région Militaire 2 Plaines des Jarres **(CIA and Gen Vang Pao)** where the Long Tieng air strip provided mission to rescue downed US Pilots who were on bombarding mission in North Vietnam. Royal Forces FAR and Thai irregular mercenaries were sent **into Skyline Ridge to rescue besieged Long Tieng.** This base was strategically important as it was used to block NVA and Pathet Lao from entering Laos Capital Vientiane in Region 5.
2/23/1971		**ARVN III Corps Commander Gen. Đỗ Cao Trí was transferred to II Corps** from Kampuchea campaign to replace failing Commander Maj Gen. Hoàng Xuân Lãm (Lam Son 719). He was killed on the way to II Corps. He was considered to be the RVN Gen. Patton.
2/21/1972		**Nixon met with Zhou Enlai in the PRC** without consulting with partner Pres. Thiệu to negotiate China to apply pressure on Hanoi
3/31/1972		**EasterTide Offensives**: NVA and VC's (8 divisions) attacked major cities in Quảng Trị, Huế, Kontum, Pleiku, Bình Long, An Lộc, and Long Xuyên. It last until 1/31/1973
4/1/1972		**The Battle of Khong Sedone** in Laos was lauched as the NVA were flooding Khong Sedone when they were mobilizing divisions to attack South Vietnam **(Easter Offensive)**

Date		Forces	Events
July and Aug. 1972		Kampuchea RepKhm 200,000 KhmRouge 70,000	J. Fonda, as well as AG Ramsey Clark and others visited North VN… visited North VN. Aug 1972, causing a major anger among those who were fighting for Freedom from Communism. Upon their return to the US, they fueled more uprising among the antiwar movement.
			ARVN started rationing ammunition and supplies as it increased military draft to reach 1.1 million
1973-74 (93th)	**Senate% 58D/40R House% 56D/43R**		
10/1/1972		RVN & Allies RVN 1,048,000 US 24,200 ROK 36,790	Nixon-Kissinger audio conversation as taped was later declassified after 1996: Nixon and Kissinger's plan to abandon the RVN. Nixon prepared for 2nd term, **Oppressive Paris negotiations by Kissinger, Pres. Thieu refused to sign treaty**
11/7/1973		Aus 130 NZ 50 Phil 50	US Congress (majority Democrat) passed the **War Power Resolution** to block the President from making war decision.
12/1/1973		Thai 40	Thanks to **Operation Line Backer** (Xmas 1972 B52 Bombing), Paris Accords Agreement was finalized. Nixon promised the US would resume bombarding the North if it attacked the South again.
			Buddhist Communist sympathizers uprising (led by Monk Thích Trí Quang) in Saigon and many major cities
1/27/1973			**Paris Agreement was signed** between all 4 sides. US Veterans returning home were abused and disgraced.

32 | Peace or Freedom - Part 2: THE WAR

Date	Political	Troops	Events
1/28/1973			**Paris Accords went into effects:** Immobilization of all troops. Return of US POW, total US troops withdrawal.
1/28/1973			NVA could keep occupied territories and continue to resupply from Soviet/China, all military aids to RVN were cut. That was oppressive to the RVN!
8/9/1974		RVN & Allies RVN 1,110,000 US 50	**Pres. Nixon resigned** as Watergate broke out. **VP Ford** took over. In South Vietnam, **Buddhist monks and followers protested and demanded Pres. Thiệu to resign.**
1975-76 (94th)	**Senate% 62D/38R** **House% 67D/33R**	Other allies 0	
12/13/1974	Pres. Gerald Ford 1974-77 Republican	VC/NVA 370,000	**Spring Total Offensive:** NVA and VC's attacked in full forces as a deluge. Nixon was not President any longer to keep his promise to resume B52 carpet bombing of North Vietnam.
3/13/1975	Defense Secr. J. Schlesinger 1974-75 D. Rumsfeld 1975-77 ChiefsStaff Don Rumsfeld 1974R Dick Cheney 1975R		**US Congress (majority Democrat) cut all aids to RVN per Paris Accords.** Pres. Thiệu ordered Gen. Ngô Quang Trưởng to abandon I Corps and withdraw to Đà Nẵng. Gen. Phạm văn Phú was also ordered to evacuate II Corps. NVA and VC's flooded both Corps and massacred evacuees. **The War Power Resolution** passed by Congress in 1973 rendered Pres. Ford powerless even if he wanted to save the South.

Peace or Freedom - Part 2: THE WAR

4/17/1975	**Pnom Penh fell.** Pres. Lon Nol surrendered to the Khmer Rouge. There were 60,000 NVA in Kampuchea in 1974 supporting the take over by Khm Rouge. In 1977, the Socialist Rep of Vietnam turned around and repelled the Khmer Rouge, protecting Vietnamese migrants from being Cap Duon (beheaded).
April 1975	US personnel, Vietnamese VIP's and commanders of civilian agencies and military units are **evacuated by American planes and helicopters**
4/30/1975	**Saigon fell.** RVN Interim President, Gen. Dương văn Minh, surrendered the RVN to NVA Col. Bùi Tín in Saigon. He ordered all ARVN Units to surrender and follow NVA's orders.
1975-1995	US Congress passed Resolutions to allow Vietnam War refugees into the US. **Over 1,000,000 Viet boat people** poured out into South China Sea and were held in containment camps in East and SE Asia. **20 to 40% of them died at sea.**
1975-1985	Discriminated by the Socialist Rep. of Vietnam SRV, **nearly 300,000 Chinese Vietnamese** decided to move to China by land and Hong Kong by boat.

Date	Event	Political
12/2/1975	**Vientiane fell** and - Laos last King Sisavang Vatthana of Laos surrendered to SRV-backed Pathet Lao. Lao royalists & Hmong (Miao before 1975) fled to Thailand. The Socialist Rep of Vietnam helped establish a Vietnam-backed government in Vientiane.	
7/2/1976	**NFL was dissolved.** A great many NLF members were discharged but never received any benefits for their contributions to he Party.	
1977-1991	Socialist Republic of Vietnam (DRV) incursion in Cambodia against its former ally Pol Pot Khmer Rouge. Defecting Khmer Rouge took power with the SRV's blessing. SRV established a Vietnam-backed government as it had done in Laos.	**Pres. Jimmy Carter** 1977-1981 Democrat
1977-78 (95th)		**Senate% 58D/41R** **House% 66D/34R**
1979	**Khmer Rouge massacred 2,700,000 Khmer** of all ages (Killing Fields). **SRV sent army to Kampuchea to anhilate the Khmer Rouge,** these having also beheaded Vietnamese immigrants there ("Cap Duon"). **China invaded the Socialist Rep of Vietnam** to punish it for repelling the Khmer Rouge in Kampuchea. Army Divisions crossed Vietnam border but were repelled by militia, while the main forces of DRV were still in Kampuchea.	

Period	Congress/President	Events
1979-80 (96th)	**Senate% 55D/44R** **House% 64D/36R**	1979-1990: China invaded Vietnam repeatedly and was repealed by mostly militia while VN main forces were in Kampuchea fighting China-backed blood-thirsty Khmer Rouge. Land in North Vietnam was taken definitively as part of the new era of China invasion of the oil/resources/fishry rich, and strategic South China Sea (1996-current)
1981-82 (97th)	**Senate% 53R/46D** **House% 56D/44R**	
6/9/1987	Pres. Ronald Reagan 1981-1989 Republican	Pres. Reagan: **"Tear down this wall…"** He and Soviet leader decided to tear down Berlin Wall: The Cold War was over at the cost of millions of lives of people who had nothing to do with it!
1983-84 (98th)	**Senate% 55R/45D** **House% 63D/37R**	
1989	Pres. George HW Bush 1989-1993 Republican	**Resistance started in Vietnam** against the new regime. Buddhism rising again, this time fighting, not for Peace, but for Freedom and Human Rights
1991-92 (102nd)	**Senate% 58D/42R** **House% 62D/38R**	A resolution by Sen. McCain allowed RVN officer POW's into the US in exchange for economic sanctions. **Orderly Departure Program** Familiy Reunification under the U.S. Immigration Laws.

1993-94 (103th)	**Senate% 53D/47R** **House% 59D/41R**	**Vietnamese Refugee Resettlement in the World**: 1.8 M have been resettled in the US as refugees per Congressional Resolutions. Also: 600,000 in Cambodia, 350,000 France, 211,000 Australia, 200,000 Taiwan, 158,000 Canada, 150,000 Germany, 147,000 Japan, 143,000 S.Korea, 83,000 Czech Rep., 70,000 Malaysia, 55,000 UK, 50,000 Poland, 30,000 Laos, and all over the world…
2000's	Pres. **George W Bush** 2001-2009 Rep Pres. **Barack Obama** 2009-2017 Dem	**SEVEN PRESIDENCIES OR ALMOST 4 DECADES LATER:** Welcome Home, Vietnam Veterans (2011 Senate Resolution by Sen Burr) Pres. Obama was the first U.S. President to visit Vietnam since the end of Vietnam War
2001-02 (107th)	**Senate% 50R/49D** **House% 51R/48D**	Sources: Wikipedia, POTUS, "The VN War: Chronology of War (Army, Navy, Air Force Foundations and USMC Assosciation - Edited by Ret. Col. Raymond Bluhm Jr.)

TABLE OF CONTENTS

TITLE PAGE 1
FOREWORD 3
ACKNOWLEDGEMENT 5
ABOUT THE AUTHORS 7
ABOUT THE "VIETNAM: PEACE OR FREEDOM" SERIES
....10
CHRONOLOGY...13
TABLE OF CONTENTS 39

PROLOGUE AND THE INDUCTIVE QUALITATIVE
RESEARCH..44

CHAPTER 1: Pedagogy of Death............................47
    The concepts of Life or Death as applied in politics and
    war...48
    Was the any political connection between the Middle
    East and Việt-nam? 55
    Killing is purposeful 58
    Horrendous death shapes the Conscience of America.61

CHAPTER 2: Fusing Beliefs into the Death Pedagogy...........66
    Buddhism and Hinduism: Karma and Reincarnation ..66
    Popular belief in Destiny 69
    Effects of Karma on politics 72
    Confucianism and its politicized infusion...................76
    Communist versus nationalist options 78
    Totalitarianism meant strength and power 80
    Humaneness and Liberalism translate into being weak...
   81
    The Meaning of Human Life via the spiritual lenses...83
    HoChiMinh-ism as a spiritual belief...........................84
    The Atheists and folk believers' viewpoints 86
    The Christians 88
    The battle between strong beliefs 89

CHAPTER 3: And the Wheel of History keeps turning 91
    Thousands of years under Chinese Domination 91
    Fighting against Chinese Domination.........................93
    Use Chinese wisdom to fight against Chinese invaders
    (Sun Tzu's The Art of War).......................................101
    Sun Tzu 105

Resisting the French Occupation 107
CHAPTER 4: The Turning Points 110
    Who was Hồ Chí Minh? 113
    Turning point: The hidden letters 121
    Turning point: the remnants of the defeated Japanese army 124
    The Songs that move rivers and mountains 126
    Hồ Chí Minh's match: Ngô Đình Diệm – Another turning point 133
    Rule by fear 134
    Diệm's dilemma 137
    Pivoting points for the Republic of Vietnam: Democracy 140
    Real leaders or "wannabes?" 141
    Shocking news in the U.S. 145
    Mysteries and Speculations about the death of President John F. Kennedy 146
    More behind the scene 152
    Turning the history page 153
    The Unites States go to war 157
    Politicizing the dissident North/Central versus the docile South 159

CHAPTER 5: The Psychology Playbook 164
    Psychology: Maslow and the Hierarchy of Needs 165
    How does PTSD fit into this? 171
    Psychology: Daniel Goleman's Reptilian versus Neocortex Brain 172
    Psychology: Snapshots and the haunting 176

CHAPTER 6: Communist versus Nationalist Brain 181
    Communist brain 182
    Nationalist brain 184

CHAPTER 7: Easter-West Incompatibility 187
    Predictability in warfare 190
    Incompatibility between U.S. and RVN policy 193
    Massacre as tool of terror or secret of winning a war? 195
    Democracy in the Eye of the Beholder 198
    The Power of Fear 200
    Instigation and Fear 201

CHAPTER 8: Let the Game Begin… The Paris Accords 202

Introducing the Players and the Sideliners 202
Minister Trần văn Lắm (1913-2001), Republic of
Vietnam Foreign Affairs 203
Mr. Lê Đức Thọ (1911-1990), People's Democratic
Republic of Vietnam primary negotiator 204
Minister Nguyễn thị Bình (1927-), South Việt-nam
Communist Provisional Revolutionary Government
(NLF or Việt-cộng) Foreign Affairs 205
Secretary of State Dr. Henry Alfred Kissinger (1923-
), Also U.S. National Security Advisor, U.S. primary
negotiator 206
United States 37th President Richard Milhous Nixon –
Republican (1913-1994) 213
Second Republic of Vietnam President Nguyễn văn
Thiệu (1923-2001) 215
Second Republic of Vietnam Vice President Nguyễn
Cao Kỳ (1930-2011) 218

CHAPTER 9: Kampuchea and Laos were left out of the Picture
....221
Republic of Kampuchea or Khmer (Cambodia, as
renamed by Pres. Lon Nol) 222
The "Secret Khmer Special Forces" that were the most
"secret" 224
Kingdom of Laos 227
First DRV Invasion of Laos 227
The Original Lao "Secret Army" 230
Gen. Vang Pao and his Miao Secret Guerilla Units in
"Région Militaire #2" 231
Clandestine Operations by the RVN "Secret Army" in
Laos..234
Like Việt-nam, Laos was politically divided 236
The Trail in Regions 3 and 4 238
What happened in Việt-nam was "trailed" from Laos
then through Kampuchea 240
Is Vietnam War a Civil War or Second Indochinese
War? Neither! 242

CHAPTER 10: In this Ball Game, whose Side was the press
media on? 246
The beer venders set fans up against their own team.250
How could the press media determine Win or Lose? .....
....254
Who else is the enemy? 266

CHAPTER 11: The Fans join in to Boo at their Home Team 272
    Democracy divides – Part 2 272
    The political divide 275
    The religious divide 277

CHAPTER 12: Paris Accords, the Twisty Treaty 283
    Death by strangulation… by our own friends! 285
    The Generals 292

CHAPTER 13: Win or Lose? 298
    Taking advantage of the rear debauch and decadence 298
    Benefitting from the allies' withdrawal 305
    Whether winners or not, they are the heroes 307
    Win Lose, Lose Win 309
    Turning the "Peace with Honor" switch on 311
    Eternal resistance "trường kỳ kháng chiến" and the anti-war movement in the US 316
    Liberate or Invade 317

CHAPTER 14: The Confusion of the American Public about the War 322
    The ARVN don't fight? 322
    Some did, some didn't 326
    They Might be Humble, but Certainly Not Lazy or Coward 340
    No one really cares about the ARVN Victory or Agony .. 344

CHAPTER 15: Maslow Revisited 346
    From "lost in the new Independence" to "lost in Freedom" 351
    Justifying "their" absence in community-at-large activities 357
    How does Maslow's theory apply to Việt-nam during the war and now? 364

CHAPTER 16: A Completely New Environment to Cope With . 369
    The Way of the Warriors: Either chose a heroic Death or obey Supreme Commander and lay down weapons 369
    Sweet re-education or death camps 372
    Share or repent 373
    The other solution: flight or fight 378

Time to wake up 380
1973 – POW exchange 382
1975 - An NVA Colonel's confession 382
1981 – Father kept uttering in his coma 384
1990 September - NVA Gen. Bùi Tín defecting from Việt-nam 386
2013 - Another defecting official, Mr. Đặng Xương Hùng, Consul to Genève, Switzerland (2008-2012) .. 388
1992-1995 – Buddhist Monk protesting against government oppression 390
2009 – Taxi driver in Hồ Chí Minh City 391
1976-2016 – Catholic priest Thadeus Nguyễn văn Lý 392
May 2016 – Pres. Barrack Obama 394
What really matters? 396

CHAPTER 17: The Awkward Truth about Winning and Losing! 398
    Win or Lose? 399
    Heroes or Villains? 402
    Like eagles, they watch their preys 410
    New Art of War: Win the invincible by hitting their Conscience 412

CHAPTER 18: We are Home, America! 419
    Major James Connelly (name was changed), MACV 420
    Tom (unknown), Long Range Reconnaissance Platoon 422
    Gene Giunta, a baby-killer 427
    Tony Lazzarini and the Forever Brotherhood 429
    PTS 430

EPILOGUE 434

GLOSSARY 448

BIBLIOGRAPHY 452

PROLOGUE

The answers to the What, How and Why-questions from a multicultural viewpoint

Knowing why Vietnamese life was so cheap, as observed by U.S. Gen. Westmoreland, explained how they would fight the war. Not understanding this would cause fear. Not understanding how the Việt Minh had defeated the French would make us unprepared. Knowing the connection between Việt-nam, Laos and Cambodia and recognizing there was a tight relationship between the wars there could have made it easier to fight the war. Ignoring the three factors for success, "Thiên Thời, Địa Lợi và Nhân Hoà (celestial socio-political timeliness, favorable location and harmony with people), would doom us to failure. Fighting the Vietnam War and ignoring the Sun Tzu's Tao of War would definitely warranty no success. Superiority in modern weapons would not necessarily warranty victory. Failure to deal with and control turmoil in the "hậu cần or rear support" had definitely affected the success at the "front line." **The War** will address the other culture and the psychology of the war that Western strategists should know in order to understand what have happened the way it has in the past, as well as in the present, as the aftermath of the war is still affecting the veterans and the refugees.

An Inductive Qualitative Research

Ten years of relentless research triangulating verifiable documentaries, actual observation of events, cross-cultural references, interviews of actual combatants, personal in-person experiences… have been rationally summed up in this analysis. The suggested conclusion at the end of "**The War** " has not been drawn from the traditional way as it is usually done by a deductive approach used in traditional quantitative research: start with existing theories, e.g. that Communism is bad, that American presidencies /government are doing their best to win the war and rescue the South Vietnamese, American GIs are baby-killers and hut-burners, RVN government is either tyrant or corrupt leaders, ARVN are busier with coups d'état than fighting the war,...

On the contrary, researchers should start with a blank page of paper, and with a neutral and non-partisan mind, because a social-scientific analysis cannot be biased, otherwise we can never learn from history and, instead, we will end up being a political one-sided activist. It cannot be just like a high school or college research paper!

Traditionally, the quantitative research[1] is used in formal scientific researches (e.g. doctoral dissertation). It starts with the researchers deciding on an unconfirmed focus (what to study), then it follows the sequential steps once it has already identified the specific questions, narrow topic down, collect facts and numerical data (%) from related participants, analyse these data using established statistics and conduct the inquiry to support, verify, and justify the original topic. The conclusion is deductive from an unconfirmed theory.

A quantitative researcher is like an angler looking for specific brown trout to catch... Meanwhile, the qualitative research, which has been newly introduced in the late XXth Century, starts without a pre-existing theory or focus. It is like an angler trolling the boat up and down the river and trying to use a variety of baits and lures to see what fish can be caught. Loop after loop on the river, trolling back and forth, he/she might catch the same type of fish (the same information might be repeated), but each time, some new fish is caught, making the loot more and more interesting (new materials are added to enrich and expand the old finds). Nevertheless, if there are more of the same crappies caught (and nothing else), he/ she will conclude in an inductive way that the area has a lot of crappies, and not many walleyes or basses (a conclusion may be drawn). With patience, the angler will definitely find out where else to catch basses versus walleyes or crappies... Qualitative researchers look at the mass of possibilities and participants, ask broad and general questions to probe possibilities, triangulate to find correlations and patterns, analyze those themes to come up with possibilities by induction. The conclusion might be subjective and biased due to small samples, and, in some cases, can be contradicting common beliefs, but the Truth will not be missed. Readers need

---

[1] https://research-methodology.net/research-methodology/research-approach/inductive-approach-2/ Qualitative Induction vs. Quantitative Deduction in Researches.

to note that qualitative process is always in a progressive mode until the very final conclusion; the algorithm keeps expanding as new information is brought into the triangulation, thus making the reasoning sometime redundant. As a result, they need to read along and do not get stuck with the flow as the ending might be something totally different.

The War is compiled in the qualitative approach to accommodate the randomness of the available data that vary from some one-sided documentaries, some unverifiable oral history, to other true-from-the-horse's-mouth audio-tapes, first-person encounters or even declassified materials… This book is an investigation of the war. It is what war researchers need to have for their projects. It is what veterans and families need to use in order to guide them through the maze of this War.

CHAPTER 1
Pedagogy of Death

(Before you read on, I need to warn you this chapter is graphic and might not be suitable for the sensitive soul. Please read at your own discretion)

(the key concept words will be capitalized)

Please note that the following belief-based analysis in this as well as the next chapter is absolutely not what-so-ever meant to compare which country, culture, religion, spiritual sect or belief is better or which political group is more or less courageous, heroic, patriotic, inhumane, or barbaric. My intent in this chapter is to make readers become more aware that the way people have been fighting in wars depends heavily on how they perceive Life and Death as applied to themselves as the ones being possibly sacrificed, as well as applied to others as the ones they kill. It explains how combatants' feeling about their own or the victims' Death can determine their readiness and devotion to their cause (willingness to die for it) and, consequently, how they view and decide on either ending others' life or preserving it in the course of action. While some may view taking lives as a mean to achieve their ultimate goal, others look at it more seriously and chose between Life and Death as, for the latter, Life matters more and, therefore, alternative courses of action should be taken instead. While some are using terrorism and intimidation to fight and win at any cost, others fight but still look at various choices of humane actions and make decision over a wide range of alternatives, between Life, Death, Magnanimity, Compassion, Pity, Mercy, Love and Peace.

The Concepts of Life or Death as applied in politics and war

Why are we talking about Death? Don't we have enough death before, during and after the war? Death means differently in each culture, so, if we want to learn about cultures, we need to research, not just about food, architecture, etc… but also about how each culture views Death as it relates to Life. In my opinion, as a person who has worked with many cultures and subcultures of many age groups, I strongly believe that how a person sees Death will determine how he/she lives, fights and dies, flies away in Fear, or just freezes up and lets Fate decide. Death as defined will also determine how he/she kills, and whether to inflict more pain by extending suffering or make it a merciful quick event. Punitive killing is a cause-and-effect action, not aiming just at making the victims "payback" for whatever wrong doing they have performed, but more for the purpose of intimidating those who are still alive, this being the case of terrorist executions and massacres.

My father had told me about how some aboriginal tribes in North Việt Nam where he grew up had treated their family members who had passed away. These dead were not buried as in most cultures, but kept standing inside either hollow walls or wooden pillars that were supporting the house ceiling. They were actually not considered dead and gone, but that they were still living and serving a good purpose in the household: protect the living. For these people, death and life were very close to each other, and the living and the dead were still living in harmony together. Perhaps they had no fear of Death, and no fear of dying, as dying is believed to be just a process of transitioning to another life, along with the living members of the family. Imagine we had to fight an army of people who were not afraid of dying…

The Vietnamese, through thousands years of foreign domination and exploitation, actually learned about death more by being the victims (rebels against invaders) than the executioners (indigenous collaborators). During more than twenty centuries of domination of Việt Nam, as a warning to potential uprising organizers and followers, the Chinese invaders had used terroristic execution methods against Vietnamese "rebels" such as the one-thousand cut sentence (cut off live victims piece by piece to the beat of a drum "tùng xẻo" until they died), using horses to tear up victims into pieces in multiple directions, having elephants crush victims alive... sentencing from one to three generations at a time in mass public executions. Indescribable creative methods of torture and mutilation were used to literally extract information from unbroken spirits. Public beheading had always been the least painful to the victims: death came fast and there would be no suffering. During the French colonization, sentences were less harsh than under the Chinese Domination, but public executions by guillotine, hanging or firing squad were still intimidating to the public. Their purpose, as I had mentioned earlier, was not just to punish the perpetrators, but more to intimidate the living ones and to give out a strict warning to potential perpetrators. Of course, insurgences were immediately eradicated, or at least delayed, while suspects were subject to terror or cruel torture. Public decapitation was also used under the Japanese occupation to shape the Vietnamese mind. Nevertheless, it had always been foreigners executing Vietnamese, until Vietnam War when the Vietnamese started crushing and massacring their own kind, perhaps copycatting from the Russian and Chinese Revolutions.

I have just discussed how death has been used in ideological and political cleansing. Death can also be used for uplifting and honorable purposes.

In the Eastern World, Japanese Samurai also had a different attitude towards Death. To die for the Emperor was a top Shinto honor in the Spirit of Bushido, the way of warriors. Even though the samurai groups had been dismantled in the late 1870s in Japan[2], their Code of Bushido was still respected by Japanese warriors and noble-spirited civilians as far as honor was concerned. "Sepukku" or "Hara Kiri[3]" had been performed as a ceremonious way of preserving one's honor whenever one was considered dishonored because of a duty or mission failure, or when one needed a desperate but honorable way of serving one's country or supreme leader. It was actually the Code of Bushido that had made "Kamikaze" pilots courageous enough to fly fighter planes filled with gasoline and bombs into the United States establishments and

---

2   https://en.wikipedia.org/wiki/Samurai , and Smithsonian Channel documentary series on Samurai Headhunters, Samurai Warrior Queens.

3   Hara Kiri consists of committing suicide by using a short knife "tanto" to disembowel self while an assistant uses a longer sword "katana" for beheading.

Navy ships during WW2. It was the Code of Bushido that a great many soldiers and family had chosen to end their life, by using blades and explosives, or by jumping off the cliff, instead of letting the U.S. Army capture them at the end of the American-Japanese war. The world was wondering then why the Japanese people, civilian included, who would not hesitate to commit suicide rather than surrender to the Americans so they could take advantage of Peace coming to restart a new life. If we had learned more about the Spirit of Bushido, we would see that it was that same Way of the Warrior that had helped Japanese fight with no fear during the war, and that had made them choose to die and preserve their honor rather than accept defeat… Preserving Honor was utmost important, more than Preserving Life, especially when it had to do with country, clan, or family. Dying and self-sacrificing for the Emperor and the "Empire of the Sun" were considered to be as light as a feather. Was that Courage? The fact of the matter was that Dignity and self-respect were the key values that had really mattered. It was a strong unbreakable **Belief! And that belief was somehow connected to Greek Divinity…, this making it even more powerful.**

    In the meantime, in the Western World, the legendary Greek King Leonidas' 300 Spartans had dared confront Persia King Xerxes' 100,000-men strong army in 480BC at the Thermopylae Pass. Even though these 300 guards were later reinforced by more troops and slaves, the mythologized high combative spirit as well as the excellent use of tactical maneuvering and topographic knowledge (location: narrow pass) ought to be recognized as among the most commemorated efforts of human kind to fight for their country sovereignty. The 300 Spartans were defeated, but their sacrifice remained unforgotten as recorded in Greek mythology[4]. The sacred drive that had made the 300 daring enough to battle against the gigantic Persian army was probably propelled by the Spartan's culturally-defined expectation from Men. The Greek rites of passage for all boys required that they be forged into strong, tactically smart and risk-taking warriors, besides being particularly pious in their worship of the Greek Gods. Men were probably not born that courageous; they were trained into an unbreakable Belief to defy death and serve their Gods,

---

4   Historical Myths: The Truth Behind The 300 Who Held Thermopylae and 300: History vs. Myth http://europeanhistory.about.com/od/ancienteurope/ a/hist-myths2.htm , http://clioseye.sfasu.edu/Archives/Main%20Archives/300%20chron.htm

their leaders, and their country… Was it Courage? What was Courage, anyway?

So many lives were sacrificed during the two-century long Crusades (1095AD-1291AD) when generations of soldiers fought against each other from generation to generation to defend their religions. Columns and columns of military personnel dragged their heavy weapons and marched on foot from afar European countries to Jerusalem in response to Pope Urban II's call to aid Christian Byzantine Emperor Alexiox I[5] against Seljuk[6] Turks' invasion (Anatolia). Such was probably how the Crusades had started a lingering conflict between Christianity, Judaism and Islam that had outlasted for an eternity. How could religion convince men of many nations to confront all the hardship and long distance, and put their life and their own descendant's on the line to serve their cause? How could the monarchs from different antagonistic cultures and sovereignties train and convince their people to categorically and uniformly follow the same path of mutual destruction throughout years since the antiquities to the present time? Was it by instilling Courage or group self-determination? Obviously, faith and beliefs played the imminent role in these wars. Unfortunately, the costs of waging war were not just military but always involved defenseless civilian lives that, in the eye of commanders, did not mean much as the end always justified the means. This could be how terrorism against common innocent people had started a very long time ago. Human life, like animal's, was cheap and meant nothing compared to **Group Belief and Cause. Again, we could see that Belief was divine, as crusaders would not hesitate to risk their life for Christianity, and the other fighters for Judaism and Islam.**

The Jews' Saga during the biblical exodus as cited in the Christian Bible, books and stories, as well as shown in one of my favorite grand movies The Ten Commandments (1956), was portrayed so well that it seemed to be engraved in even a Buddhist boy's mind like mine the idea that people should be free even at the price of their life. Charlton Heston was then known by the Vietnamese, many who were Buddhist and/or

---

5   https://en.wikipedia.org/wiki/Crusades the seven Crusades 1095AD-1291AD

6   http://www.fsmitha.com/h3/casia01b.htm Seljuk Turks and the first Crusades

Confucian like me (I was both), as if he was the actual Moses[7] who had led nearly two million refugees (the first political refugees in the world?) out of their natal Egypt in search for Freedom[8]. Yul Brunner played the role of Pharaoh Rameses so authentically I even hated his face. This film hit Sài-gòn right after the big exodus of 1954 from North Việt Nam, when as many as a million refugees had left the North and headed South in their Journey to Freedom (actual name of Geneva Conference evacuation). A great many of them were Catholic from the villages of Bùi Chu and Phát Diệm. The theme of Freedom in the film could be the reason why the movie had become really popular among viewers regardless of what their religion was. I was wondering if it would incite even more refugees to leave North Việt Nam if the film had been released much earlier, and shown in North Vietnam... As the Jews (Israelites) did when Pharaoh Ramesses II (Ramses II) chased them from their homeland in the 1300s BC, heading for the impassable Red Sea to escape for Freedom, most of the Vietnamese Northerners had chosen water escape in 1954, and so did the Southerners 21 years later in 1975 when Vietnam War ended, both via South China Sea... Was it Courage that made them run away for safety? Or was it just fright, then flight to save your life? Or was it a strong Belief in Freedom?

    The 1st Century AD myth of Masada, Israel, had made the world wonder about the 960 of diverse communities of Jews called Zacharias who would rather die in a mass suicide than live and become Roman slaves. Even though identified by Jewish Historian and Roman collaborator Flavius Josephus (author of The Jewish War) as rebels and terrorists (they killed innocent people, including Romans and other Jews because they were oppressed by these groups), and in spite of its controversy with Judaist rules against suicide, besides no evidence was found to prove this myth was true (negative results from archaeological remain searches), this group of Jewish women, men and children, led by Ben Ya'ir[9],

---

7   http://www.bbc.co.uk/religion/religions/judaism/history/moses_1.shtml Life and Times of Moses, https://en.wikipedia.org/wiki/Ramesses_II Pharaoh Ramses II, http://www.huffingtonpost.com/sr-hewitt/the-ten-commandments-movie

8   http://www.jewishhistory.org/the-exodus/ and https://en.wikipedia.org/wiki/The_Ten_Commandments_(1956_film)

9   2015 Smithsonian Channel documentary <u>Siege of Masada and CBS historical movie Dove Keepers based on New York Times best seller Dove Keepers written by Alice Hoffman. Archaeological search revealed only remains of two adults and one child; no other remains of the 960 could be found, assuming they were all cremated according to Roman tradition.</u>

had exemplified the unshakable mind of freedom lovers and oppression fighters. The Masada legend had become Israel symbol of National Unity. This was an example of Courage being used to instill *more Courage...*

More Jews' Saga was portrayed in contemporary history. Jewish Holocaust survivors, after having overcome WW2 Nazi genocidal persecution in their own homeland, gathered themselves from different corners of the World and refugee camps (e.g. Cyprus) and found their way to resettle in Palestine where they proclaimed their State of Israel (Proclamation of Independence) in 1948 in the two-state arrangement by the United Nations. In spite of the Palestinians fighting back, with tremendous united efforts and persistence, the Israeli seemed to have outgrown the former's forces. They had succeeded in building up an army, in spite of a serious lack of weapons at the beginning, from the initial stage of fighting with rudimentary weapons to gradually expanding with better and better equipped units. Small groups of barely armed militants from everywhere in Europe, many not having even time to recover from concentration and refugee camps, had recovered their biblically-proclaimed homeland into a powerful and technologically competitive Israel.

Some clandestine and/or covert American assistance in the 1950s had without any doubt helped strengthen Israeli position[10]; nevertheless, without patriotic self-determination, hope, mutual trust, exemplary sacrifice and unity, the people of Israel would certainly not have succeeded in reclaiming their home. Being united and having an unbreakable Belief had undoubtedly allowed Israel to build up a country capable of defending itself against neighboring countries while surviving an extremely harsh desert and political environment. Most had accepted the same cause, which was fueled by one single Spiritual Belief: to recover biblical ancestral homeland.

While the Jews were reclaiming Biblical Israel in Palestine, the Palestinians felt their very own homeland being invaded. Time went by, and, like a chunk of cheese gradually eaten by a mouse, between 1947 and to the present, more and more Palestinian land was chewed up by Israeli occupation.

---

10   U.S. Army Col. David "Mickey" Marcus, promoted Israel Aluf (general), helped reorganizing Israel Mahal (volunteer) army and initiated U.S. support t Israel in the late 1940s. 1966 motion picture Cast of a Giant Shadow and http://www.zionism-israel.com/bio/Mickey_Marcus.htm Zionism and Israel- Biographies: Mickey Marcus

Small local Palestinian defense groups started popping up in small villages to counteract Israeli expansion, armed with a strong belief that their ancestral land, also biblically-defined, and was being invaded by people from outside Europe. The war against the new Israeli State grew from small skirmishes to bigger local battles and soon spread out, and turned into a perpetual regional international conflict when neighboring Islamic countries got involved to defend each other, in the name of Allah. It was a strategically inevitable explosive situation when three major religions with centuries of history of conflicts were trying to co-exist around the same mount, the Temple Mount in Jerusalem[11]. More so, any efforts from any side to politicize and to make efforts to "appropriate" Jerusalem would ignite this cauldron of fire.

Considering from the major atrocities of Nazi genocide against Judaism, to the political turmoil in the Middle-East between Islam and Judaism, or the disastrous Buddhist Crisis in Việt-nam in the 1960-70s, or the bloody rise of Islamic States of Iraq and Syria (ISIS)…[12] all of which being more and less religiously instigated and manipulating the people using religion, one can see that religious beliefs, if mixed with politics, can lead people into repression or senseless bloodshed. Chaotic situation usually creates havoc among the people who do not have high critical thinking skills but have to rely on emotional judgments to interact (Reptilian Brain Theory to be discussed later – Daniel Goleman). That is the only reason why advanced political systems avoid mixing Religions and Politics, as the former, the Divine Determination, interferes in the latter process, the Human Interaction. That is also why Communism chooses to eradicate religions (the mind control), as the Party knows they are the "People's Opium" that corrupts them and intervenes with their "desired" politicization (the other "Mind Control): only the supreme Party leader, not any God or Buddha, can have total control of the people. On the other hand, as I will further discuss in chapters to come, the Communist Party has applied reverse psychology, and take advantage of religions to control the people. I do believe the lesson learned in Vietnam War, if one is conscious about learning a good lesson, should tell us to be careful with the leaders, or by the same token the followers,

---

11   PBS Rick Steves The Holy Land: Israelis and Palestinians Today. March 2017.
12   https://www.ibtimes.com/isil-isis-islamic-state-daesh-whats-difference-2187131

who can artfully instigate their people to manipulate them for the sake of mind controlling. Downstream in history, we will see those politicians in action causing havoc and never-ending war, while fanning hatred to maintain followers, so that the flame of the war will keep bursting. Of course, the followers are so emotionally encroached they can never see get out of the vicious circle of mind control (to be further discussed in chapters to come).

Was there any political connection between the Middle East and Việt-nam?

Politically worldwide, Israel occupying Palestine had become more and more a serious matter, causing the rebellion of local inhabitants and the uprising of the Palestinian Liberation Organization (PLO) in 1964. Perhaps, more or less, the PLO had been influenced by many colonized countries' efforts in regaining their independence, such as the Vietnamese experiences against China, France, and Japan. Many Vietnamese political analysts in the 1960s believed Hồ Chí Minh's efforts in defeating the French in 1954 and in "fighting Americans" in the 1960s (National Liberation Front of Vietnam or NLF) had greatly influenced Yasser Arafat in forming the PLO[13] and fighting against Israeli's occupation and for Palestine Independence.

I kept wondering for years how the Palestinian people felt about having to give up their land to occupation, while Indochina, Africa and India were fighting to successfully take their land back. A two-state solution (which made sense as both Palestinian and Israeli had originally lived in Palestine, along with the Christians, of course) would definitely not work any longer when one country overpowered the other one... Why did the PLO movement fail? Was it because PLO did not have enough support from any other world power and neighboring countries that shared the same belief? Was it because Israel was supported by the U.S.? Was it because the Arab block was not united? ... because the PLO belief in Independence was not strong enough? ... because it lacked trust in the U.S. facilitating Peace and believing that Israel would not make Peace? Probably, with bitterness, hopelessness and anger, the Palestinians had to watch powerlessly the Palestine map shrink

---

13 https://en.wikipedia.org/wiki/ Palestine–Vietnam_relations

more and more between 1947 and the 2010s. In the similar way, France had been eating up Cambodia, Việt Nam and Laos like cancer for a hundred years. How could France, which was much better-equipped and well-backed by the U.S.[14], be defeated in 1954 by the ill-supported Việt Minh, more or less single-handed?

My father had often talked with me (even though I was too young and had no clue about what he was saying) about Yasser Arafat's Liberation efforts, besides, other international celebrities, such as Fidel Castro's revolution in Cuba and fighting against the U.S., or Egyptian Pres. Gamal Abdel Nasser's dilemma in the Six Day War against Israel in 1967. He had been following up on news updates by secretly listening to Hà-nội short wave radio broadcasting or reading Vietnamese newspaper international reports. He seemed to have a discrete admiration for both Independence fighters, Hồ Chí Minh against the French and Arafat against the Israeli. Inspired by Hồ, Arafat had visited Communist Việt Nam at least 10 times. Hồ Chí Minh might have sided more with him as they had a common goal, waging unconventional war to regain their beloved Ancestral Land from the invaders. North Việt Nam was one of the first countries that had long recognized the Palestine Liberation Organization (PLO in 1976, then the Palestine State in 1988), while it did not establish relations with Israel until 1993[15].

In some way, both the biblical plight of the Jews taking Palestine back, and the contemporary struggle of the Palestinian people trying to regain their historically-proven ownership of the ancestral land, were similar to the Vietnamese case which, for over a thousand years, had lost their ancestral land and had been trying, at any cost, to get it back. "Patria above all" had been their motto. Patriotism places Ancestral Land above All: above self, above family, and above death, indeed! At any cost, using whatever resources available, either through world networking or political lobbying, the Jews had more or less successfully fought for Israel, since the 1948 rebirth of Israel, to the 1967 Six-Day War then afterwards, the 1973 Yom Kippur War, and so on, just like a "spreading drop of oil" (as the Vietnamese would say, "vết dầu loang").

Israel, as it had become more and more an ally of the

---

14   The U.S. gave France at least 1 Million dollars in military aid for fighting the Việt Minh in the 1950s.

15   https://en.wikipedia.org/wiki/ Israel–Vietnam_relations

U.S., might have been known more to the South Vietnamese (nationalist) press media than the PLO. Later on, Israeli Gen. Moshe Dayan, then a colonel, took a special trip to Việt Nam to study the guerilla war: it had made some politically informed South Vietnamese side more with the Jews because of their tenacity when they had to sacrifice all and stayed focused on their goal of reclaiming their biblical long lost home. What had helped the Jews stay unbroken and pursue their ultimate final goal of reclaiming Israel? Was it Courage or religious drive (Belief)? Or just plain Self-Determination and Patriotism! What had helped them face death with no fear while confronting opposing countries that surrounded them? Had their afterlife belief in paradise, reincarnation[16] (or no belief at all), as some sources of information had described, influenced their action? Why did the PLO fail to secure Palestine for the Palestinians[17]? Why did the neighboring countries not intervene in the Two-State efforts? It would be hard for an outside observer like me and, I am sure, a great many other researchers to understand the intricate intertwining complication of differences in beliefs that lead to different outcomes in history…

…Like in a high school group conflict, more and more friends coming and joining in a conflict to support friends always makes the conflict grow bigger and complicated, to the extent problem solving (or peace negotiating) and diplomacy might not resolve anything. The war has grown from PLO fighting Israel, to a multi-national Islamic conflict against Israel ... and, maybe by extension, hatred against the U.S., whether the Americans want it or not, because of its covert intervention in support for Israel. As an outside observer who have somewhat experienced it, or, at least, who have grown up in a similar conflict, I believe each side is seeing the other side occupying its territory. The U.S. has been viewed by many European and Middle Eastern countries as a strong supporter for the State of Israel. Is this the reason why many Middle Easterners hate the American governments?

---

16   https://uk.answers.yahoo.com/question What do Jews believe happens after death? Kabalistic or Hassidic Jews believe in Reincarnation, while Orthodox and Messianic Jews in Gan Eden - Paradise

17   Comparative maps of Palestine between 1947, 1967, 1973 and 2005 or later showed Palestine being eaten up as if by terminal cancers. A typical map is Palestinian Lost of Land 1946-2005 http://www.whatreallyhappened.com/WRHAR-TICLES/mapstellstory.html

Killing is purposeful

Most recently, the Jihadist movement, and especially its massacring innocent bystanders all over the world, had woken and shaken up the whole world and made people wonder why and how human beings could resort to such atrocity. How could radical Islamic terrorists convince young Middle Easterners (or even Westerners) to join in and "punish non-believers." Most of the victims were *defenseless and innocent civilians and children, Muslims included, oftentimes executed when they were tied up and folded up at their knees,* in the most gruesome and inhumane way... How could they convince terrorists to strap bomb vests around their torso and be ready to do the unbelievable things such as murdering innocent defenseless people, or, if needed, killing themselves? Was the purpose to punish those whom they considered as enemy or intimidate the survivors?

It was all about the eye of the beholder: what the world saw as innocent civilians could somehow be actually seen by perpetrators as the guiltiest of all. Not like some other killers who actually believed that the ones they were going to kill would have a chance to reincarnate for a better life (Hindu, Buddhist...) or to go to Hell/Heaven (Christian...)..., these killers did not even care what would happen to those whom they killed in their afterlife. They might even enjoy the idea that their victims would go to hell... Definitely, we could conclude that, what the executioners wanted was, not just to punish the victims, but also to inflict horrific images, fear or pain among the survivors. On the one hand, they killed thinking they were protecting their beliefs and punishing the non-believers or "infidels". On the other hand, they strongly believed they would be rewarded with the ultimate Martyrdom as well as be with Allah if they were to be killed. In an ego-centric sense, self-sacrificing for their religion would actually elevate them to Heaven. Such would be the benefit. Was it Courage? Or was it a strong Belief that fulfilled its sole purpose of intimidating the world, as seen as corrupted? A Belief impregnated with religious interpretations[18].

At the beginning, I thought Jihadist movement was to respond to Palestinian people being oppressed by Judaist Israel and the Christian West (especially the U.S., United Kingdom, France and Germany). I gradually realized it had to

---
18   Interpretations can be re-interpreted to accommodate desired effects.

*do with something very different that I and, perhaps, a great many others could not understand, no matter what sort of critical thinking skills and knowledge of socio-political cultural psychology and methods we could come up with to analyze it. Even more confusing and inconceivable, the majority of the victims had been Muslim people who were minding their own daily business, the "innocent" as seen by Westerners. Cable News Network CNN television host Fareed Zakaria was trying to shed some lights to this matter in his May 2016 analytic television show "Why do they hate us?" Definitely, we saw nothing proving that the Jihadists, Al Qaeda, Taliban or Islam State ISIL were avenging Israel's oppressing the Palestinians: so far, those groups had not done much to attack any single Israeli site. Not one single time, even though Israel was just across the border from the major turmoil zone! On the contrary, they had been destroying their own world-venerated Islamic sites (because they did not believe in worshipping statues?) and annihilating more Muslim people than anyone else. Could a **belief** lead people to commit unacceptable crime? Mr. Zakaria pointed out that many Muslim radicals and Islamic State of Iraq and the Levant ISIL[19] militants were usually not religious themselves, but rather disenfranchised or dysfunctional members of various societies. He concluded that the converted terrorists had joined because radicalization gave them a sense of purpose and accomplishment. This was also a matter of extremely strong Belief that drove actions.*

Along the same line of thought, people concerned about terrorism had been wondering: "why were non Islam-related people also committing mass massacres, this being expanded almost world-wide?" Could non-political terrorism in the U.S. be viewed in the same fashion: non-sense indiscriminate mass shooting or massacre of innocent people in churches, movie theaters, schools or at entertainment events had been occurring there! Whether by radicalization or because of mental health problems or repressed racial supremacy, they were all murdering innocent and defenseless people. The truth of the matter was that, even though the world did not pay much attention to it (or that, for some reason, it was not exposed by the press media, except for the case of Mỹ Lai), the same horrendous act had spread out widely in Việt-nam during the

---

19   https://en.wikipedia.org/wiki/Islamic_State_of_Iraq_and_the_Levant , Islamic State of Iraq and the Levant ISIL (in Arabic DAESH) is the original of ISIS, a political and military ideology originated from Iraq and expanding into Syria in the mid-2010s towards the end of the American invasion of this country. ISIL/ ISIS organization is classified by the UN as terrorist.

war, mainly committed by the Việt-cộng VC (Southerners converted to Communism - NLF[20]) and NVA North Vietnam Army (main force "bộ đội", also PAVN People's Army of Vietnam) or Khmer Rouge in Kampuchea against hundreds of thousands of innocent at a time (adding up to millions). These defenseless victims were of all ages and both genders. These massacres were not committed because of religious beliefs, or insanity, or racial discrimination. Masses of defenseless people, little children or elders, male or female, some were still sleeping, others were tied up like sausages, or bundled up with telephone cords like grapes, begging for mercy… were exploded with missiles/ explosives, or had their skull cracked open with steel shovels or bats then thrown into mass graves. These were Vietnamese-on-Vietnamese or Khmer-on-Khmer genocides!

    These attacks were absolutely not acts of courage, but were rather acts of cowardness and inhumanity using weapons ranging from old-technology machetes to modern-time mortar shells, B40 RPG's, grenades, Surface to surface missiles or plain undetectable[21] improvised explosives devices. The executioners' job was as easy as killing a helpless hen or a fish immobilized on a chopping board: victims were tied up and had no ability to defend themselves. Their course of action had to do with either substantial brainwashing and mind control, or simply insanity. They killed like a mindless machine, with an unshakable Belief it would be right with the Communist Party or proletarian equality to do so; besides, as proven by history of the rise of Communism, it was acceptable and encouraged during the Russian Bolshevik or Chinese revolutions that millions be got rid of in the name of the new order. Massacres were purposeful! Purge! Exemplary sacrifice!

    What mattered to the terrorists were the effects their actions could cause by killing innocent civilians: to inflict terror with the intention to pressure and drive others into submission. Killing others would be, not really to get rid of them, or just to punish the individuals for opposing them, but rather to use death as a strict warning and a serious threat to the rest of the living ones. The more gruesome death appeared, the more terror would be inflicted onto the living. For that reason,

---

20   Many RVN strongly believe the VC (National Liberation Front) are bloody-cruel than the NVA. VC usually seek out victims for revenge and terroristic punishment. They do not operate in larger units as the NVAs do.

21   They were made of natural non-metallic elements that cannot be detected by mine detectors.

the Jihadists would not hesitate to behead[22], bury alive[23], hang to dry[24], dismember, disembowel, eat victims organs… these would create horrendous visuals (snapshots) that would terrify their enemies, and, for sure, uplift their supporters (in Vietnamese, make them thirsty for blood "làm cho chúng hăng máu"). It had become, not just a physical attack on the individual victims, but also a mental assault on the surviving mass. Believe it or not, the VC[25] had done the same to innocent civilians and Republic of Việt-nam RVN (also South Vietnam) partisans, of their own race and creed, even of their own family.

In order to understand more deeply how death could impact the killers as well as the victims, we would need to look more carefully into the spiritual belief systems. Ideologies did not just get indulged by words of mouth or examples; they would be more easily indoctrinated when they were based on the faith of the believers.

Horrendous death shapes the "Conscience of America"

It would be hard for anyone to imagine how people could bear the image of innocent civilians being tied up like sausages, kneeling on their knees and bending their head down awaiting a sharp or hard object to strike them and end their life. It would be even harder to understand how the executioners could perform such an act of taking defenseless people's life, most of whom were elders, women and children. This had happened in European countries during medieval time, in China under imperial dynasties, in Japan under Shogun's rule, in the Middle East countries under different rulers as well as terroristic oppression, in Việt-nam during the communist aggression. These group massacres had been used for political and warfare purposes.

---

22   Note that beheading is a very common capital punishment method in many traditional countries, as dated back to the medieval eras. It is a horrendous image to Westerners because they are not used to this idea any longer.
23   Tết Offensive of 1968
24   U.S.A lynching
25   Many of us ARVN soldiers believed that VC and small NVA units were much more inhumane than the main corps NVA (divisions) who abode more by conventional war and had more professional leadership.

What had been puzzling me the most was why American people were so upset about the 1969 massacre by American troops of the 504 villagers in a small village of Mỹ Lai (aftermath of 1968 Tết Offensive), when they had totally ignored the 7,800 South Vietnamese civilians massacred in just one city of Huế by the communist during the Tết Offensive. Much more had been killed indiscriminately on a daily basis throughout the war in terroristic attacks and yet, such slaughters were not even mentioned on the television, news or English internet. So was the fate of hundreds of thousands more who died in re-education camps or at sea (boat people), silently and unnoticed. The three million innocent Khmers massacred in a genocidal killing spree in Kampuchea also went silent. Why did they pay more attention only to the victims killed by American troops and their allies ARVN, and not the multitude of the others by the enemy? Or was it because this news had all been hidden or ignored by the U.S. press media?

Bloody visuals[26] as such had actually affected hundreds of millions of television viewers when the latter could see with their own eyes Vietnam War victim's cadavers left out to dry. When they linked those snapshots they could see, and even deeply feel, from newspapers or television to the images of American soldiers burning huts or bodies of babies and women or elders, they would lose their mind imagining their own husband or sons having committed inacceptable crime against humanity. Would they go even more "berserk" if they were told there had been much more, a great many more innocent people killed by the communists? Nevertheless, their self-inflicted guilt would take them to the conclusion that, if their husband and sons had not been sent to Việt-nam to wage war, those massacred by the communist would not have been killed. Either way, mea culpa was probably the most common reaction that American people would claim taking the blame on self and one's own, rather than blaming it on the enemy. It did not help when they were unfamiliar with what was going in Việt-nam, and uninformed about the history and culture of Việt-nam.

What would Americans think if the newsreels of the old wars (e.g. Holocaust victims and military dead in WW2, or soldiers' bodies in WW1 trenches…) had shown them

---

26   As portrayed by iconic photos of burned down huts, Mỹ Lai massacre babies, the "Napalm" baby girl, the execution of VC Lém by an RVN Police Colonel, Mỹ Lai massacre… These snapshot photos, with limited, if none, accompanying information have mislead viewers with sensitive heart into explosive reactions against the war

soldiers' dead bodies, or, even worse, piles of corpses of mostly civilians? Would they react against the U.S. involvement in those wars? The non-governmental news media were probably not allowed to see and film those atrocious pictures to show to viewers, and, consequentially, the citizens back home could not react against it. Nevertheless, it was totally different during the Vietnam War. Reporters from hundreds of newsprint and television networks were allowed to embed with the American fighting units and to report what they had witnessed directly back to their agencies without going through any censorship (freedom of press).

People have learned from past experience that unverified and misleading news, "fake news," have caused, not just public confusion, but catastrophic panicking and disturbance (civil disobedience) that have led to major chaos. More and more people want the truth and the social-media is proving that such is a fashionable trend among the younger generation. As a result, in order to attract readers and viewers, network companies that want to advance are also "digging out" more and more truth, as it pertains to Democracy and Freedom of Press. Of course, the truth can be deceiving! Nowadays, most television news media networks are relying on fact-checking technology and methods to verify their sources. This is not true during the Vietnam War. More on discrepancies among news media coverage will be discussed later.

While visiting his troops in Việt-nam in the late 1960s, U.S. supreme commander Gen. Westmoreland had commented that "Asian lives were cheap, really cheap…" perhaps as a heads-up to his troops. Obviously, coming from peaceful America, they had never witnessed such carnage as they could now in Nam. The general had actually admitted how easily Vietnamese people's life had been wasted, in mass, everyday anywhere in the South. Life really meant nothing! The Tết Offensive had exemplified what the general said: To kill or be killed happened as if it was just a new normal lifestyle one had to accept. We would find out why really soon.

Not like those who grew up with the concepts of Karma and Destiny, the Westerners who were not impregnated with those concepts felt totally differently about Death. Compassion and humanity put the latter in a mold that would restrict possibilities that only Asian combatants in the Vietnam War could afford to have: take lives for the main cause, if needed, to win. The truth of the matter was that the U.S. government had strictly demanded that RVN fight the war the American democratic way in exchange for U.S. military aids. It had banned the RVN from using oppressive measures (execution, torture…) against the enemy and collaborators, or the sympathizers who would most likely turn communist because of NVA ruthless blackmailing and oppression.

It appears that there is always a conditional status when one needed to receive aids. I am wondering whether the present Democratic Republic of Vietnam DRV is in the same situation… Is it true Hồ Chí Minh and Phạm Văn Đồng have to agree giving up the two islands Spratly and Paracel in the South China Sea to China in exchange for military aids as it is rumored? That might be the reason why Việt-nam dares not take China out to La Hague International Court as the Philippines have to dispute its invasion of these islands. The following archival photo of the letter September 14, 1958 sent by DRV Premier Phạm Văn Đồng to China Chou En-lai confirming the current 1958 agreement by Hồ's government on border between China and North Vietnam. 2010s activists allege that it confirms the two islands have been relinquished to China. It has to be noted that the islands are under the Republic of Vietnam RVN control until Sài-gòn falls on April 30, 1975.

Interestingly, as admitted by communist sympathizers, collaborators and VC or NVA, the most successful combat units were the Republic of Korea ROK divisions: their soldiers, besides being effective and brave, would not hesitate to use the Asian way (execution of rebels as I have already discussed, and now applied even with those suspected for collaborating with the enemy). The very simple reason was that the ROK army was not subject to American rules, consequently, ROK troops were allowed to use the Asian way, which could be considered as cruel by the American liberal standards! In the meantime, the enemy was able to achieve their superiority over the Vietnamese nationalists as they could use those atrocious measures (mainly torture or even execution/ massacre) to force their "persons of interest" or their beloved ones to cooperate with them like docile sheep against the RVN. Any means could be used to serve the ultimate end goal: use the civilians to overwhelm the nationalists (to be discussed further in later chapters).

When growing up, I had heard people mention the word "con thiêu thân" referring to little insects that kept flying into hot light bulbs in bundles and fell to their death being roasted by the heat. Couldn't they realize that those alluring bulbs were the cause of their friends', family members' and their own death? They did not mind and continued to head for that mysterious murderous source of energy. The "con thiêu thân" actually made me think of those multitude of Vietnamese, whether they were friends or foes, armed or unarmed, purposeful or inadvertent killers, who would day after day head straight towards their own death, year after year. Common people would not really understand why they did that, and neither would the Vietnamese. Was it because of their spiritual belief, their peer pressure, their surrendering attitude towards destiny, their courage or foolishness? With those questions in mind, let's explore more how they conceived Death.

CHAPTER 2
Fusing Beliefs into the Death Pedagogy
(the key concept words will be capitalized)

By examining Death, one can understand why combatants have no fear when confronting their own Death, or any hesitance when killing others. To counter balance the horror caused by Death, glimpses at life after-death will lighten up the fear of being killed while fighting in the war or the hesitation of killing others to make examples. Combining notions of death with Destiny and Karma gives oneself a self-consolation when thinking about one's own death, or motivates self to kill as one has been reassured that the victims will be given a better chance in their next life, and that it is their Destiny that has determined it is time for them to be reincarnated. Like learning about Death, learning about what happens after death varies depending on individual and group beliefs.

Buddhism and Hinduism: Karma[27] and Reincarnation

These two major religions teach their followers the fundamentals of reincarnation which make them believe that after Death, they will come back to life in a different form depending on how they have lived their previous life. Usually, the expectations are that they would fulfill good citizenship prior to death in order to be reincarnated in a better form of life, such as another human in different social classes, hopefully higher, or an animal or plant with different conditions and lifespan if they have not cumulated enough good deeds... Karma is then instituted as the cause and effect mechanism that will govern how the individuals ought to behave if they wish to gain a rewarding future in the near (close future) or far future (next life).

Normally, Christians might be very concerned about

---

27   Karma is defined in http://hinduism.about.com/od/basics/a/karma.htm by aboutReligion, http://berkleycenter.georgetown.edu/essays/karma-hinduism by Georgetown University, and by Wikipedia https://en.wikipedia.org/wiki/Taoism and https://en.wikipedia.org/wiki/Karma_in_Buddhism

Hell or Paradise when thinking about possibility of dying; however, interestingly enough, non-Christian Vietnamese people do not seem to worry much about Hell, even though Hell, as it has vastly appeared in a great many myths as reflected by popular stories, including children's ones, is described to be incredibly horrendous. The most common version of Hell based on the Vietnamese Buddhist belief seems to differ much from the original one Naraka as described in Buddhism/ Sikhism/ Jainism[28]. These myths might have been infused in other non-Christian ones in Việt Nam and are used to educate citizens to be honest, kind and have servitude towards others. It is believed that all souls, whether they are sinful or not, will have to go through ten Gates of Hell, each governed by a separate King called "Diêm Vương[29]," before they are allowed to be reincarnated. The soul of the newly dead person is to be escorted by horse and buffalo-faced guards to the King for judgments and redemption. Each of the unacceptable and inhumane sins committed during her/his lifetime will determine how long the soul will stay at that gate, and how it will be punished: be deep fried in a boiling oil vat, attached to a grilling tube, torn up and eaten by wild animals, repeatedly disemboweled, and skinned, or swallowing molted copper over and over until all sins are dissolved. Punishment may be a combination of the sentences above. Good souls usually clear through Hell fast, oftentimes without sentencing. Upon completion, the souls are escorted to a non-ending river where they are all served a bowl of "forgetful" porridge "cháo lú" before they reincarnate. This porridge will make them forget all of their past so they will be reborn as a new "something" based on their Karma from their recent life.

Can it be possible that the belief that everyone has to go to Hell makes the non-Christian Vietnamese not much concerned about Hell? Everyone has to go to Hell, so why bother? They will just need to behave and do good deeds in their current life so they can avoid suffering in Hell and have a good Karma for their next life! Such is the way Hinduism and Buddhism teach disciples to be good exemplary citizens and not to fear Death...

This deep behavior-realigning process in reincarnation and causality-infused indoctrination, in my opinion as a

---

28   http://buddhism.about.com/od/thesixrealms/fl/The-Buddhist-Hell-Realms.htm Original Buddhist Hell

29   https://vi.wikipedia.org/wiki/Diêm_vương in Vietnamese: Vietnamese Buddhist Hell

professional educator with expanded experience in cross-cultural sociology and psychology, has throughout centuries shaped many docile Confucian[30], pious Hindu/ Buddhist-influenced[31] Vietnamese communities to abide by societal requirements, and therefore, by expansion, by diverse political expectations. Political or military torture might be therefore interpreted as "pre-hell" redemption, therefore "bearable," and Karma can be used as causality in the political life... Many of these ideologies have become traditionalized into secular societal customs as they are observed even by non-Buddhist/ Hindu people.

    While Buddhism, the dominant established religion in Việt Nam, has not encouraged self-sacrificing for others' wellbeing, a few monks have self-immolated to protest for Peace or against the RVN government in an utmost level of self-sacrifice for religious rights. Martyrdom, as far as I know, has never been encouraged or condoned in Buddhism either, as committing suicide might be judged as "Karma avoidance."

    The Buddhist main goal has always been to follow Buddha's teaching[32] and self-correction in order to perpetuate the Buddha's beliefs of Dharma in order to have a beneficial Karma. The ultimate goal is to escape the Reincarnation cycle and achieve Nirvana eternally; nevertheless, this goal is too far to reach and seems not to help the combatants much, in my opinion. It is mainly the possibility of being reincarnated into a better life that has helped soldiers be at peace with their possibly fatal Destiny. I know that is my case and my Buddhist men's in my platoon: we all hope, by serving our Ancestral Land and, in case we have to sacrifice, our reward will be to reincarnate as a person with a better status. We are also certain our family will be at peace with this arrangement. That is why, whenever mourning for a soul that has just left Earth, we

---

30   Confucianism, observed by a vast majority of Vietnamese regardless of other religions, is a philosophy of Life, not a religion, teaching citizens to become compliant and conform to the law/ order and obligations of the society

31   Including the 12-18.4% Buddhism, 6% Hoà Hảo and Cao Đài which are Buddhist-influenced religions. The other groups are 70-75% folk religion family worshiping and Paganism, and 1.5-2% Christianism. Adjusted from https://en.wikipedia.org/wiki/Religion_in_Vietnam#History

32   About Religion: Buddhism.about.com. Buddha, "awaken one" in ancient Sanskrit language, in expanded definition, refers to a person who has attained enlightenment which release him/her from the cycle of reincarnation and all sufferings, desires,... Buddha then is seen as a teacher for all Buddhist believers. Siddartha Gautama was the founder of Buddhism and the first Buddha. In some temples, there can be many statues of Buddha's, including male and female, but the main one is always Siddartha Gautama sitting in the main chamber alter.

always pray "I pray that his/her soul will soon float towards the serene Peace of the West" ("Xin Cầu cho linh hồn người sớm phiêu diêu cõi Tây Phương cực lạc[33]")

Popular belief in Destiny

Another aspect of superstition has occupied most Vietnamese's mind during the war: their Belief in Destiny. This belief has become somewhat pagan (regardless of religions) as it consoles the person who is about to commit his/her life to a dangerous political or military cause, or who has failed in some aspect of life (business, relationship...) that s/he has no choice or way to avoid the outcome of events. It has been predetermined! It encourages the soldiers to move forward at the front line even at the cost of their tragic sacrifice. Fate has already been determined[34], so there is no way to resist or avoid it. They have to accept their Fate and will execute an assigned mission no matter how the outcome might turn out. There is no question to ask and there should be no hesitation! Their Fate has already been set up as soon as they come to life. Is that Courage? Or is it just a total surrendering and submission? Definitely, it is based on a very strong Belief, the one that is rendered divine, whether it is rational or not. They have to let go of everything, their loved ones, their properties, their mental assets..., do their best at the battle field and let Destiny decide? No matter what happens, they are content.

It has become a normal lifestyle for people, younger or older, to consult astrologists or to read the horoscope predictions, especially during New Year festivities, as they are anxious to know what the future has lined up for them. When are they going to start a family? Have children? How is Health? How is Wealth? Will they have Longevity? What kind of death are they subject to? Will there be loved ones at their death bed or funeral? Therefore, New Year or wedding wishes must always reflect these concerns: longevity, health, wealth, lineage... This lifestyle has become a routine for the atheist as well as the religious ones, as it is part of the ancestral culture. The future might be dim, but a quick peek at it, whether it is accurate or not, still gives them some reassurance

---

33   YouTube in Vietnamese Diệu Âm Cõi Tây Phương Cực Lạc Funeral Guestbook http://www.legacy.com/guestbook/DignityMemorial/guestbook
34   Our non-Christian Asian allies might share these same beliefs.

as they somehow have an acknowledgement of their Destiny. Predicting the future might have originated from Hindu-Buddhist Karma-influenced or even pagan shamanistic beliefs: I notice that most amulets, including the ones granted to me during my boot camp time, are issued by Theravada[35] monks and shamans. Its influence has been so deeply rooted in peasants' lifestyles that it has subtly infiltrated paganism as well as religious sects. Theoretically, Karma can be adjusted by improving behavior in order to be awarded with a good remaining life. Destiny cannot be altered, as it has already been determined. Consequently, destiny is referred to when individuals need to make peace with possible catastrophic results of events, e.g. permanent loss of dear ones.

>...I had almost forgotten that my father had my astrology chart drawn when I was a young kid. I suffered for a long time from a strange spleen disease, to the extent my skin had turned literarily "as yellow as turmeric." As I had mentioned in Peace or Freedom – Part 1: My Memoir, I was so sick that French medicine could not even cure. This chart had laid out my whole future and gave me a glimpse at the path I would go through: my Destiny. It said that I would not die of a disease when a kid, but that I would lose everything I owned, and die away from Ancestral Land, in a foreign country far away. It said:

>"Tiến vi Quan, Thối vi Sư,
>Một tiếng hô, Vạn người hưởng ứng"
>...meaning:
>"If I advanced in career, I would become a "mandarin" (either civilian leader or military commander), but if I retreated, I would become a teacher, One holler from me, ten thousand would respond."

>As of now, most of these predictions had come true: I lost everything I owned in Việt Nam, and came to the U.S. to resettle. I was a military leader there, and when I came to the U.S. after the war, I became a teacher then school

---

[35]   Small Vehicle branch originating from Ceylon, Thailand, Laos, Cambodia, Myanmar https://www.accesstoinsight.org/theravada.html Theravada monks wear saffron, maroon and/or yellow robes and are not allowed to cook, but receive food of any style of cooking from believers. They do not have to be vegan vegetarian as the monks of the Big Vehicle (China, Japan…) who wear grey robes.

administrator. A great many students/ parents and staff worked well with me... Could it be that these uplifting predictions had actually foretold my future, or had they given me a guiding path to follow (made me work for it?) to achieve the success I now had? Or was it just a coincidence my life had shaped exactly to that prediction? There would be no way to tell!

    At some point after my escape from Việt Nam and my successful resettlement in Minnesota, when considering how successful my career had been, I realized that the tremendous efforts I had put into high achievement in community development and educational, as well as professional advancement was not really by relying on my predicted Destiny. I did work very hard for my success! It was not a totally blind belief in fate that had encouraged me to move on. The two essential "talismans" I had relied on the most were the untold[36] expectations from my parents and my Strong Belief in Đức Tả Quân Lê văn Duyệt's Divine Intervention and Benediction on my behalf. He was a sanctified general whose temple I was taken to for spiritual adoption when I was gravely ill. It was my strong and unbroken will that had made me flee from Việt-nam, made me keep trying consistently, urged me to move on, to study on, to do my best at what I was doing, helping others, and to achieve the top I could ever do. "If you think and tell yourself you can, you can" had always been my motto. And, yes, definitely, it was strong 9-shot a day of good espresso coffee that I had been drinking during my career ascension in Minnesota. If I had to repeat this incredible path, I believed I might not be able to.

    For most Vietnamese, this second belief can be as important as Karma, if not more, as it can reach beyond Buddhist and Hindu Karma to influence other believers. It has dominated among many Vietnamese, mostly of the Buddhist and Pagan/ Folk-belief groups, and probably a great many others including Christian: Destiny or Fate. Didn't the French have a song by the name of "Qu'est-ce sera, sera" (What will be, will be). It is described by a great many terms in Vietnamese "số kiếp, số phận, duyên nợ, duyên kiếp, số mệnh, duyên phận, duyên số," as well as derivative phrases such as

---

[36] As I had said before, they never told me what they had expected me to do or become; it was mostly the children stories they bought for me that had engraved in my brain they might like to see me achieve high, in both military training and education areas as described in the Vietnamese Confucian rhetoric: "Văn Võ Song Toàn."

"an phận, an phận thủ thừa, referring to submission to fate, or số phận hẩm hiu, số kiếp bẻ bang," consoling oneself of one's poor destiny that we have to abide by, "chấp nhận duyên phận phủ phàng," accept our fate no matter how disappointing it can be. This rich vocabulary just for one notion of destiny shows how deeply it is engraved in the Vietnamese mentality of compliance to their fate. Destiny may be considered, in one way or another, unmovable, and, if so, the people will have no choice but to follow the path and submit to it. In order to cope with this self-indoctrinating belief, the Vietnamese people must tame themselves to accept whatever has been stored for them, and not to resist or question it. This attitude might have made a great many Vietnamese even more obedient and submissive to whoever has taken control of the situation (such as authority figures) as well as governed their life. It also gives the one with authorities the power over the followers, causing the latter to become disempowered and even more submissive.

Effects of Karma on politics

This might also be the reason why horoscope, astrology, prediction, fortune telling, and even prophecies have played an important role in all aspects of in the Vietnamese daily life, from individual to group/ family decision, in business or politics. People want to know, even if it might not come true or if it involves tragedies or atrocities, what can be in the offing in their future, and might try to find ways to mildly circumvent predictions (by praying, changing lifestyles to prepare for better Karma, converting to a new belief…) or at least to soften the impact in case they cannot possibly avoid any mishappening. Eventually, they will comply with their fate when they cannot make changes, or, if they are able to deflect from their predictions, they will say their destiny has determined such a new course of action.

Due to this submissive "accept-fate" mentality, the common Vietnamese tend to chose to maintain a status quo and are not usually daring enough to stand up to take the leadership initiatives and instill systemic changes (e.g. rebel against an oppressive establishment). The most they dare do is to choose to be a follower and to pick some entity that shows potential power and leadership. This attitude by a great many followers might have given ideas to many power-hungry individuals to show off that they have leadership potentials and take advantage of the peasants' naivety. This might explain why Việt Nam, from the first days of Independence from the French through both Republics, has been swarmed with conflicting rebel-lords and "wannabe's" who have on many occasions incited civil disobedience and political as well as religious uprising. Non-stop "coups d'état" launched by insatiable rival generals between 1954 (against PM and 1st Republic Pres. Ngô Đình Diệm) and 1967 (against 2nd Republic Pres. Nguyễn văn Thiệu[37]) have caused distrust among religious, military and political leaders. This significantly unstable political turmoil has especially made the American government (Presidents Kennedy, Johnson, and Nixon[38] followed by Sec. McNamara's substantial reports) strongly believe the U.S. cannot support the South and precipitated an urge to withdraw troops. This topic will be discussed in a later chapter[39].

Besides trying to pacify the country and fight against Communism, true leaders have to battle hard against the "wannabe's" who are doing their best to polarize and divide up the docile Destiny-dependent population[40]. It takes these true leaders a great deal of tolerance, inquisitive and reactive knowledge, resourceful skills and unconventional out-of-the-box thinking and adventurism in order to free oneself from one's own "colonized mentality" and to lead the submissive people into self-determination. Perhaps these leaders have no choice when they must resort to using intense pressure and possible oppressive measures on the compromised population that is getting out of control… This will certainly contradict the American democratic policy imposed on the RVN (against what is perceived as dictatorship).

The question still remains: in a confused and confusing socio-political arena (a country that has just received the Independence it has never had before), how do people know who a true leader is, especially when "everyone" thinks s/he is a better leader? What do they know about their newly acquired Freedom and Democracy? Tribalism in a post-colonial and imperialistic context made it absolutely confusing for both

---

37   There have been "coups" since the South became the State of Vietnam until the end of the first republic, when Thiệu became president (1st term).
38   Sec. Robert McNamara
39   Section The Political Divide in Chapter 10.
40   During my forty some years living in the U.S., I have not heard or witnessed Destiny playing such an important role in people's life as much as among the Vietnamese people.

the leader "wannabe's" and the submissive semi illiterate followers. They are definitely lost in the jungle of their new Independence where choices/ responsibilities versus power/ benefits are confusing them. Probably fate gives the common people a hope they are following the right leader? The worst thing they can do, if they have chosen the wrong leader, is to blame their mistakes of Fate…

> "*Mã đề dương cước anh hùng tận, Niên (Thân) dậu đương lai kiến thái bình*[41]," *predicting that, from the beginning of the year of the Horse until end of the succeeding year of the Goat, all heroes would die, then that the years of the Monkey and Rooster would bring back Peace*… was a well-known and very popular prophecy by veneered scholar-prophet Trạng Trình Nguyễn Bỉnh Khiêm[42], a high-ranking wise mandarin during the unstable Vietnamese imperial Mạc Dynasty of the late 1530s. During this period of time, major concerns about the ravaging civil war between the Trịnh (North An Nam) and the Nguyễn (South An Nam) Lords might have urged the mandarin to issue a series of prophecies to predict far and near future. Just like other prophecies, this one could not be verified until the events happened, but it had given the Vietnamese people a glimpse at Hope that there would be light at the end of the tunnel.

  By extension of this prophecy, the actual Vietnam War of the 1966-1967 (years of the Horse and Goat) gradually grew into the hero-decimating Offensive of Tết 1968 (Monkey), but the Peace everyone had expected never came. Nowadays, nationalist Vietnamese were still referring to the prophecy with the hope that 2017 (Year of the Monkey) would end communist control in Việt Nam. After near half a century since the end of war, people, especially those who used to be obedient and submissive to the communist regime, and even those who had once supported it, were all becoming weary as they had realized that Democracy in Việt-nam was nothing but an oppressing "fake" one[43]. Nevertheless, families that had lost loved ones during the war could still console themselves that their loss was inevitable as it had been predestined, and, for

---

41 https://vn.answers.yahoo.com/question/Ma De Duong Cuoc Anh Hung Tan, http://quanlambao.blogspot.com/2014/05/ma-e-duong-cuoc-anh-hung-tan.html and http://bacaytruc.com/index.php?option=com_content&id=1489:ma--dng-cc-anh-hung-tn

42 https://en.wikipedia.org/wiki/Nguy%E1%BB%85n_B%E1%BB%89nh_Khi%C3%AAm

43 To be discussed later.

many, their departed ones had already been long reincarnated, with a good Karma. When all failed, they would resort to similar consolation, this time referring to another of the wise man Trạng Trình Nguyễn Bỉnh Khiêm's prophecy, that *"Mười phần, chết Bảy còn Ba, Chết Hai, còn Một, mới ra Thái Bình..."* when, out of Ten, Seven would die, then Two more would die, leaving One surviving, then there would be Peace...

"Ancestral Land above all" ("Tổ Quốc Trên Hết") is the motto, calling on heroes to serve the people and the Ancestral Land! That has been our destiny already, as soon as we are born. Good citizens are born carrying their expected responsibility. This has become an ethno-centric tradition incorporated in the family ancestral respect and belief. At any cost, they are to pursue the plan, as this is their mission, even if they die succeeding or failing it. It is all because our destiny has determined so. As soon as we are born, we have already accepted that mission, and, also, this acceptance of one's fate makes it an obligation that one had to strictly comply with the demands of the mission.

After all, worse for worse, if killed in action, the communist soldiers know they will be worshipped by the living ones left behind as martyrs, while proceeding to their next well-deserved life. On the other hand, in the same way of thinking, the Republic of Việt-nam RVN civil servants or military personnel serve the common good and patriotic causes accordingly as established and required by the systemic political entity. If needed, they will not hesitate to sacrifice themselves to fulfill the Karma and to serve their country as good Confucians.

By applying these speculations, I believe that Buddhist Vietnamese combatants (as discussed earlier regarding afterlife beliefs[44]) from both sides communist and nationalist strongly believe they will be in good compliance in terms of reincarnation and Karma as well as Confucian servitude and duties if they have to face Death. If the mission is impossible, they will take it anyway, as it has already been dictated by their destiny and governed by Karma "Luật Nhân Quả." The communists, however, might have no hesitation accepting their fatal destiny compared to the nationalists who might be a little more hesitant. The former have no choice as they are brainwashed by "Uncle Hồ's bible" (similar to China Mao

---

44 http://www.vietspring.org/religion/religioninvn.html

Tse-tung's Red Book) to sacrifice for Martyrdom serving the People and Ancestral Land Tổ Quốc, while the latter have the freedom and choices between matters of life and death and alternative options.

Buddhism, even though it has great influence on the Vietnamese culture, is actually not a dominant belief. The one that has affected the most is not a religion, but the "law and order" philosophy imported from China after many centuries of Northern domination: Confucianism. This social value-driven belief has actually the deepest influence on the majority of the Vietnamese regardless of their religious inclination.

Confucianism and its politicized infusion

Confucius was a philosopher and extremely influential socio-political teacher during "Thời kỳ Xuân Thu or Spring-Autumn Era" in China history 501-479BC. His teaching was "religiously" followed by the Chinese in all social classes from the Imperial family down to the bottom echelon in the society through dynasties. It preached a secular duty-bound moral principle and set up mutually supporting rules and regulations between age-gender groups in the all aspects and hierarchies of life. Divinity was not the focus. Consequently, it was easily adopted by and fused into religious beliefs such as Buddhism, or even Christianity in Việt-nam since the Chinese domination through contemporary time. Confucian influences in the making of the Vietnamese societal sovereignty were substantial as the Vietnamese imperial dynasties had relied on them to strengthen their authorities as well as the people's Nationalism. Men and women, from childhood until they died, were bound by separate sets of values and duties that reinforced their roles in the family, among friends, in the work place, in the society and national structure.

During its development in Đại-Việt (Việt Nam), Confucianism has evolved and morphed into a more typically Vietnamized version that seems to somehow differ from its original form from the Chinese one as it is referred to in various websites[45]. What I am going to describe below is what I have been trained in since childhood via family education,

---

45 https://vi.wikipedia.org/wiki/Nho_gi%c3%a1o#Tư_tưởng_về_Thế_giới_đại_đồng Confucianism in Vietnamese, http://www.advite.com/Tam_Cuong_Ngu_Thuong.htm "Tam Cương, Ngũ Thường" in Vietnamese

human behavior, children stories, adult stories, mythologies, and textbooks.

Vietnamese men had to strictly abide to the "Tam Cương" three harnesses (loyal obligations to the leader "Quân," the teacher "Sư" and the father "Phụ") and the "Ngũ Thường" five virtues (humaneness "Nhân," loyalty "Nghĩa," respect "Lễ," wisdom "Trí," and trust "Tín"). The Confucian men "người Quân Tử," more or less the equivalent of the British "Gentleman," had to always fulfill all the "Tam Cương, Ngũ Thường" at all point in time and space in order to be respected. Respect to his family name (honor), which was of utmost importance, determined how he decided and how he behaved in all situations: at any cost, he had to protect family Honor. It was perhaps this same family name honor that urged them to do anything to uphold it, thus creating the pejorative connotation of "quân tử Tàu" (Chinese gentlemen).

In the meantime, women had to faithfully fulfill the "Tứ Đức" four virtues (household duties "Công," beauty "Dung," speech "Ngôn," and grace "Hạnh") and the "Tam Tòng" three submissions (to the father when he was alive "tòng Phụ," to the husband when married "tòng Phu," and to the eldest son when a widow "tòng Tử "). The women's submissions made it a subsequent requirement that they be married and reproduce at least a son to carry on respect to the husband's name.

Everyone was trained to know his/her place in the society and family. It boiled down to a one-way hierarchy: submit to and obey whoever was above and had control over your life, a leader, commander, father, eldest son, colonizer, someone with higher knowledge, supervisor...

This indoctrination had made the Vietnamese people extremely good followers who would obey at any cost. This docile behavior had perhaps encouraged the aspiring leaders to show off their "bags of tricks" to entice and recruit followers. The more followers (and territories) the latter had, the more power they would think they had.

This might explain how "wannabe" military leaders had taken advantage of people's obedience to leaders to launch their "coups d'état" and fulfill their wish for power, leading to a series of "coups d'état" from 1954 until the end of the second Republic (under Pres. Nguyễn văn Thiệu). As we went along with this analysis and logical flow, we would realize how the French colonizers had succeeded in recruiting their collaborators and "servants," and, by the same token, how the Communist Party was able to find people to sacrifice for its ultimate goal. Rival tribalist leaders and "wannabes" caused an internal division between the people. We would also understand why the "little" ARVN soldiers were confused between obeying the direct order given by their Vietnamese commanders and following the conflicting American advisors' recommendations. We would see why the civilians were so easily caught in between when having to obey overpowering NVA and VC and following the ARVN or allies' instructions. Even though the bottom line was to serve their Ancestral Land, the different factions from all war participating sides had their own agenda using the Confucian mentality of the people. More discussion on this topic will follow later on in this book.

Confucianism could blend well with Destiny and Fate as the latter, being encrusted in people's mind, could explain and justify why the individual should be satisfied with the end result that fate had determined for them to comply their societal demanded virtues and obligations. That was why, even knowing they were heading to their own death, the Northerner youth did not hesitate to join southbound forces (the National Liberation Front or Việt-cộng) to liberate the South when they were told the almighty Americans were going to occupy and colonize it with the French. For the same reason, like other docile Southerners, I, as well as many other young ones, no matter how patriotic[46] we were, had also joined the ARVN to repeal the invaders from the North and the VC rebels. We all did it mainly because of that **Confucian sense of duties** that was engrained in our DNA!

Communist versus Nationalist options

While the nationalists had diverse and perhaps contradicting attitudes towards Life or Death, to Save Life or to Kill, the communist soldiers were "monovalent," and homogeneously determined: they seemed to have no hesitation to kill no matter the victims were fierce or helpless, related to them or strangers. This could have made their job easier

---

46   Patriotism has to be incited by politicized rallies and propagandas. Most of us young men were not directly politicized.

as there was no reasoning needed: rationalizing required balancing between opposing values and emotions. The enemy seemed to have grown out of the same China-influenced (Chairman Mao's) mold created by "Uncle Hồ" who was their only God, and the Communist Party, which had always been discretely manipulated by Commissar Lê Duẫn[47], the strategist who masterminded most of Uncle Hồ's moves during the war[48] after having successfully initiated and organized the National Liberation Front in South Vietnam (1940s-1957). Hồ's "bible" (like Mao's Red Book) was the sole guidance and directorate. Since childhood, Uncle Hồ's youth had been proudly wearing a red scarf around their neck to differ them from the others who were still "unworthy." These totally submissive disciples of "Uncle Hồ" ("con cháu Bác Hồ") had only one choice, to Believe and Obey, as their family name mattered and had to be visibly honored. Some children were even raised by the state[49] so they could be shaped up more perfectly, and their parents were nothing but their comrades. The party indoctrination had made them disavow all family ties, religious or pagan belief. When they were grown up (16 and older), they would join the main force "bộ đội", the People's Army of Vietnam (PAVN, also called in the South the North Vietnam Army NVA) and were trained to make their Party cause the sole ultimate goal. All had to live up to the party expectations, otherwise they would be subject to "strict consequences." This would serve as an example to the others. No one would want National Shame to the Family Name!

Personal preferences and individual fates would not matter any longer as the Party goal had replaced them with only one mission: Defeat the Americans/allies and their "gangs of servants" (the RVN people). Only the end would justify all the means. It was a total belief in the final purpose of their mission, to "liberate the South from the imperialist

---

47   Vietnam War: A Memoir of the Unsung Heroes – Janette Marin (Facebook, Sept. 2018) and https://alphahistory.com/vietnamwar/le-duan/

48   "Uncle Hồ" became much more popular than Commissar Lê Duẫn, even though, according to a great many commentators, Duẫn had made most strategic plans during the North war against the U.S., probably thanks to Hồ's extended international activities promoting Communism abroad. Duẫn had instituted political commissariat in most if not all facets of Party movements, civilian as well as military, perhaps based on the Soviet model. This explains why the enemy excels in psychological warfare. At some point in 1960, Duẫn was the Party leader, leaving Hồ in the figure head position. His direct intervention in the South had helped strengthen newly formed National Liberation Front cells, this making Hồ and Duẫn rivals. https://en.wikipedia.org/wiki/Lê_Duẫn

49   **Children were raised by the government and would not see their parents often as the latter had to work and totally focus on war efforts.**

Americans," this ultimate goal being forged from the main propaganda that the U.S. was in Việt Nam to take it over from Americans' ally, the French colonials, and that the RVN system had betrayed to Ancestors' Land and the People being servants to invaders. Their choice could not be simpler than that: self-sacrifice! As a reward, they would become Martyrs (a national honor to the Family Name) if they died, and forever be worshipped as the People's heroes. To refuse to fulfill this fate (to die like a hero) would be to die like a coward: be executed by their superior, and, as a result, bring disrespect to their Family Name and ancestors.

By the same token, their loved ones would be expected to do the same thing: family members, friends or neighbors would cooperate with them by agreeing to provide Intel information on the nationalists or allies, supply food and medication, participate in the Party war efforts, etc. Refusing to collaborate with Uncle Hồ's children would equate to betraying the sacred Ancestral Land and its People, meaning "deserved to be executed and would serve as an example to the others." This would explain the reason behind all genocidal massacres they had committed from 1940s (land reforms) through 1990s (offensives, re-education camps) for defiance, dissidence, resistance, questioning, failing to comply. It was all about the illiterate people using terrorism to govern and dictate!

Totalitarianism meant Strength and Power:

This explained why vindictive and punitive executions had been widely used to subjugate submissive peasants to full cooperation. Common soldiers (most did not finish elementary school, if they had gone to school at all) already had the "go ahead" and could decide on their own to execute or not. Of course, the latter would rather obey the oppressive enemy in order to stay alive and keep loved ones alive, than cooperate with the kind and compassionate nationalists.

On the contrary, the nationalists (including the allies) had much more responsible choices along with moral restrictions to think about, to juggle with and decide; of course, the process would certainly be confusing, and even freak them out, and therefore make them hesitate[50]. Unfortunately, in a state of

---

50   As I am saying this, I am sure many readers, who have not been trained in modern psychology, might object to this statement.

war, we could not afford to have doubts, be confused and/or hesitate. Consequently, we were showing to our Foes and their collaborators we were weak, and to our own people we had no cause in the efforts. Having mercy, compassion, or humanity meant "weak."

Humaneness and Liberalism translate into Being Weak:

Any RVN decision had to be subjected to others' approval (especially by American advisors) along the line of command in compliance with the U.S. aid agreement. All decisions, no matter how strategic they could be (e.g. intelligence gathering, avoid civilian villages even if they refused to stay within their strategic hamlets, these being created to help protect them[51]), had to be filtered through Geneva Convention guidelines before they could be executed.

Shots are coming from the village! Snipers are killing our soldiers from there... tunnels... These peasants were supposed to live in the strategic hamlets "ấp chiến lược", not here. Should we shoot back? Stored rice found from buried containers... Are these enemy supplies? Should we burn them? These huts are not allowed here. Should we burn them? Civilians living in unauthorized areas and suspected for collaborating with the enemy had to be released since they all were actually wearing civilian peasant garbs. Anyway, they would come back, even stronger, probably with "Intel" they found from being captured! Nobody at home (USA) would understand any of these, nevertheless, they were the ones judging how we had fought the War and deciding on its outcome, as well as the life and death of millions of lives!

Of course the enemies knew about this nationalist weakness so they could take advantage of it and even exceed what they were allowed to do (torture, exemplary individual to mass execution), not caring much about how they would be judged. After all, they were just lawless "rebels:" they had no convention to submit to, but just follow the path like good obedient Confucian followers! As a result, while allies and ARVN soldiers, also good followers, had to abide to all

---

[51] As seen in the movie The Green Berets. Also, https://thevietnamwar.info/strategic-hamlet-program/

kinds of guidelines, the enemy did not have to, because no one paid attention to them, even when they executed our men when captured as prisoners (1960s-1972, instead of keeping prisoners of war POW's alive and protecting them humanely as per Geneva Convention)[52]. Of course, no one at home knew or even cared about civilians being used by the enemy as human shields and were massacred as a strategy of intimidation against the RVN government and the people. How many knew about the 506 peasants killed by U.S. troops in Mỹ Lai massacre (March 1968[53])? How many knew or cared about the 7,800 civilians massacred also during the Tết Offensive of 1968 in Huế (January-March 1968[54])? Some massacre was more tragic than other!

    In my opinion, from generation to generation, during their training, our enemies had been heavily subjected to communist propaganda that resulted in horrible dehumanizing metamorphoses that had somehow brainwashed them into completely heartless creatures. This was a very common practice used by the Communist Block during the Cold War to change and command total subjugation of people's will. Many, who had grown up in this peer-pressured and party-controlled robotization process, like in the case of my step brothers from North Việt Nam, had adopted the party as if it was their only religion. They had no choice but give their soul away. I was sure about it! After the war ended, my civilian brother in law #4, as well as many of my other ARVN friends, had also been subjected to the same brain-washing treatment when they were "doing time" in concentration camps. Most resisted authorities for a short time then gave up because of torture. Those who counteracted had been executed in public (intimidation) or simply disappeared. This had been accounted for by survivors.

    Yes indeed, the process of brainwashing disciples since their childhood in "Uncle Hồ's" Youth propaganda program (similar to Hitler's or Mao Tse-tung's Youth, and most recently, Islam State of Iraq and Syria ISIS Youth in the Middle East) would turn normal people into senseless unshaken and unbroken fanatics. A great many NVA soldiers even had their

---

52   Probably, as the Paris Accords allowed VC and NVA to keep territories they occupied, starting Spring Offensive of 1972, after capture, more allied and ARVN soldiers were kept prisoners. The post Paris Accords POW exchange showed there were at least 6 to 8 times more communist POW's than ARVN and allies being exchanged.

53   https://www.history.com/topics/vietnam-war/my-lai-massacre-1

1. 54   www.nytimes.com/2018/02/20/opinion/hue-massacre

body tattooed with the vow to "Sinh Bắc, Tử Nam" (be born in the North, and die in the South). Their Love for Ancestral Land "Yêu Nước, Yêu Tổ Quốc" was unequivocally equated to their new destiny and translated into "doing whatever the party had prescribed and dictated," including killing anyone or indiscriminately massacring groups they considered as "traitors[55]." That might be the only choice they were told they had! It was perhaps for this reason, during the multiple attacks, especially during the two "offensives" of Tết Mậu Thân of 1968 and Summer Easter Tide of 1972, the communists had massacred and buried in mass graves thousands of defenseless innocent nationalist civilian elders, women and children, unarmed civil servants and military personnel. Thousands and thousands of civilian families had been shelled upon on the National Route called "the Highway of Terror" when they were trying to escape from war zones by moving southbound towards Sài-gòn. The same tragic situation happened during the 1974-75 ARVN Army of the Republic of Việt-nam "strategic evacuation" when I and II Army Corps in Central Việt Nam were abandoned by the ill-supported RVN. The final Spring Offensive of April 30th, 1975 arrived with millions more casualties[56]...

The Meaning of Human Life via the spiritual lenses

The terms "terrorism" might not have been familiar to the ears of many people in the U.S. and Europe until the 2000s, but it had already happened when the Nazis committed the horrendous "Holocaust extermination" against millions of innocent unarmed Jews in WW2 concentration or labor camps. Mass executions had also been performed in the two Russian Revolutions of 1905 and of 1917, and the three Chinese Revolutions, including the People's Revolution of 1911-13, the Communist Revolution of 1927-1949 and the Cultural Revolution of 1966-76. During these events, millions of lives, mostly defenseless and innocent civilians, had been wasted. For some reason, it might be possible that genocides had been so unbelievable the term "terrorism" could not even

---

[55] Who had cooperated with the "American invaders" or worked for the "Puppet Regime of Sài-gòn."

[56] For fear of mass massacres, approximately one million terrorized Southerners ran away for Freedom, and, so sadly, another million of RVN civil servants and military personnel who could not escape were incarcerated and murdered while in captivity

be connected until it started pounding heavily on people's mind with the Middle East Wars and World Terrorism in the Twentieth Century.

HoChiMinh-ism as a spiritual belief

To follow their big brother comrades, the Vietnamese communists had chosen terrorism and massacres as their preferred tools of trade to intimidate those who dared stand up against them, who refused to collaborate with them, or were related to some nationalist partisans, or who just happened to be in their way[57]! The fact of the matter was that the enemy wanted to make the living watch the victims die and suffer, become horrified, think about what would happen to them if they disobeyed and would not bend into total submission. Stories retold by survivors would indirectly serve as intimidating threats to the others (psychological terrorism), resulting in more compliance to and cooperation with the enemy. Only a distorted mind of brainwashed terrorists could come up with such savage methods of domination.

Along the same line of thoughts of sacrificing lives for the common cause, the VC and NVA had also adopted without hesitation the Chinese old tactics of "Human Ocean" attack (the Army of the Republic of Việt-nam ARVN called "Tấn Công Biển Người"), which meant "attack in numbers as large as an "ocean" while disregarding how many of their own lives might be sacrificed. This also explained why many VC NVA troops had, at the expense of their own Death, used their own body to block cannons rolling downhill out of control on the Hồ Chí Minh trail. Others had been martyred for having "heroically" used their body to carpet their troops' way across mine fields or over razor (post 1965) or barbed (pre-1965) wire fences... Some determined that it was Heroism and Patriotism. Others believed it was pure insanity or effects of peer pressure.

During the last months of the war, the Army of the Republic of Việt-nam ARVN had discovered VC bodies chained to their sniper positions on tree tops where they were hanging down from. Autopsies performed by ARVN medical

---

[57] Article in Vietnamese Chiến-tranh, Bản chất và Mục Đích : Phần 2: Kinh Nghiệm Lịch-sử (War, Characteristics and Objectives - Part 2: History Experience written by Researcher Trần Chính Trung published in August 2009 Vietnamese magazine Chiến-sĩ Cộng-hoà (The Soldiers of the Republic)

corps had proven these VC had been given hormone shots to help them become fearless. The communist casualties[58] seemed to always be three to four times higher than the nationalist's (ARVN and allies); nevertheless, the Party always found new troops from their docile pool to replenish their exhausted units, as if they had a bottomless supply of them. These people appeared to not really mind losing their life for the Communist Party for the very simple reason they strongly believed in its propaganda and that their heavenly afterlife, or reincarnation "for a better life," had already been guaranteed; besides, they knew they would receive the honor of Martyrdom "Anh Hùng Liệt Sỹ," as well as be worshipped nation-wide by, not just their loved ones, but most importantly, by the Ancestral Land, this being a greatest Honor for them and their family. That might have been their only wish, a "Death-with-Honor wish of the Hero of the Ancestral Land!" This was actually the only choice the Communist Party had allowed them to make. All of their peer society had elevated this choice to a supreme belief, in the name of "Heroes against American Invaders" and "Liberators of our land and waters" [59]! Eventually, the Communist Indoctrination, with the leadership of many political cadres, e.g. Lê Duẩn, Trường Chinh[60]... who had actually master minded the Communist Party behind the scene, was able to create **a new religion to replace all existing ones**, a religion that all members must be part of: HồChíMinh-ism. By infusing HồChíMinh-ism into a spiritual belief for the VC and NVA, the communist party has actually transformed the Vietnam War into a **spiritual war**, the war of the fanatics.

In the March 17, 1975 meeting at the White House with a congressional delegation that had just come back from a visit in Việt Nam, Sec. Kissinger stated that "they (Lê Đức Thọ, delegate of the People's Democratic Republic of Việt Nam PDRV, Democratic Republic of Vietnam DRV or North Việt Nam, Kissinger's counterpart in Paris Negotiations,

---

58 According to Wikipedia 2009, there were up to 1.1 Million Communist dead compared to 313,000 ARVN. Approximately 2 Million civilians were killed. https://en.wikipedia.org/wiki/Vietnam_War_casualties Note that these numbers were estimates and might change with new data updates. The ratio between NVA-VC vs. ARVN was approximately 3,1:1. However, some ARVN unofficial source believed that the ARVN casualties were higher than reported on the website: around 1 Million, making casualties comparable between Communist and nationalist sides.

59 "Anh hùng chống giặc Mỹ xâm lăng,""Giải-phóng Đất Nước"

60 Lê Duẩn was a founding member of the Indochina Communist Party (the future Communist Party of Vietnam). https://en.wikipedia.org/wiki/Le_Duan. Trường Chinh played an important role in shaping the politics of the Democratic Republic of Vietnam (DRV) and creating the socialist structure of the new Vietnam https://en.wikipedia.org/wiki/Truong_Chinh

and his team) are the most devoted, single-minded abrasive communists I have ever seen...He (Thọ) is a dedicated revolutionary. They are hard cases and in some ways rather admirable. Lê Đức Thọ and all the others have fought all their lives...because of their history, it was natural for the Vietnamese to think that all foreigners are treacherous to them...[61] *'If our generation cannot win, then our sons and nephews will continue. We will sacrifice everything, but we will not again have slavery. This is our iron will. We have been fighting for 25 years, the French and you. You wanted to quench our spirit with bombs and shells. But they cannot force us to submit.'"*

In a PBS Globe Trekker show on Railroad in Việt Nam Tough Trains, during his interview, a former North Vietnamese Army NVA member, a professional "Reunification Express" train conductor, had told Globe Trekker host Zay Harding about one of his missions carrying North Vietnamese Army NVA troops, weapons and China-Soviet Union military aids from Hà-nội area to Vinh where troops and supplies would cross into South Việt Nam via the Hồ Chí Minh Trail that started at the passes[62]. "When the American bombers started to hit our train, the only thing I thought about was to accomplish the mission Uncle had assigned to my crew: to take our brave soldiers and supplies to Hồ Chí Minh Trail safely so we could achieve our sacred goal of liberating the South... a few of my men had already been killed by shrapnel, but I stayed focused on my mission even though seriously wounded...," he said, obviously still emotional even though it had already been nearly fifty years since the days the events happened. He had extremely unbroken faith in his mission! Most NVA men believed in "uncle Hồ's" preaching that the South really needed them to liberate.

The Atheists and Folk believers' viewpoints

We have looked into the mind of the religious combatants. What about the majority others? Those who are Folk-belief and non-religious Vietnamese (73%) are pagan or

---

61 *No Peace, No Honor Chapter 3 of Prologue by Larry Berman.*
62 Mụ Giạ Pass in Quảng Bình Province to Ban Karai Pass in Hà Tĩnh Province

Taoist[63] naturalist. A great many observe ancestral worshipping, and respect nature and spirits. They might be more or less influenced by their friends or relatives' reincarnation belief via Ancestral worship; by extension, they respect the cause and effect law (Buddhist Karma relates also to cause and effect), and, might view Death as either a change in life form, a transition into the next life, or a "return to sand and dust" ("Trở về với Cát Bụi"), which refers to the sacred cycle "Return to Nature[64]..." Nevertheless, they resort to carrying shamanist potions, amulets and/or charms which will, as they believe, emanate the folkloric magic power that will keep them safe and protect them from hitting an explosive device or being struck by bullets or shrapnel. These are pieces of jewelry used to circumvent maledicting Destinies. Most Vietnamese combatants, regardless of religion, might have relied on more than one protection device.

Many had relied on some sort of luck charms, or at least some memento of their family for good luck. Besides believing in reincarnation and partially in Destiny, I used to carry, in addition to my two dog tags, two amulets around my neck at all times since the day I had joined the ARVN until I arrived in Minnesota as a refugee in 1975: one set of rolled up 24K gold leaf Hindu "bullet-proof" amulets issued by a popular Vietnamese shaman, and a Buddha charm issued by a Buddhist monk. In addition to amulets, before heading to the front line, I had gone with Mười to the Hindu Champa Tháp Bà Temple for a sacred blessing in Nha Trang where my boot camp ("Quân Trường Đồng Đế") training took place in 1973. I had then promised I would return to pay respect if I could survive at the war. In 2009, I returned to Tháp Bà to offer my thanks and to recognize the blessing I had received from the Hindu goddess many years ago. During a time when life and death were battling each other, one could not afford not to be spiritual. Whether the charms had protected me or not, I was not sure, but I had indeed survived the war, safe and sound, without a single scratch.

My students in the public schools in Minnesota where I

---

63   Like Confucianism, Taoism is a philosophy of Life, not a religion, that teaches people who do not want to conform with the societal laws to go away into nature

64   Many pagan Vietnamese also believed that everyone, without exception, would have to go through the Ten Gates of Hell, be purged by torture to redeem the sins committed in their past life, before they could be released back to Earth to reincarnate. Afterwards, they were fed a bowl of porridge "cháo lú" that would help make their soul forget what they had just gone through.

was an administrator were very curious whether I was injured when fighting during the war or whether I had any scars. I did show them one set of amulets I used to carry, the one from the Shaman in Việt Nam. Sometimes, to make it sensational, I had shown them a long scar across my neck and told them "I almost got decapitated by the enemy when they tried to cut my throat with their fierce and inhumane machete, the Mã Tấu." Actually, it was just a scar left after a benign nodule removal procedure a few years ago. That was the only story I made up to sensationalize the Vietnam War stories I told some male students; the rest was all true...

The Christians

What about the Vietnamese Christians? Statistical figures had been changing after the war, but the percentage might have stayed unchanged. The 7.3% Christian population might openly display more their religious devotion in the South (RVN) than in the North where "Religions were considered the People's Opium[65]" had been indoctrinated. Nevertheless, probably the majority of devoted Catholics in the North had already migrated to the South after the French defeat in 1954 (Passage to Freedom).

As for the Christian Northerners who have chosen to stay, I speculate that their commitment to the Party political cause has been as strong, if not dominant, as their belief that they will go to Heaven, if they die serving "Tổ Quốc" Ancestral Country. As a matter of fact, Patriotism and Sacred Sacrificing for the Ancestral Land being encrusted in all Vietnamese for many centuries makes it much easier for the Communist Party to convince them to sacrifice themselves when fighting against the Americans and the RVN "traitors of the People." Nevertheless, they have to abandon their Christianity in order to survive.

My friend recently married a Northerner whose three generation family was Catholic. The grandfather decided to stay in the North instead of taking advantage of the 10 months of Passage to Freedom to migrate to the South. He chose to stay to protect his parish church in the North so the congregation had a place for worshipping. He was executed

---
[65] Karl Marx.

by the Việt Minh. His son, a Catholic priest, was also executed. That was strong respectable Belief!

The Battle between strong Beliefs

As described in Find your Fate: Life after Death[66], Christianity has historically taught that everyone has only a single life on earth. After death, an eternal life awaits everyone either in Heaven or Hell. There is no suffering in Heaven; only joy. For those sent to Hell, torture seems to be eternal as if there were no hope of cessation. The overwhelming majority of mainstream Christian denominations rejects the notion of reincarnation and considers that this theory challenges basic tenets of their beliefs. The Vietnamese,[67] American, Australian, and New Zealander Christian combatants would definitely view death in a way very different from the non-Christians: Paradise for those deserving ones is obviously preferred to Hell for the non-deserving. How would you know for sure your soul will be going to Heaven if you die in this war? It might become a more difficult choice when they have to decide in an instant whether to sacrifice oneself for Democracy and die, kill others and let them die, or avoid the conflict and retreat from it. Contrary to the communists (the fatalists) whose only choice is to sacrifice, the nationalists, including the allies, *have the conscious freedom of choice* as I have mentioned in a previous paragraph: it can be a difficult choice as one believes no one should die!

This freedom of choice leans more towards Life in the Christian's eyes. Life is extremely precious as *one has only one life to live:* it ought to be valued and respected, but not wasted! Back home in the U.S. (and perhaps in other allied countries), it hits the Christians' heart even more heavily when hearing about and seeing in the news life being taken away, whether from a civilian or military, whether they are good or bad.

---

66   http://death.findyourfate.com/life-after-death/christianity.html
67   A great many of ARVN soldiers were descendants of the 1954 North Vietnamese refugees from the Catholic villages of Bùi Chu and Phát Diệm. Also, many newer protestant denominations, including ethnic minority groups such as Mnong Nùng, E De, Ja Rai.., had been started by Americans and French missionaries: French Reformed Church, Anglican–Episcopalian, Christian and Missionary Alliance, Baptists, Churches of Christ, Worldwide Evangelization Crusade, and Seventh-day Adventists. https://en.wikipedia.org/wiki/Protestantism_in_Vietnam and http://www.sacred-texts.com/asia/rsv/rsv11.htm

Americans back home unequivocally refuse to let others kill or their own people to be killed. For these people of conscience, it is unacceptable that the Vietnamese population be treated inhumanely, especially children, women and elders, whether they are enemy or not. They cannot tolerate it that their own Christian Americans are the ones killing those "innocent ones." It is not a Christian thing to take anyone's life. The war is therefore expected to be fought with Humaneness and Compassion by the allies and the ARVN. It is a very strong Belief, too;

  As of 1970, even the American combatants started to question themselves "it is not right to kill for Peace[68]!" while the North Vietnamese Army and Việt-cộng continue to escalate their killing spree. Massacre after massacre, which have rarely been covered by the television media (where are they when we need them?), the communists have taken advantage of their superiority in causing terror to ravage the South and to "win" peasants' support (out of fear, of course). We have here a struggle between two strong Beliefs: One belief is to kill indiscriminately, regardless of whether an individual was an enemy or not, while the other belief is to spare the lives of all in providing a second chance for them. Which side will survive? It gets worse when the malevolent ones were considered the heroes, while the life saviors the demons.

  Now that we have examined the mind of the combatants through their viewpoints on life and death beliefs, we still need to understand their past history, which unequivocally will influence their readiness to fight as well as the way they interact with their enemy while waging war. Even though I grow up with the war, from the French establishment through the first and second republics, to this day, I still remain astounded by how intricate it is.

---

[68] Walter Conkrite's The Vietnam War (1985) – The World of Charlie.

CHAPTER 3
And the Wheel of History keeps turning

It has been said: "history not learned will be repeated." In order to understand more and dissect this war successfully, we ought to look further into the past to explore the root of the Vietnamese mind. With that in mind, let's go back a few thousand years…

Thousands of years under Chinese Domination

In order to more fully understand the Vietnam War and the resultant outcomes, we need to understand the Vietnamese attitude and psychology towards war by carefully examining Việt-nam History. This could be the deep learning part of "Know our Foe, Know ourselves…" the U.S. might have needed to know this before entering their Việt Nam involvement. As mentioned earlier, Việt Nam has an incredible history of over three thousand years of tributary subjugation to and War against the invasions and colonization by China. Wars have been waged in multiple uprisings by different community leaders, through hundreds of generations of Vietnamese painfully struggling and heroically sacrificing for the Independence of their Ancestral Land "Tô Quốc."

Ever since 2879BC, Xích Quỷ (the origin of Việt-nam), Văn Lang, Âu Lạc, Lạc Việt" through 1427AD, Giao Chỉ, Lĩnh Nam, Giao Châu, Vạn Xuân… Đại Việt[69] had been enduring, struggling and combating against China tributary protectorate and domination. Its oppressed people who longed for Independence had raised armies to try to liberate themselves, especially during some 1,500 years between 111BC and 1427AD when major open wars had taken place. Heroes of both genders had risen up one after another to fight back. Among these included the two well-revered Trưng Sisters [70], or the Vietnamese Joan of Arc Lady Triệu[71], and Vietnamese imperial dynasties of Lý, Ngô, Đinh, Lê, Lý, Trần,

---

69 https://en.wikipedia.org/wiki/An Nam_(Chinese_province) From a Chinese province under Tang Dynasty 679AD to independence
70 https://en.wikipedia.org/wiki/Trưng_Sisters Sisters Trưng
71 https://en.wikipedia.org/wiki/Lady_Triệu Lady Triệu, Vietnam Joan of Arc

Hồ, Hậu Lê[72]...who had dared stand up and defy the almighty Chinese imperial dynasties of Qin, Chu, Han, Liang, Sui, Tang, through Ming. Many of these Annamite Emperors had adopted the Chinese way of life, culture, military warfare as well as literary wisdom, and used them to fight back. This was an implementation of the phrase "dùng gậy ông, đập lưng ông" (use their stick to strike them on their own back).

  A large portion of the Annamite population was raised to become obedient Confucian (as observed by many Western commentators), loyal and multi-talented in the Chinese Confucian way to serve in either the Chinese (during occupation) or the Annamite (during tributary subjugation) system. These Annamite also tended to be very loyal to their past, no matter how insignificant it could be, meaning that the past was much more important than the future, especially when what was coming seemed to be unpredictable and futile, if not detrimental to existence. Consequently, they would usually refer to past experiences and traditions when trying to solve a current situation or planning for future interventions. Luckily, Chinese systems seemed to also be conservative and, consequently, might be predictable. Since then, hatred against foreign invaders kept building and had been suppressed into the Annamite DNA, making them even more attached to their past, and urged them to DISTRUST FOREIGNERS AND HATE "INVADERS." As history also showed, this hatred became intergenerational and eternal.

  Collaboration with China was at that time insignificant, if non-existent. As the Chinese establishment would not trust indigenous people, so not many Annamite were used by the system. On the one hand, they learned to bend and be submissive in order to maintain a status quo that would allow them to buy time and wait for the ripe moment to turn around and strike back. The common people, mostly illiterate, had no other choice but accept their destiny. On the other hand, those striving for advancement would study at their best with the hope they could be employed by the indigenous Annamite emperors, and, if lucky, they would achieve mandarin's (officials in the government) status and serve the Emperor against the invaders. Destiny would guide leaders towards a form of subdued collaboration: They passively submitted themselves to invaders' power and accepted to make their annual tributary offerings, while waiting for the "celestial

---

[72] https://en.wikipedia.org/wiki/History_of_Vietnam

time to rebel" (we will discuss the three elements for success later). Their submission would allow them to have Peace and security for their extended family. Success or failure always affected their whole family (everyone bearing the same name). Resistance would be futile as Annamite rebelling leaders were usually executed by the Chinese occupiers, along with their three generations of close or distant family members, all beheaded, torn apart by horses, cut into pieces ("death by thousand cuts") or crushed by elephants in public mass executions. This could be the original medieval "standard" terroristic punishment in history, as adopted by many countries in the world: execute or massacre in public to warn the living not to up rise against the dominating establishment.

Fighting against Chinese domination

After tens of centuries dominated by the Empire of China, the little Dragon woke up: many leaders of Đại Việt made attempts to repel the North in order to regain Independence and Sovereignty. Sino-Annamite wars followed one after another, and failure after failure on the Đại Việt part. Nevertheless, with perseverance, the Việt, in spite of its small army, had defeated China expeditionary armies on many occasions.

The first Việt victory was at the spectacular naval battle of Bạch Đằng Giang (near Hạ Long Bay) in 938AD, where Giao Châu (another old name for Việt-nam) Commander Ngô Quyền defeated the Southern Han Dynasty ruler Liu Yan's Armada. Tall spikes (hidden in the Descending Dragon Bay cave which nowadays people can visit in Hạ Long Bay) were planted on the river bottom to incapacitate the enemy naval ships allowing the Annamite army to use "hỏa công" incendiary tactics[73] to destroy the whole Chinese fleet. China Prince Liu Hong Cao was killed, and Liu Yan took the retreat back to China.

Việt Gen. Ngô Quyền proclaimed himself as King of Giao Châu, which Southern Han Dynasty had no choice but to recognize. For the first time, the Annamite had a sweet taste of Independence which lasted for over 300 years.

---

73   http://viettouch.com/hist/ngoquyen/ Ngô Quyền defeated Southern Han at Bạch Đằng River

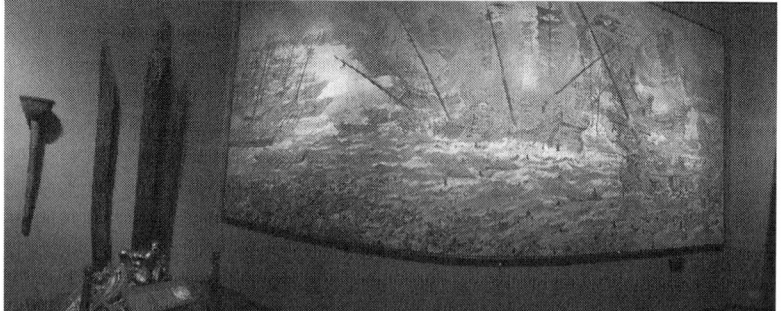

The Two Battles of Bạch Đằng Giang[74] victories in 938 and 1288
Source: Author's photos taken at the History Museum in Sài-gòn
(2019)

During this détente, they started new traditions and laws to accommodate this independence and to reaffirm their sovereignty until the Yuan Dynasty invaded again with three consecutive major attacks.

Yuan Dynasty Emperor Ku Blai Khan (grandson of

---

[74] The Chinese Army was defeated at the Battles of Bạch Đằng Giang by Đại Việt Gen. Ngô Quyền (938) https://vi.wikipedia.org/wiki/Tr%E1%BA%ADn_B%E1%BA%A1ch_%C4%90%E1%BA%B1ng_(938) and Gen. Trần Hưng Đạo (1288), in both cases using the same tactics of spikes planted on the bottom of the river (actual spikes on the left of this photo). https://en.wikipedia.org/wiki/Battle_of_B%E1%BA%A1ch_%C4%90%E1%BA%B1ng_(1288)

Genghis Khan and founder of China Yuan Dynasty) sent his army south of the border in three salvoes of attack to invade this weak tiny country in 1258, 1285 and 1287-88 AD[75]. At the end of the first invasion of 1258, their massive army had subdued Đại Việt (new name for An Nam) into a few years of vassal submission; however, Ku Blai Khan's forces encountered major disasters when they invaded the third time...

Just before China Emperor Yuan's second invasion, the historical Hội-nghị Diên Hồng People's Convention was summoned by Đại Việt Emperor Trần Thánh Tông in 1284AD. The latter, Father of succeeding Emperor Trần Nhân Tông, called for a referendum to seek advices from the wise elders on how to deal with the Yuan Emperor Ku Blai Khan's plan of invading their Ancestral Land[76].

This early-day democratic event was depicted in the XXth Century lyrics of a patriotic song "Hội-nghị Diên Hồng" written by North Vietnamese Army NVA politician and composer Lưu Hữu Phước [77]:

"Toàn dân nghe chăng?
Sơn Hà nguy biến..."

<To the People of Đại Việt:>
"Hear ye, Hear ye,
Mountains and Rivers (Ancestral land) are at risk!
Hatred piling up! Borders are quaking.
Stumping on our Mountains and Rivers, (enemy) cavalry is thundering
Causing Hatred throughout thousand years,
All people, children of Angel and Dragon! Our Mountains and Rivers are at risk!
Hatred piling up! Peace or War?
....
Watching Yuan's army destroy our Homeland,
Seize our forts and stump on our sacred temples and mausoleum

---

75 https://en.wikipedia.org/wiki/Mongol_invasions_of_Vietnam Mongol invasions of Vietnam.

76 https://vi.wikipedia.org/wiki/Hội_nghị_Diên_Hồng Diên Hồng Convention of 1284AD

77 https://vi.wikipedia.org/wiki/Lưu_Hữu_Phước Composer of Hội Nghị Diên Hồng song, anti-French anti-collaboration revolutionary, Việt Minh active politician used music to move people's will.

*Seeing so many of Thoát's (Gen. Thoát Hoan of the
Yuan's Dynasty) brigades flood our land
All over, our Mountains and Rivers are trembling as the
people cry!
Ask: Facing national shame, should we make Peace or
War?
Answer: We are determined to fight!
,,,
Fight forever!
Save our land,
Pursue our Pride to be Heroes!
Ask: If Homeland is poor, what do we fight with?
Answer: Sacrifice!
We swear to risk our life for our Mountains and Rivers
Through decades of Glory..!"*

Translation by H. Tuong

Even though this song was probably written in the 1940s, for communist propaganda instilling patriotism against the French and Americans, it also became popular among many nationalists themselves after the 1960-70s war as it really demonstrated how patriotic and unbroken the Vietnamese were against foreign invasions. Like the Jews using their 1st Century AD myth of Masada to raise Israeli's patriotism in the 1940s, the Việt Minh had more or less successfully used myths of heroes and symbolic means, e.g. songs, to incite Vietnamese Nationalism in order to defeat the French. Later on, many, even among the nationalists, had recognized how uplifting this spiritual weapon was when used again by the Communist Party to ready up its army, the NVA in the North, as well as raise the National Liberation Front NLF army[78] in the South, but this time against the South...

Based on the myth that Emperor Trần Thánh Tông had used Hội-nghị Diên Hồng to incite Đại Việt people to up rise and defeat the almighty Chinese invaders, Lưu Hữu Phước believed his song would have the same effects against the Americans. What was left for the Communist Party to do was to convince the docile peasants that the U.S. was there to take

---

78 *When withdrawing their main battle forces back to the North in 1954 per Geneva Conference (Indochina), the Việt Minh had secretly implanted Communist seeds in the South to start an guerilla-warfare insurgence against the new State of Vietnam, the Việt-cộng VC and their National Front of Liberation of the South (Mặt Trận Giải Phóng Miền Nam).*

Việt Nam over from the French colonials.

Yuan's second punitive invasion in 1285AD was repelled as its cavalry got caught in muddy Red River. The whole army was pushed back to China. Two years later, Emperor Yuan sent his mighty army again to invade Đại Việt for the third time in 1287AD! Even though reinforced with tribal auxiliaries, Ku Blai Khan's near half a million-men strong army, commanded by fierce generals, was disastrously defeated by young Đại Việt venerated commander in chief Gen. Trần Hưng Đạo[79] (Prince Trần Quốc Tuấn) and his generals at, once again, the same Battle of Bạch Đằng, followed by the final Battle of Vân Đồn[80]. Chinese wiseman Sun Tzu's Art of War tactics were well used by the Đại Việt strategists: "chase the dog, but always leave an escape route so it will not bite back, then catch it at the dead end, when it's weakened by the chase…" Ku Blai Khan's army remnants were chased to and defeated at Vân Đồn. Even though he won the battle, Đại Việt Emperor Trần Nhân Tông[81], in order to appease the Yuan Emperor and to show humble submission to him, had voluntarily offered continued tributary submission, thus allowing the Vietnamese to have a short half-century period of Peace and partial Independence.

The people of Đại Việt might be longing for Peace at that time, but in their mind, they had actually started to realize Independence from China was what they really wanted. Unfortunately, this short period of Peace was abruptly interrupted by the return of the Chinese army, this time under the Ming Dynasty. This fourth and last invasion and occupation by China last only 21 years. It was extinguished in 1427AD as the Ming dynasty great army was defeated by An Nam Emperor Lê Lợi (thereafter came the legend of the Golden Turtle Kim Quy and the Return of the Sword[82] at Hồ Hoàn

[79] Among the top 100 military commanders in the world at all time, https://www.thetoptens.com/top-military-generals/ ranked Trần Hưng Đạo third, after David IV of Georgia (1st) and Alexander the Great (2nd) and before Napoleon Bonaparte (4th). NVA Gen. Võ Nguyên Giáp placed (5th), before Genghis Khan (6th)),and Nguyễn Huệ (10th) after Julius Cesar (8th). Others ranked were Union Gen. Robert Lee (11th), Nazi tank Gen. Erwin Rommel (12th), Pres. George Washington (13th), Soviet Gen. Georgy Zhukov (15th), U.S. tank commander Gen. George Patton (17th), China strategist Sun Tzu (18th), Israel King David (20th), U.S. Gen. Ulysses Grant (24th), U.S. Pres. "Ike" Dwight Eisenhower (29th),… based on 35,000 votes.

[80] https://en.wikipedia.org/wiki/Mongol_invasions_of_Vietnam
[81] https://en.wikipedia.org/wiki/Trần_Nhân_Tông Emperor Trần Nhân Tông
[82] https://prezi.com/v9s1vobejl9a/the-legend-of-le-loi-and-the-giant-golden-

Kiếm Lake, downtown Hà-nội) in a bloody ten year long campaign[83]. During this final war, the first form of guerilla tactics using suicide attacks[84] (by Gen. Lê Lai), as well as infiltration insurgent warfare, were born. Both An Nam and the Ming Dynasty had used ethnic forces from Ai Lao, now Laos, in their troops. This was also the first time ethnic tribes Thái, Tày and Mèo or Miao from the China side of the border had joined in with the An Nam army to repel the almighty Ming's army. Later on, some of the Thái, Tày and Mèo/ Miao (later changed to Hmong in the 1970s, meaning Free people) tribes had fled South-bound to seek refuge in North Vietnam[85], settling in nowadays Cao Bằng and Lạng Sơn Provinces, North Việt Nam, besides in Northern Laos (highland).

An Nam's new customs reflecting Freedom, Peace and Independence that were developed during the short period of Independence in the late 900s, but were later abolished by the Ming's invasion in 1406AD, were all reinstated after the Ming's army was totally expelled out of An Nam in 1427AD. What followed were four centuries of Independence and some Peace for An Nam, except for some small but continuous wars against Siam (Champa), and a civil war between Lord Trịnh (North An Nam) and the Lord Nguyễn (South An Nam) that literally divided An Nam into two separate countries. This separation might have made it easier for the French to invade this long country, and, later on, caused the South to consider Vietnam War as a war of aggression by the communist North, instead of Civil War between two parts of the same country.

Eventually, An Nam, which was at that time nowadays Northern and part of Central Vietnam, expanded South into Champa territory[86] via three imperial marriage alliances in

---

turtle-kim-qui/ Emperor Lê Lợi and the legend of the giant golden turtle
83   https://en.wikipedia.org/wiki/L%C3%AA_dynasty#.C3.AA_Th.C3.A1i_T.C3.B4ng Đại Việt and the Lê Dynasty
84   https://en.wikipedia.org/wiki/L%C3%AA_L%E1%BB%A3i Emperor Lê Lợi and the new guerrilla warfare.
85   https://en.wikipedia.org/wiki/List_of_ethnic_groups_in_Vietnam
86   This annexation had conflicting background history. According to the Khmer, from un-updated source https://en.wikipedia.org/wiki/Khmer_Krom , Khmer King Chey Chettha allowed the Annamites to operate a custom house - fishing village in Mid 1600s, which expanded into a port, then was "unofficially" annexed to An Nam to become nowadays South Việt Nam. Vietnamese Wikipedia https://vi/wikipedia.org/wiki/Cong_nu_Ngoc_Van clarified that Vietnamese Lady Ngọc Vạn, daughter of Lord Nguyễn Phúc Nguyên (1613-1635), married Khmer King Chey Chettha and part of Cambodia was annexed to South An Nam. Lord Trịnh was ruling in North An Nam then. Perhaps the Khmer people in other areas of Cambodia were not informed of this alliance, or they were but ignored it. Cambodia

1306, 1620 and 1631. In 1306, Princess Huyền Trân Công Chúa, daughter of Emperor Trần Nhân Tông who had defeated the Yuan Dynasty army, was promised and given in marriage to Champa King Jaya Sinhavarman III. Queen Paramecvariin's dowry was the two territories of Ô and Lý that were eventually annexed to Đại Việt to form the current provinces of Quảng Bình, Quảng Trị and Thừa Thiên. Later, Lord Nguyễn Phúc Nguyên of the South gave in marriage his daughter Princess Nguyễn Thị Ngọc Vạn to the Khmer King Chey Chetta II in 1620, this opening up Mô Xoài territory (now Bà Rịa province) to the Vietnamese. Lord Nguyễn betrothed his other daughter, princess Nguyễn Phúc Ngọc Khoa, to King Pôrômê of Champa in 1631 allowing the Vietnamese to migrate to the Southern part of the nowadays[87] Việt Nam.

It has become more and more obvious to this writer that many Cambodian (or Khmer) leaders have omitted or ignored this part of history shared between the two countries, and have propagated hatred among the Cambodian, hence promoting among the peace-loving people of Cambodia the genocidal practice of "Cáp Duon" against defenseless Vietnamese immigrants in Cambodia. The campaign, which is usually led by Khmer leaders, accuses the Vietnamese for having brutally invaded in the far past history and occupied their land to create what is called the Mekong Delta nowadays. Thousands and thousands of Vietnamese immigrants of all ages and both genders have been beheaded and their head thrown into the Mekong River to float down to the delta in Việt-nam. More information will follow.

---

during that time was divided because of internal turmoil and was under threat by the Siamese Empire. This alliance also helped Cambodia receive protection by An Nam against Siam/Cham. Wikipedia https://en.wikipedia.org/wiki/History_of_the_Cham%E2%80%93Vietnamese_wars related to constant wars between Siam and An Nam from 982AD to 1835AD, most of these wars were won by An Nam.

87   http://www.asiafinest.com/forum Princesses Princess Huyền Trân, Ngọc Hân and Ngọc Khoa marrying Champa Khmer Kings. These two events might have been omitted in Cambodia's history, causing the Khmer people to hate Vietnamese and massacre them in the early 1970s. See earlier section in Chapter 8 regarding "Cap Duon" behead Vietnamese.

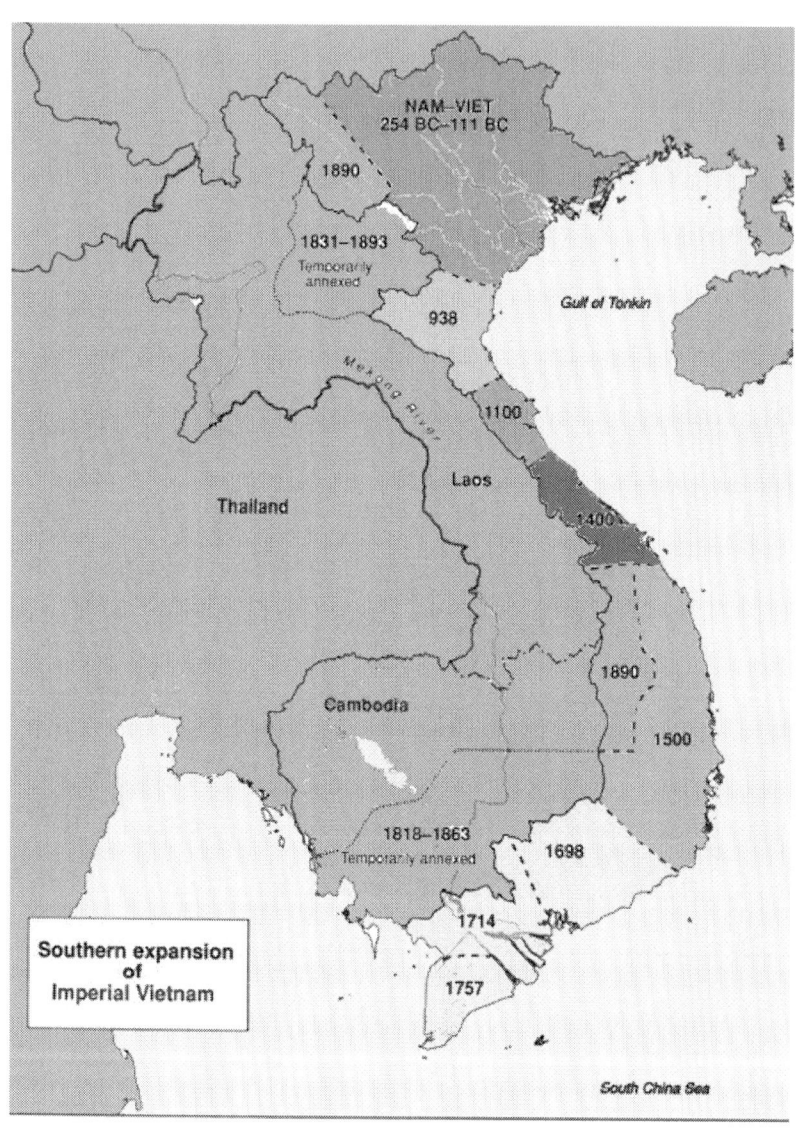

Đại Việt Expansion through Lê Dynasty (1428-1788AD)
Source: www.globalsecurity.org

Use Chinese wisdom to fight against Chinese invaders (Sun Tzu's The Art of War)

Although my father had never discussed politics or war, either at home, or with any friends whenever they had an eating gathering (usually Tết New Year celebration or ancestors' death anniversary), I could tell he was much interested in international politics and news. He liked to entertain himself by listening to the news on BBC and on Hà-nội channels (illegally) on the short wave radio, every night in the late 1950s and the early 1960s. I had picked up this habit from him and listened to VOA and BBC[88] in the mid-1970s to keep updated on the war. As a part time, I watched television news on networks such as CNN, NBC, ABC, FOX, Al Jazeera and Russia Today and, most recently, Paris 24, Deutsche Welle DW (Berlin) ... almost every day to following up on the interesting political era of Middle East Wars as well as the Trump presidency.

As my father never talked about his past, my siblings and I never knew what political inclination or military (most men at that time were in some branch of the ARVN) experiences he had. What was my father doing before he migrated to the South? He was definitely not part of "Uncle Hồ's clan," otherwise he would have remained in the North to fight the French. He might have found out about "Uncle Hồ's plan to purge the Việt Minh of nationalist patriots. He migrated to the South, not in the 1954 Passage to Freedom program for refugees. The one thing I was sure about was that he did move to the South from North Việt Nam in the 1940s and started a new life with our family, leaving his first family behind (unknown to us until after the war). Anyway, one day, I discovered one thing that sparked many questions in my mind: he had a French parachutist field jacket secretly hidden in his storage area. Feeling that he would not be inclined in sharing his story, I never asked him whether he had served in the French allied Airborne (the well-praised all-Vietnamese 5th Airborne Battalion which I had just found out while doing research for this book) against Hồ Chí Minh's Việt Minh.

Was there something political in his past he was trying to hide and forget? Was there something that led him to meet my French godfather who later hired him? Up until then, it had been a total secret. Most of the elders in our family in the South

---

88   Voice of America and British Broadcasting Corporation

who might be able to shed some shreds of light about father had already passed away. It became more puzzling when I found out my application for post high school study in the U.S. was rejected by the Internal Affairs Department (equivalent of the U.S. State Department) for an unknown reason that my mother hinted that "he was blacklisted for something he had done with my French godfather when they were traveling in Laos in the late 1940s..."

      My father liked to quote, not French, but Sino-Vietnamese[89] phrases, when he wanted we children to remember some wisdom he thought would help us succeed in life. When he was young, he did attend Sino-Vietnamese schools in the country-side in Hưng Yên Province, in North Vietnam where he was born, but he did not know any of the spoken Chinese as used in China such as Cantonese or Mandarin. The phrase I heard the most from him was "Thiên thời, Địa lợi, Nhân hoà[90]," which was a Vietnamese adaptation from The Art of War written by the great 5th Century BC Chinese strategist Sun Tzu. The latter always conceived that one had to study carefully three factors which people could rarely control, but which always affected people's life: celestial timeliness (or socio-political trends) "Thiên thời," favorable location "Địa lợi", and harmony with people "Nhân hoà." One should always make plans accordingly based on these three conditions before starting any project. Neglecting one would cause definite failure. Sun Tzu's teaching seemed to have influenced father, as well as other Vietnamese: they grew up reading story books about the Art of War wisdom Annamite political leaders had used to deal with Chinese conquerors. Another quote he sometimes referred to was "Biết người, Biết ta, Trăm trận, Trăm thắng" or "Know our Foe, Know Ourselves: One hundred Battles, One hundred Victories." Of course, story books also taught them to be good Confucian, obey the supreme leader (Emperor) and protect their Ancestors

---

[89]   In the early periods of An Nam history, Chinese characters Chữ Hán were taught and learned by Annamite using Annamite pronunciation, so Chinese speakers of ancient Chinese might not recognize the spoken part, but could understand it in writing. Sino-Annamite schools for Vietnamese children existed in small towns, probably more in North Vietnam, as a classical program (similarly to learning Latin in French schools) until the 1930s.

[90]   Even though Sun Tzu emphasized 5 elements all army generals must master before waging war, Moral Law (harmony between the ruler and the people), Heaven (Timeliness, weather, elements...), Earth (Location, positions), the commander (quality of the commander) and Method/ Discipline (organization, strategies, training...), the Vietnamese paid attention the most to Heaven, Earth and People's Harmony (Moral law)

Land and serve the people... I apparently had followed that track faithfully!

This teaching by the well-known Chinese strategist Sun Tzu was focused more on army intelligence and strategies than on life wisdom, "Know our Foe, Know Ourselves, One hundred Battles, One hundred Victories." As there was no one on my side of the family in the military, I had never heard about war experiences besides watching RVN government news cinematographic reports or reading from Sài-gòn newspapers (more or less controlled by the government). The long history of many thousand years fighting against the Chinese had developed among the Vietnamese a keen interest in Chinese strategies. They loved to study Chinese epics of ancient wars and how that wisdom had eventually been used to fight back against China.

I remembered reading stories in children and adult story books (or even cartoons) about "silent wisdom battles" that I had mentioned earlier between the Chinese strategists sent to An Nam to test the Annamite mandarins (Emperor's cabinet) in a battle of the mind. Overtly and covertly, the former had oftentimes challenged the latter[91] into actual or virtual battles in both military might and literary/ strategic wisdom battles. Their purpose was to research An Nam strengths and weaknesses, and evaluate feasibility whether or not they would demand an increase in tributary submission (if An Nam showed signs being mighty) or attack An Nam (if there were signs of weaknesses). This might be the reason why Vietnamese rarely trusted foreigners! "Why are they here?" and "Are they here to test us out like the Chinese did?" were the usual probing questions that would tickle their brain whenever they could probe foreigner's interest in any kind of involvement. "Know our Foe, Know Ourselves..."

Generation after generation being harshly colonized by China and France, the Annamite, and later on Vietnamese, had been engrained in their genes so that they would purposefully embrace the occupiers' or foreigners' culture, language, technological advancement, history, and even their way of thinking with the sole purpose of using the enemy's knowledge to fight back[92]. They went even further to the extent of living,

---

91   Emperor's court high-ranking officials were divided into two branches. The martial mandarins were the generals who planned and conducted war, while the literary ones dealt with the laws and management

92   To my astonishment, I found out, when researching for my doctoral

impregnating themselves with that wisdom so they could feel/ act them out as if they were the foreign invaders themselves. As a result, the Vietnamese could wage long and subtle (guerrilla) insurgency wars.

In the meantime, many others had chosen a different route and used their eagerly acquired knowledge to serve the colonizers as collaborators. The latter would use their knowledge about their own people and adapt it to the new culture to serve the dominating master. In order to market their skills and knowledge, they would not hesitate to put others down so they could stand out and therefore elevate themselves and be noticed, so that the colonizers would choose them for socio-economic or political advancement opportunities in the system. This could probably be true more with the Southerners as the French were more interested in the more fertile and resourceful South. This political arena had caused more divisiveness between North and Southerners, and, in particular, a major distrust between the nationalist activists and the potential collaborators, and, therefore, caused the Southerners to be divided into opposing factions. Psychologically speaking, this polarizing phenomenon had probably expanded through time up even to the present time. Bullying others (between Vietnamese) became a nuisance behavior when they were showing off to prove they were worthier or superior, even if they were incompetent or lacked knowledge and common sense… "Know our Foe, Know Ourselves…"

Political and patriotic learning had been infused in all forms of Vietnamese literacy from children's stories and cartoons to adult books, and proverbs such as "nuôi ong tay áo" (raise bees in one's sleeves), "gậy ông đập lưng ông" (use their stick to beat them up on the back), "điệu hổ ly sơn" (lure the tiger away from its den so an attack can be performed successfully on it), "nội công, ngoại kích" (striking inside-out while attacking out-side-in)… which had been incorporated into their daily life, then had sprung out whenever any situation came up requiring strategic planning. All along in Việt Nam history, it had always been "learn from the enemy to defeat them." The key to military success was undoubtedly to study

---

dissertation, that, on the contrary, many from the African American community and culture do their best to boycott, refuse to participate in, or just choose to do less in the established education system (or government, corporate business) because they perceive it is a White system. The African American who succeed in it were classified as "sold-out." My dissertation as published via University of St. Thomas, St. Paul Minnesota: *Discourse of Respect Among African American High School Students.*

about self and the enemy, and match our strength and strategies with theirs. "Know our Foe, Know Ourselves…"

…I grew up practicing different styles of martial arts and had been deeply exposed to strategies used while confronting an opponent: what should I do to block his/her attack and, in a quick second-split reaction, must know what move to use to counter attack, quickly but with precision. Know all of their vital points: they are our targets. Know all parts of our body: they are our weapons. Then, having spent almost three years in the ARVN, I strongly believe that, if we "know the enemy (intelligence), and self (self-evaluation), one can win one hundred battles out of one hundred[93]." As a matter of fact, one should always out think and outwit one's enemy, and act faster than they could in order to have superiority in the battle. Consequently, one's eyes should always remain open, scrutinizing from top to bottom the opposition, and taking advantage of any occasion to "tuỳ cơ ứng biến" attack or counter attack accordingly. You blink, you die! People might be wondering how fighters could do all of those simultaneously and swiftly… It was all about training and practicing with perseverance what one had learned in a combination of elements…

Sun Tzu

As Sun Tzu[94], a 5th century BC Chinese army general, military strategist, and philosopher who lived in the Age of Spring and Autumn [95] ("thời Xuân Thu") of ancient China, had said: "If you know the enemy and know yourself, you need not fear the result of a hundred battles. If you know yourself but not the enemy, for every victory gained, you will also suffer a defeat. If you know neither the enemy nor yourself, you will succumb in every battle.[96]" He believed that "major military strength is not as critical in winning a war as outwitting the enemy. Study their weaknesses and strength, avoid confronting where you are weak and attack where they would not expect

---

93   In Vietnamese: "Biết Người, Biết Ta, Trăm Trận, Trăm Thắng."
94   https://en.wikipedia.org/wiki/Sun_Tzu
95   https://en.wikipedia.org/wiki/Spring_and_Autumn_period was Chinese medieval period associated with Confucian teaching promoting imperial law and order
96   Sun Tzu's Quotes, http://www.goodreads.com/quotes/17976-if-you-know-the-enemy-and-know-yourself-you-need

the most they would be attack..." Use Soft to conquer Strong and use their own strength to overpower them. Even though most of his strategies and teaching are now outdated and inapplicable, some had actually evolved and proved to still be very applicable towards successful campaigns, like the ones mentioned earlier. What was necessary to a campaign success was beyond army intelligence and researching about the enemy or weapon superiority. It also required serious and true learning about self: self- reliance, power, sustainability, one-to-one comparability between self and the enemy, the vital factors that will lead to success in planning strategies and tactics before, during and after confronting the enemy. These strategies would not apply just to army campaigns, but had also proven to be effective in planning and running a business, operating a company or even starting a family.

Source for both photos: France archival – Indochina
Many battalions of soldiers from Cochinchina were preparing to fight for France in WW1 in 1916. They were trained in boot camps in Saint Raphael, Southern France.

Resisting the French Occupation

After having gained some Independence from China, the Annamite began to really appreciate what it felt like to be free from a foreign invasion, until 1880s when the French came to invade and colonize their Ancestral Land. The next nearly three fourths of a century turned out bloody again. In the same way annual tributary offerings[97] "Triều Cống" was paid to China (e.g. spices, precious delicacy food, precious gemstones, China ceramics, silk and brocade, clothes, and even concubines) during its Domination for over a thousand years, goods exploited in Việt Nam were

---

97   https://en.wikipedia.org/wiki/History_of_Vietnam , https://en.wikipedia.org/wiki/List_of_tributaries_of_Imperial_China#List_of_tributaries ,

now taken to France (e.g. rubber for Michelin industry, spices, exotic fabrics, clothes, tea, rice...), and human power was passed from hand to hand in slave open markets serving the "Bạch Quỷ White Devils."

Some electronic photos recovered from 1916 French archives shows An Nam soldiers in training in French military camps in St Raphael boot camps in the Mediterranean Southern part of France (Provence-Alpes-Cote d'Azur). These are among the many Annamite[98] infantry battalions that have fought for "mother-land" France in WW1 during its utmost decisive phase. These are also the first Vietnamese used in the French colonial army (not French Legionnaires which consist of foreign fighters from various countries)... These photos have really justified my earlier statement that Vietnamese are obedient and loyal, and that, if they have decided to collaborate, they will serve the enemy-rulers faithfully even if they have to risk losing their life...

The French-made movie "Indochine[99]," even though fictive, has actually reflected well relationship between the French colonials and Vietnamese indigenous (Annamites) during the early 1900s. The film shows that collaborators are treated better than others at different levels, while different factions of rebels are organizing insurgences. Many ambitious Vietnamese who want to collaborate have done their best to attract the colonizers' attention so they can be chosen to serve, by smashing down the competition and elevate themselves.

The French colonial government divided Việt Nam into three parts, with their own government run by a separate Annamite governor Đốc Phủ Sứ, the one in the South being probably the most prestigious one, because it was the most fertile region: this was where the first French colonial central administrative Capital (of the whole Indochina) was located. The strategy was to actuate a plan to divide and conquer[100].

---

98   As the Annamite were called by the French.

99   Indochine is a French movie released in 1992.It is based on a fictive story of a rich French lady (actress Catherine Deneuve), a rubber plantation owner, who has adopted an imperial princess orphan. The story developed into the rise of communist insurgence that ended with the French defeat and the Geneva Peace Conference.

100   This plan was somewhat successful as the Vietnamese started discriminating against each other in one way or another, based on whether they were from the North (the poorest agricultural part of Vietnam) , Central (the Imperialist part) or South (the agriculturally richest, as exploited and perhaps the most favored by the French).

These three parts of Việt Nam, Tonkin (North), An Nam (Central) and Cochinchina (South), were combined with Laos and Cambodia to form the five French Indochinese parts that were run more or less as separate countries, but all attached to motherland France and her central colonial administration[101]. The "Gouverneur General" ruled all five at an administration center rotated between the capitals Sài-gòn (1887), to Hà-nội (1902), then Đà Lạt (1939), and finally back to Hà-nội (1945)[102]. As Tonkin was agriculturally much less fertile and, perhaps, politically more rebellious, the French were more interested in exploiting An Nam and, in particular, Cochinchina, where crops from the luscious rice fields, and tea and rubber plantations were harvested and shipped back to France.

Not like the British in India, according to many history analysts, the French were much more brutal and bloodier, as total exploitation, not just politics, was the real driving factor that drove the French into colonizing Indochina[103]. Colonial officials and French companies converted Việt-nam's thriving economy towards serving the French Motherland: a total exploitation by divide and conquer. "Annamite[104]" patriotic scholars and fighters had tried to stand up to regain Independence, but their movements had been defeated for the next many decades. That sentiment of longing for Independence, which oftentimes less educated Annamite had confused with longing for Peace, came back ruminating among the hopeless and tamed. What we have lost, we appreciate more!

---

101 Imagine these were five states in a federal government of the United States.
102 https://en.wikipedia.org/wiki/French_Indochina the capital of French Indochina
103 Chinese colonization before that was both political and economical, as harsh as opposition was completely eradicated by mass execution (up to three generations).
104 As the French called the Vietnamese at that time.

CHAPTER 4
The Turning Points

Millions of indigenous people were working diligently to build up power and resources for their French overlords[105]. The former's obedient, loyal, and devoted Confucian Chinese-influenced rearing had kept them in status quo so they could enjoy whatever grace they were receiving from their masters. In exchange for being submissive, loyal to their new master, they would get treated more or less kindly than those who had chosen not to collaborate, this perhaps encouraging and boosting collaboration with the French more than under the Chinese domination (also see Democracy and the Great Divide in Chapter 11). Those who had been politicized realized their Ancestral Land needed to be liberated, while the collaborators enjoyed their more advanced status, socially, politically as well as economically. As most Annamites and Cochinchinois cooperated or passively resisted colonization[106], others had secretly or openly opposed the regime.

The first anti-French insurrection movements, such as the Đông Du[107], were started to rally patriotic Annamites into an anti-French coalition, to punish collaborators and to attack the French establishment. Those who had appreciated a taste of Independence from Chinese colonization were now discretely rebelling and instigating secret rebellion groups, such as the groups led by Đề Thám, Phan Đình Phùng, Việt Nam Quốc Dân Đảng, Prince Cường Để's monarchist "Phục Quốc" (Restoration Party in 1938[108]), Phan Bội Châu (the Vietnamese pioneer of Nationalism who formed a group called the "Vietnamese Restoration League," or *Việt Nam Quang Phục Hội*, modeled after Chinese Sun Yat Sen's republican party), Hồ Chí Minh..., or movements strategizing new political directions, such as Văn Thân, Cần Vương, Phong Trào Đông Du. The latter were

---

105   http://alphahistory.com/vietnam/french-colonialism-in-vietnam/ French colonization of Vietnam
106   https://en.wikipedia.org/wiki/History_of_Vietnam_during_World_War_I
107   Vietnam and the French website by Sanderson Beck http://www.san.beck.org/20-10-VietnamandFrench.html
108   https://en.wikipedia.org/wiki/Cường_Để.htm Prince Cường Để's anti-French movement

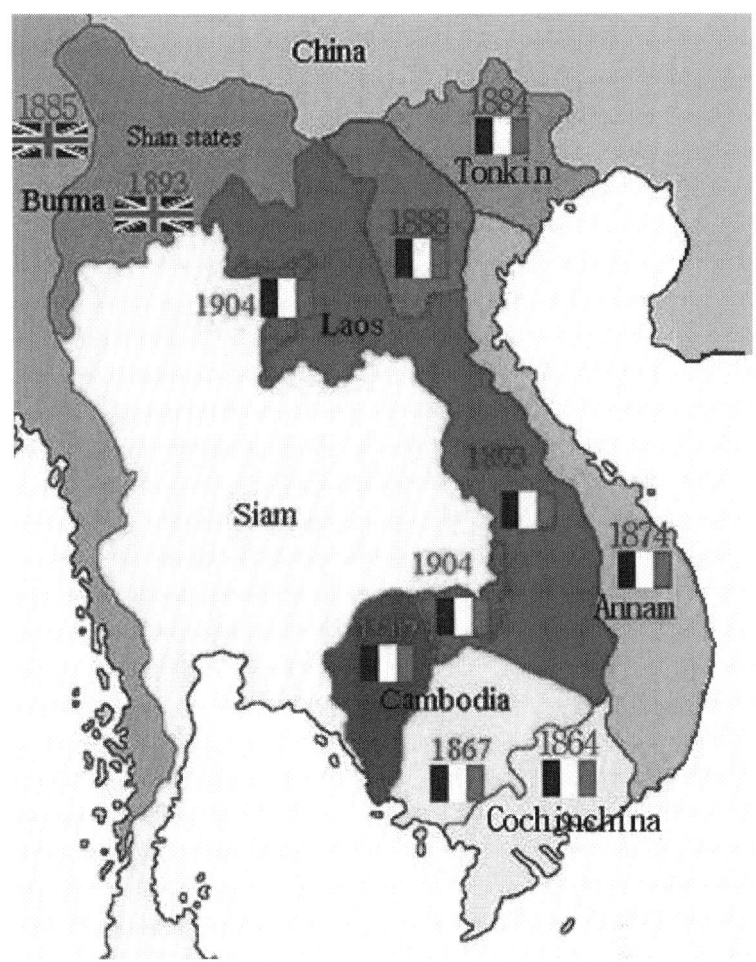

Map Source: www.lahistoriaconmapas.com   Wikimedia.org

seeking an alliance with the East, in particular Japan and even the old enemy China, these two countries experiencing the same cataclysm: the invasion of the "White Devils." Others, such as the An Nam Imperial Court movement, were religiously motivated, trying to curb Catholic expansion.

Later, when France was invaded by the Nazi, the Vichy Government was created by French collaborators in Southern France to govern occupied France administratively.

When Japan expanded its occupation in the Pacific Ocean[109] during WW2, it took over Việt Nam militarily in 1940. The established French Colonial Government was allowed by the occupying Japanese armies to continue running the five Indochina's administratively (the general administration was in Đà Lạt, Cochinchina). Japan had, by analogy deduction, considered Vichy France as its ally as it was Germany-occupied at that time with a collaborating government in Vichy, Central France. The U.S. secretly supported Hồ Chí Minh and Gen. Võ Nguyên Giáp's Việt Minh[110] (communist and nationalist united in Việt Nam Độc Lập Đồng Minh Hội[111], or Vietnamese League for Independence) by having the Office of Strategic Services OSS (the origin of the current CIA Central Intelligence Agency, DEER Team) train their troops[112] [113]to start an insurgence war against the Japanese. At the same time, Vietnamese anti-Japanese movements[114] [115]rose significantly with the Việt Minh being the most significant group.

Who was Hồ Chí Minh?

There had been many legends about Hồ Chí Minh, as many as he had pseudo-names, to the extent that many analysts believed "those Hồ Chí Minh's were not actually the same man, and that they could not even be the real Hồ. To make it even more complicated, China had released a certain Hồ who had served in the People's Army in the rank of major... married to a Chinese woman... Who was he? Or, to be more intriguing, which one was Hồ? Or which one had done what in the War?

---

109   http://www.globalsecurity.org/military/world/vietnam/hist-wwii-jap.htm Japanese occupation of Vietnam
110   http://nova.wpunj.edu/newpolitics/issue23/goldne23.htm anti-Colonial movement in Vietnam
111   A great many nationalist Vietnamese extremists believe the Việt Minh equated Communist; This was actually true only after French defeat at Điện Biên Phủ. The American OSS (pre-CIA) helped train the Việt Minh (nationalist and communist) in 1945 when the latter were fighting the Japanese.
112   Office of Strategic Services https://en.wikipedia.org/wiki/French_Indochina_in_World_War_II
113   Office of Strategic Services http://www.historynet.com/ho-chi-minh-and-the-oss.htm
114   http://www.ibiblio.org/pha/timeline/410801awp.html
115   Vietnamese Blog https://www.wattpad.com/345858- cac-anh-hung-trong-lich-su-danh-giac-viet-nam-cac

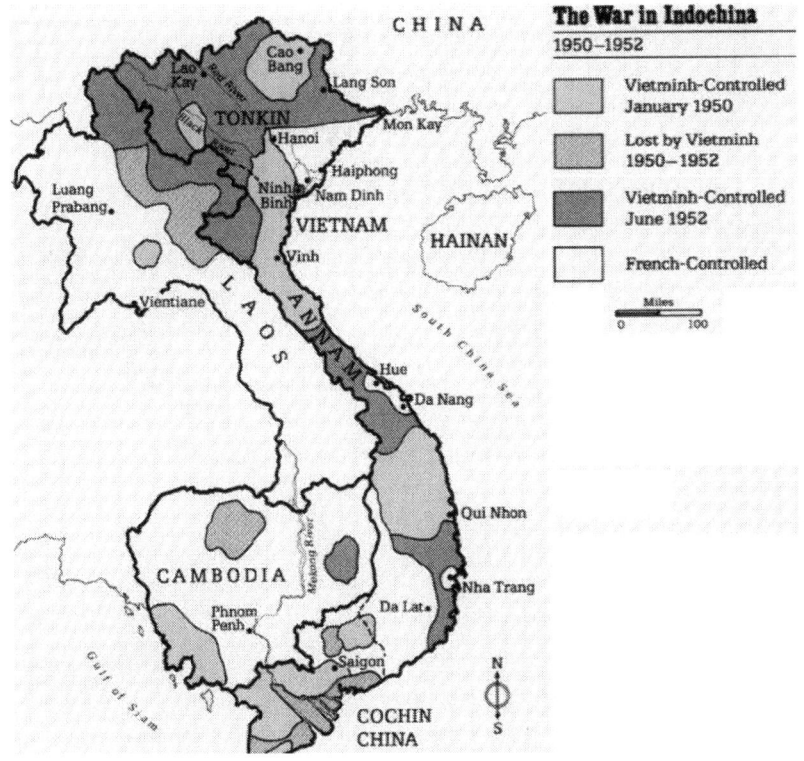

Source: users.humboldt.edu
Vietnam.html
(Note that the Việt Minh was mixed communist and nationalist at the beginning and was not purged into Communism until after 1954)

One version alleged that Hồ had worked with the OSS, and that this was not the first time Hồ had some experience with the U.S. He was a well-traveled man in France, working as a kitchen helper on ships under different pseudo-names and aliases, from birth name Nguyễn Sinh Cung to Nguyễn Tất Thành, Văn Ba[116]... The U.S. was probably where he spent more time on mainland, particularly in New York City and Harlem 1912-13, where he had left long-lasting friendly impression among the African Americans because of his deep

---

116   https://en.wikipedia.org/wiki/Ho_Chi_Minh#In_the_United_States Hồ wondering in the U.S., Europe and Asia https://abibitumikasa.com/forums/show-thread.php/133630-Ho-Chi-Minh-attended-Garvey-Lectures-in-Harlem

compassion for their being racially oppressed[117][118], and, later on, in Boston (Parker House Hotel baker). Then his travel path took him to the United Kingdom where he spent even more time (1913-1919). He was perhaps in search of his own identity and an inspiration or a possible enlightenment for a Free Việt Nam which at the time was under French colonization. It was possible that he had witnessed the poor in the U.S, mainly African American, being racially oppressed. Could he relate well to what the African Americans were going through because such oppression was very close the socio-political calamity the Annamites (as the French used to call the Vietnamese in the three Indochinese parts of Việt-nam) were experiencing? The Civil Rights movement was ignited and going stronger and stronger in the U.S. at that time, especially with Rev. Martin Luther King's devotion. Anyway, the sympathy the African Americans had for Hồ had made King suspected for being a communist or a sympathizer. I could still remember having read about this in Sài-gòn newspapers.

  Upon his return to France where he settled down for an even longer time (1919-1923), he started getting involved in politics hanging out with the Socialist Party of France and the Vietnamese nationalist movement in France. He petitioned for civil rights for the Vietnamese in French Indochina, under different pseudo-name Nguyễn Ái Quốc. He kept pushing for more networking and founded the "Parti Communiste Français" which he used to attract French citizens to the plight of indigenous in all French colonies. Seeing that his efforts in France were in vain, Quốc switched to the powers in the East: the Soviet Union and China. He left Paris to go to Moscow under the name of Chen Vang to study at a communist university, and became an active participant in the Fifth Communist International (the 5th "ComIntern") Congress before going to Canton (Guangzhou, China), where he married a Chinese woman, Zeng Xueming (in Vietnamese phonetics Tăng Tuyết Minh) and taught seminars on Socialism. In 1930, he united different Vietnamese communist factions into the

---

117 http://hyphenmagazine.com/blog/2013/12/9/december-lit-ho-chi-minh-harlem-phong-nguyen Ho Chi Minh in Harlem described "Bác Hồ" as a charismatic friend whose alternative life might be totally different if, according to Ho's African American lifelong friend Ed Winston, he had chosen to become an American citizen instead of returning to Việt Nam. When asked about how he felt about America, Ho, at that time nicknamed "the Mute of Harlem" and a pastry chef at Hotel Theresa in Harlem, said "America is fine…but Harlem is my home."

118 https://abibitumikasa.com/forums/showthread.php/133630-Ho-Chi-Minh-attended-Garvey-Lectures-in-Harlem Hồ Chí Minh attended Garvey Lectures in Harlem written by Hồ Chí Minh's anti-Communist biographer William Duiker

Communist Party of Việt Nam, and kept soaring when he returned to Moscow to teach at Lenin Institute (1933). In 1938, Quốc went back to China to serve as an advisor to the Chinese communist armed forces. Recent Chinese declassified archives showed pictures of him being Maj. Hồ Quang of the Chinese People's Army. In 1941, he returned to Việt Nam as Hồ Chí Minh to lead the patriotic Việt Minh independence movement (comprising both nationalists and communists) against the French.

At this point on the timeline, a great many sources and television news documentaries showed the Office of Strategic Services[119] officers of the "Special Operation Deer Team #13" training Hồ Chí Minh's troops in guerrilla warfare and readying them to fight against the Japanese. Hồ had admitted to U.S. Army Major Allison Thomas, Deer Team leader[120], at their last "mission accomplished" dinner (Japanese surrendering) that he was communist; this was the end of American aids to Hồ as what followed would be more complicated: the U.S. would not help him fight against its ally, France… for sure!

After the Japanese surrendered in August 1945, the French returned to An Nam and Cochinchina and resumed its full power in the newly-regrouped Indochinese Federation. Taking advantage of the French-Japanese transition, Hồ seized Tonkin from the retrieving Japanese and declared it independent, as the new People's Democratic Republic of Vietnam. Hồ Chí Minh also pressured An Nam last Emperor Bảo Đại to abdicate during his "Cách Mạng Tháng Tám" (August Revolution)[121]. Following was his September 2 Declaration of Independence[122], which precipitated the preparation of his army for the "First Indochinese War" against the French.

The Vietnam Forum on Human Rights, Democracy and Freedom on Paltalk ("Diễn Đàn Việt Nam về Nhân Quyền, Dân Chủ, và Tự Do trên Paltalk") [123] had identified in a blog

---

119   http://www.historynet.com/ho-chi-minh-and-the-oss.htm HistoryNet,
120   https://truehochiminh.wordpress.com/2012/08/02/ho-chi-minh-and-the-oss/ Hồ and the OSS
121   https://en.wikipedia.org/wiki/August_Revolution August General Uprising *Tổng Khởi nghĩa tháng Tám*
122   some believed it was based on the American Declaration of Independence as George Washington was said to be Hồ's role model
123   Wordpress.com Nhã Thanh Sử's blog about the five Hồ Chí Minh posted in May 2017 had been deleted within a few months later.

written by commentator/blogger Nhã Thanh Sử five different Hồ Chí Minh's, based on five different personalities, political ideologies, achievements, allegiances, relationships with partisans, alliances with super powers... One, a Vietnamese man, was real and four, Chinese, were fake. Nhã believed the real one was longing for a real Democracy and true Human Rights and that he was influenced by America: he was the one who delivered the Declaration of Independence for Tonkin and who had dissolved the Vietnamese Communist Party while pushing for a united government with multiple parties. He was the one elected as Chair of the Parliament. He was also taken away by the International Communist Party which had obliterated all of his legacies and images from circulation. His replacement was a clone whose image had been "photoshopped" onto the history of the current Hồ.

    This was a conspiracy theory, but it did make sense as it could explain why the Hồ whom Harlem African American community had nicknamed the "Mute of Harlem appeared to be a modest and reserved person, completely different from the out-going and charismatic revolutionary one we all knew about nowadays. The original one was, according to the African Americans who had known him, pleasant, patient and hardworking, but shy and quiet: a man who loved America and whose hero was George Washington! Along the same note, many other researchers and historians[124] commented that Hồ could have been a Hero for having liberated Việt Nam from the French and made it a modern and democratic country; but, instead, he aimed at making his Ancestral Land a part of the Communist International and subjugating his people to a fake Democracy.

    History Professor Nguyễn Thế Anh of the Institute of Vietnamese Studies[125] pointed out, using Communist Party members' own writing, including Hồ's and his closest disciples' such as Phạm văn Đồng's, that, by using constantly changing

---

[124]   Such as French commentator author Jean-François Revel and author Olivier Tod, Prof. Ralph B. Smith (University of London), Bùi Xuân Quang (founder of Đường Mới Group), Prof. Nguyễn Thế Anh (former Huế University), Prof. Tôn Thất Thiện (Québec University), Prof. Lâm Thanh Liêm (former Liberal Arts Văn Khoa College in Sài-gòn), Prof. Đinh Trọng Hiếu (University of Paris VII), Prof. Nguyễn Ngọc Huy (researcher of Harvard University) https://ongvove.wordpress.com/2009/08/21/nhom-duong-moi *Hồ Chí Minh, Sự Thật về Thân Thế và Sự Nghiệp (Ho Chi Minh, the Truth about his life and career) in Vietnamese by Nhóm Đường Mới (ongvove.wordpress.com) published by Nam Á Publishing Company, Paris, France (1990).*

[125]   Article was deleted in July 2017. http://www.viethoc.com/Ban-Ging-Hun/nguyen-the-anh

identities and made up background stories, Hồ had actually, not just slipped under the French surveillance and participated in many major communist conferences (e.g. the 5th Congress of the Communist International in 1924), but also taken advantage of his membership in the Việt Minh (Việt Nam Độc Lập Đồng Minh Hội) to start purging it (1941) into an eventually pure communist movement. The main question remained: Why did he switch identity like days and nights? Was it he who did it, or someone else? What the people of Việt-nam found out a few decades after "the most current Hồ" passed away was even more horrendous: Did he sell out Ancestral Land or part of it, Paracel and Spratly Islands, to China, causing a major turn up in the South China Sea? Was he China's People's Liberation Army Major Hồ Quang? Did he commit all of those crimes as Vietnamese nationalist groups had alleged[126]? If there was smoke, there should be fire, somewhere…

Recently, Catholic priest Nguyễn văn Lý, a spiritual leader considered dissident by the DRV (they had imprisoned him the 2010s without due process), had revealed some convincing evidence about Hồ Chí Minh's identity in a video posted on YouTube[127]. Father Lý asserted that the original Nguyễn Ái Quốc had died of tuberculosis in 1933 when he was on his way from Hong Kong to Moscow to find a cure. This timeline coincided with his going to Moscow in 1933 to teach at Lenin Institute as discussed earlier. The news media in the UK, France and Hong Kong had announced his death. There was a memorial organized by exchange students in his honor in the Soviet. Afterwards, the International Communism, the Soviet as well as China started a campaign looking for a Nguyễn Ái Quốc look-alike. China found a Taiwanese by the name of Hồ Cẩm Chương, who looked almost like Quốc, except that he was significantly taller by 10cm (1m72 instead of 1m62). Both Hồ had met earlier and were compatible in their communist ideology. Chương was therefore chosen to replace late Quốc. He stayed in Hồ Nam where he was assigned the new identity of Major Hồ Quang serving in the Chinese

---

126   *Tôi Ác Hồ Chí Minh in Vietnamese on YouTube, and Documentary film in Vietnamese: Tội Ác Công Sản, Sự Thật về Hồ Chí Minh Communist Cruelty, The Truth about Hồ Chí Minh (July, 2009), available on YouTube.*

127   https://video.search.yahoo.com/yhs/search?fr=yhs-iry-fully-hosted_003&hsimp=yhs_fullyhosted_003&hspart=iry&p=youtube+nguyen+van+ly+ho+chi+minh#id=52&vid=8d7d3d92a6ec0cd6962fea444fcb980f&action=click in Vietnamese. Father Nguyễn văn Lý announced the proofs that Hồ Chí Minh was a Chinaman 100%, but not Nguyễn Ái Quốc LM Nguyễn Văn Lý công bố bằng chứng Hồ Chí Minh là người Tàu 100%.

People's Army as discussed earlier. During his long stay in China, Quốc, who knew French and English, had learned Russian and, especially, the Nghệ An dialect of the Vietnamese language. Around 1940, he was released into the field in the Northern part of the DRV, Cao Bằng, to take over the newly formed Việt Minh organized by a few comrades, one of whom had also the last name of Hồ (Hồ Học Lạc, or Hồ Vinh Sơn, or Hồ Chí Minh) who just passed away because of asthma. Hồ Quang's meeting in 1942 with Zhou Enlai, China's future First Premier, was his first official event with the new identity of "Hồ Chí Minh." This explained well why there was such a big difference in personality between the original and the last Hồ Chí Minh, and why an atrocity policy (inhumane treatment of civilians and prisoners, massacres) was adopted by the NVA and VC when dealing with South Vietnamese. Questions were still pending on the original Minh's wife, a Chinese woman, Zeng Xueming (Tăng Tuyết Minh). Wouldn't she know the truth about her late husband being replaced? Father Nguyễn văn Lý further unveiled that the fake HCM (Hồ Quang, who had uneven ear lobes) had two different doubles who had balanced ear lobes[128]...

If this conspiracy theory were true, the "fake Hồ Chí Minh," the Chinese Maj. Hồ Quang who had replaced HCM when the latter died in Hong Kong Prison in 1932-33, would have to fit in a plausible explanation of the following facts that:

1. He was the author of six letters sent to U.S. Pres. Truman, one of them dated February 28, 1946, a few years after he had supposedly died in a Hong Kong prison... or did he in Moscow?
2. He died when in Moscow in 1933: he was supposed to teach at Lenin Institute there and was looking for a tuberculosis cure. Moscow confirmed his death with an obituary.
3. Another source said he was caught by Hong Kong police and had died of tuberculosis in 1932 in a prison there. Some other sources said he was assassinated in a Hong Kong prison. His ashes (under one of his pseudonames Nguyễn Tất Thắng) were kept in Kuntsevo

---

128   https://video.search.yahoo.com/yhs/search?fr=yhs-iry-fully-hosted_003&hsimp=yhs-fullyhosted_003&hspart=iry&p=youtube+nguy-en+van+ly+ho+chi+minh#action=view&id=51&vid=4c47cf69ffda65fd01f5c15a-b913610a in Vietnamese: Father Nguyễn Văn Lý discussing the three HCM impersonators. LM Nguyễn Văn Lý Kể Chuyện 3 Người Đóng Giả Hồ Chí Minh

Cemetery in Moscow[129].

```
VIỆT-NAM DÂN CHỦ CỘNG HÒA                              YKB-3739-1
CHÍNH PHỦ LÂM THỜI
BO NGOẠI GIAO                          HANOI FEBRUARY 28 1946

                       TELEGRAM                MAR 11 RECD

PRESIDENT HOCHIMINH VIETNAM DEMOCRATIC REPUBLIC HANOI
TO THE PRESIDENT OF THE UNITED STATES OF AMERICA WASHINGTON D.C.

        ON BEHALF OF VIETNAM GOVERNMENT AND PEOPLE I BEG TO INFORM YOU
THAT IN COURSE OF CONVERSATIONS BETWEEN VIETNAM GOVERNMENT AND FRENCH
REPRESENTATIVES THE LATTER REQUIRE THE SECESSION OF COCHINCHINA AND THE
RETURN OF FRENCH TROOPS IN HANOI STOP MEANWHILE FRENCH POPULATION AND
TROOPS ARE MAKING ACTIVE PREPARATIONS FOR A COUP DE MAIN IN HANOI AND
FOR MILITARY AGGRESSION STOP I THEREFORE MOST EARNESTLY APPEAL TO YOU
PERSONALLY AND TO THE AMERICAN PEOPLE TO INTERFERE URGENTLY IN SUPPORT
OF OUR INDEPENDENCE AND HELP MAKING THE NEGOTIATIONS MORE IN KEEPING WITH
THE PRINCIPLES OF THE ATLANTIC AND SAN FRANCISCO CHARTERS
        RESPECTFULLY

                                           HOCHIMINH
                                           Hochiminh
```

Among others, Feb 28, 1946 Letter from Hồ Chí Minh
to U.S. Pres. Harry Truman
Source: www.cvce.eu

---

129   The Third Republic of Vietnam Facebook post

4. He had worked with the OSS agents (Now CIA) during the early 1940's Japanese occupation. This was confirmed by French-born U.S. Maj. René J. Défourneaux[130] who was one of the OSS case officers who trained Hồ's troops in guerilla warfare against the Japanese forces.
5. He was patriotic enough as a Vietnamese to mastermind the last victorious battle of Điện Biên Phủ against the French in 1954.

    The pieces of information above are so conflicting that the triangulation process cannot lead to a plausible conclusion about the identity of Hồ Chí Minh, leaving us to be puzzled: who has been the real "Uncle Hồ" throughout the war until he dies in 1969? The only clue is that the dead one is 10 to 12cm taller than the original Nguyễn Ái Quốc or Nguyễn Tất Thắng... The truth of the matter is Hồ Chí Minh has been a public figure representing the Vietnamese Communist Party and its atrocious agenda. The real brain that has drawn and executed all the bloody plans for invading and swallowing South Vietnam is the three musketeers Lê Duẩn and Trường Chinh, with the loyal Gen. Võ Nguyên Giáp. Whether Hồ has died in the 1930s or not, the war machine has always been in top gear, as the invisible brain is still aiming at the end result, no matter what needs to be done, and how many lives it will cost.

    In the meantime, as the new President of the Democratic Republic of Việt-nam (DRV, former Tonkin), Hồ Chí Minh had sent a great many letters to Pres. Harry Truman, his Sec. of States James Byrnes (1945, 1946, 1948), then to the Governments of many countries including China, the U.K., the Soviet Union (1946) proving that France (Vichy) had betrayed the U.S. by surrendering to the Japanese and requesting help to bar France from returning to Việt-nam[131].

---

130   Author of <u>The Winking Fox: Twenty Two Years in Military Intelligence and The Tracks of the Fox – Uncovering Secrets of Wartime Covert Operation, accounts of OSS activities he had participated in</u>, in WW2 and in French Indochina.

131   http://www.historyisaweapon.com/defcon2/hochiminh/ Letters from Hồ Chí Minh to America between 1945 (Declaration of Independence of Tonkin, and Pres. Truman) and 1969 (Pres. Nixon)

Turning point: the hidden letters

As mentioned earlier, after having supported Hồ Chí Minh's Việt Minh to fight the Japanese, the U.S. declined to continue their support against the French. Instead, it would fully back up the French colonial government with substantial military aids and equipment to maintain the

Why?

- During World War II, while working directly with American agents to rescue downed U.S. pilots, Ho Chi Minh sent six letters to the U.S. government asking for support and stating that the Vietnamese wished to pattern their constitution after America's. Only after America *refused* to recognize and support their freedom, and instead supported the French suppression of their freedom, were the Vietnamese forced to turn to China and the Soviet Union.
- British colonial officer, Major-General Douglas Gracey, took surrender of the Japanese in September 1945, in Saigon, but "immediately rearmed them and ordered them to put down the Vietminh, who had already formed an administration in the South. Like the Vietminh of the North, they were a popular movement of Catholics, Buddhists, small businessmen, communists and farmers who looked to Ho Chi Minh as the 'father of the nation'."
- By 1947, thanks to the British and Gracey, the French were back in power in Saigon.
- Needing to get the French out of their country, Ho Chi Minh was still hoping for a U.S. alliance and he appealed again to President Truman insisting that he was "not a communist in the American sense". Although he had lived and worked in Moscow, Ho considered himself a free agent; but he warned that he "would have to find allies if any were to be found; otherwise the Vietnamese would have to go it alone." And alone they went until 1950 when Ho Chi Minh believed he could no longer delay accepting the formal ties and material assistance under offer from the Soviet Union and especially from China. It was the success of the Chinese revolution in 1949 that was to give the Vietminh the means to defeat the French: military training, arms and sanctuary across an open frontier.

Việt-nam Other Issues regarding Hồ Chí Minh's letters to Pres. Truman that were "hidden"
Source: slideplayer.com

French establishment in Cochinchina, the last of Indochina. The U.S. had refused to help Hồ Chí Minh for two simple reasons: Hồ had officially declared himself communist (even though the Việt Minh at that time still included both communists and nationalists[132]) and France was a dear ally of the U.S. Hồ Chí Minh's letters (one of them is attached) to

---
132 http://indochine54.free.fr/hist/begin.html

Pres. Truman requesting the latter to help intervene had been held back, supposedly by the CIA (or Joint Chiefs of Staff?). On the previous page was a list of questions analysts had asked.

Among declassified documents were many letters to Pres. Truman from Hồ. They were held up by the CIA (?). What could have happened if Pres. Truman had read these letters and agreed to mediate France and Việt-nam? Anyway, France had started to give back independence to Laos, Cambodia, Tonkin and An Nam, and, eventually, Cochinchina before it was defeated at Điện Biên Phủ

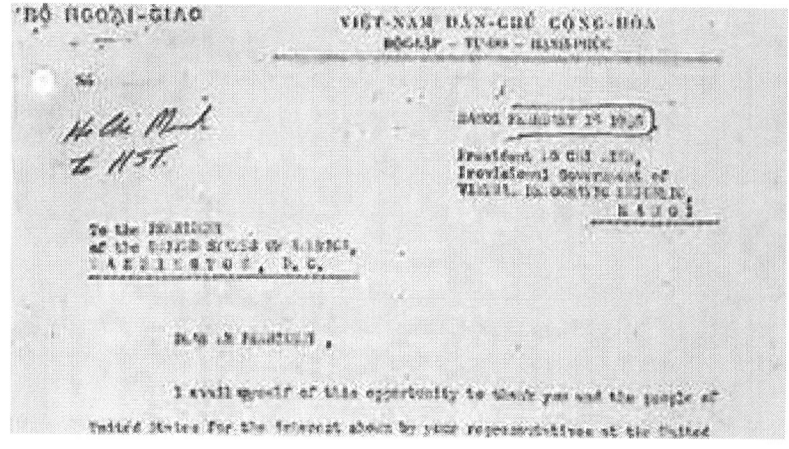

Another of Hồ Chí Minh's letters to U.S. Pres. Truman (p1)
Source: Vietnamnet.vn

a few years later... Why? Was it because someone wanted a war of attrition? These letters could have helped prevent the war from happening and save millions of lives. At that time, Hồ was still in good terms with the U.S....

On July 3, 1953, France declared it was ready to give Independence to all of Indochina as it had realized its military might had been deteriorating. By then, it had given Independence to Tonkin (3/9/45), An Nam (1948), Laos (10/22/53), Cambodia (11/9/53) and next, perhaps Cochinchina... If Pres. Truman had seen those letters and intervened, could the world have avoided the war?

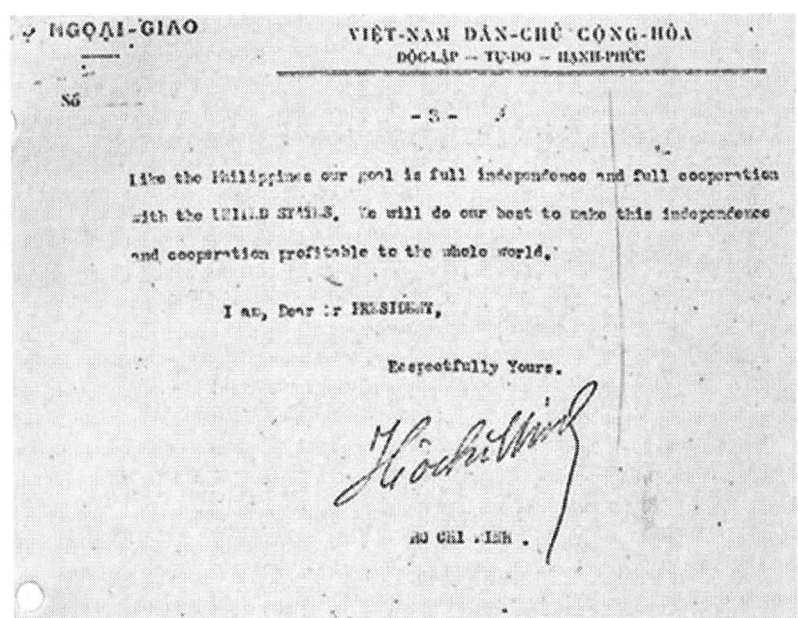

Hồ Chí Minh's partial letter to U.S. Pres. Truman
(re. the Philippines – p3 undated)
Source: dantri.com

The last communication between Hồ and the U.S. was probably a phone conversation with the U.S. Consul in Sài-gòn, George Abbott, who reported to the U.S. Ambassador in his Memorandum of September 12, 1946, that Hồ denied being communist and wished there would be Independence for Việt Nam as the Philippines (US assisted) and India had just received.

Turning Point: the remnants of the defeated Japanese army

Since the U.S. had turned Hồ down, the rogue Japanese Imperial garrisons, American former enemies, which were left behind after Emperor Hiro Hito surrendered, started training and arming Gen. Giáp's guerillas[133] to fight against the French and Americans. For Hồ, the Americans had officially entered the Indochinese theater against him as soon as they had started their Mutual Defense Assistance Program providing over a billion of U.S. dollars in military aids to France to fight the Việt Minh.

Hồ also explained that the French had not accepted Viet-Nam's demand that "democratic liberties" be restored in Cochinchina and reiterated that he would resist the French continuation of their former policy of economic monopoly in Indochina, and, if needed, he would approach other countries friendly to his plight: China and the Soviet Union. This final threat of allying with the Communist Block had probably shut down all sympathy, if any, for Hồ. The U.S. was then determined to gear up for an open Cold War with China and the Soviet Union, and a war of attrition against the DRV.

This was definitely the major turning point that had led Hồ Chí Minh to totally convert his party to Communism (Lenin-influenced) and to turn to the Communist Socialist big brothers, China and the Soviet, for assistance. Whether his Việt Minh comrades had understood what Communism was or not, whether they knew what he was going to totally turn communist or not, Hồ started purging the Việt Minh of nationalists while it was preparing for the final battle against the French.

This could be the main reason why my father had fled the Việt Minh to defect to the South in the late 1940s. My family's efforts in using my father's 1940s Việt Minh identification and membership cards (they found them after the war ended) failed. With proofs of his past membership, my father was hoping the Party would grant him leniency and spare him from the fate other Southerners would be subjected

---

133   VTV4: *Những Mảnh Ghép Của Cuộc Sống - Nhà Yêu Nước Phan Bội Châu - Tập 3 Vietnam TV History program on patriotic Phan Bội Châu, Vol. 3 March 16, 2016.*

to. Whether Father was a member or not did not matter much to the newly established Communist Party in the South (under the direction of political cadre Lê Duẩn). My Father was a defector to the South, a traitor!

1954 came into Việt-nam's history with the total defeat of the occupying French by the Việt Minh at the decisive Điện Biên Phủ Battle. This defeat marked an important turning point for Hồ Chí Minh in history as it was interpreted by him and presented to the patriotic "invaders-hating" peasants as the French passing the torch of colonization to the American, as proven by the latter's massive presence in body as well as weapons in South Việt Nam. Hồ started implanting into his strategy and his troops mind a new conspiracy against the Americans, using concrete facts observed during this last battle: the abundant military aids given to France by the U.S. to fight the Việt Minh had obviously proven that Americans were taking Việt Nam over from their retrieving French allies. The remaining armament not used by the French was passed on to the new State of Vietnam. Hà-nội used this arrangement to demonstrate to its followers that the South had chosen to serve the incoming Americans, the ally of France, as 'servants."

Hồ had also moved aggressively towards totally purging the Việt Minh fellowship of its non-communist members[134], thus making the latter search for refuge in the South during the 10 months allowed by the Geneva Conference (Accords) for Passage to Freedom (August 1954 – May 1955). He cleverly took advantage of the Vietnamese people's illiteracy and historical obedience towards authority, their hatred against foreign invaders and their total confusion about the new local leaders' and "wannabe's" political agendas to start his major campaign against what he called "Mỹ Ngụy" (American rebels) and their "Chính-phủ Sài-gòn, Tay Sai của Đế Quốc Mỹ" (the Sài-gòn Government, puppet servants of the Imperialist Americans). The Second Indochinese[135] War started to take shape as the propaganda was taken dead-seriously by a new guerila "underground," as it had vehemently been absorbed into the mind of naïve Southerners.

---

134   Britannica.com: Viet Minh Vietnamese revolutionary organization.
135   The term Indochina subtly evokes a relationship to the French Indochina and insinuates a new American Indochina (colonization of Viet Nam by the U.S.). "Second Indochina War" would then be interpreted as an "Invasion by Americans," therefore justifying the National Liberation Front to rise and fight back in self-defense. This is the reason why the NVA always carries the NLF flags in battle, even when it enters the Presidential Palace in Sài-gòn on April 30, 1975.

Recalling the Northerners' exodus to the South...*My friend Ngọc Hương was just a 2 year-old little infant when her family of 8 had chosen Freedom by fleeing Hà-nội to migrate to Sài-gòn as political refugees in 1954. As recounted by her eldest siblings, her mother had to carry her while herding the other children through hills and rivers by foot (sometimes with bare feet, trying to protect them from the cold with thatches), with no food or temporary lodging all along the road. On many occasions, they had to evade Việt Minh's checkpoints set up to push refugees back to the North. For days, the parents stayed hungry so the children could survive on whatever edible grass or cassava roots they could find, smashed in water from the creeks.*

The songs that move rivers and mountains

The road was long and painful as winter was coming, but Liberty was priceless and time was much restricted. Even though the Geneva Conference (Accords) allowed almost 10 months for refugees from the North to take advantage of "Passage to Freedom," an operation supported by American, French and other allies who provided them with evacuation transportation, the Việt Minh had actually blocked all the roads to the airport and harbors for fear of losing population to the South[136]. *Despising thorns and brushes tearing up their feet and legs, they walked and crawled until they passed the border, the 17th Parallel*[137], *to Freedom. It was almost like a death march a family of two adults and 6 children had to endure.*

Antiwar re- converted Republic of Việt-nam singer Khánh Ly's voice resonated Phạm Duy's moving song "in 1954, Father abandoned his birth place...in 1975, children left behind their homeland" reflecting the major exodus of North Vietnamese fleeing Communism in 1954 from DRV and from occupied South Vietnam in 1975:

---

136 Wikipedia: Operation Passage to Freedom - Communist prevention of emigration.
137 The 17th Parallel was chosen as the new divide line between North and South, between Communism and Freedom, and was the DMZ De-Militarized Zone.

Một ngày năm bốn, cha bỏ quê xa
Chốn đã chôn nhau, cắt rốn bao nhiêu đời
Một ngày năm bốn, cha bỏ phương trời
Một miền Bắc tối tăm mưa phùn rơi
Một ngày năm bốn, cha bỏ Sơn Tây
Dắt díu con thơ, vô sống nơi Biên Hòa
Dù là xa đó, vẫn là quê nhà
Và miền nắng soi vui gia đình ta!

"One day in 54, father left his homeland
The place where many generations were born (had buried
placenta),
One day in 54, father had left behind his sky
A Northern place darkened by tempest rain,
One day in 54, father abandoned Sơn Tây Province
And led your innocent children to resettle in Biên Hoà,
Even though far from home, this is still home
Where our family could enjoy the sun shine...[138]"
Translation by Dr. H. Tuong

By the time Việt Minh forces entered and took over Hà-nội from the defeated French in 1954, one million refugees were already on their way to Freedom in the South...The second part of this song about the 1975 exodus.

Một ngày bảy lăm, con bỏ nước ra đi
Hai mươi năm là hai lần ta biệt xứ
Giờ cha lưu đày ở ngay trên đất ta
Và giờ con lưu đày ở đây trên xứ lạ!
Một ngày năm bốn, cha lùi quê hương
Lánh Bắc vô Nam, cha muốn xa bạo cường
Một ngày bảy lăm, đứng ở cuối đường
Loài quỷ dữ xua con ra đại dương!

One day in 75, I abandoned my country
In 25 years, we had left our homeland twice
Now you were in exile in our land
While I am wandering in exile in a foreign place!
One day in 54, father retreated back in perfume land,
Avoiding North, entering South, as you wanted to be far

---

[138] https://www.youtube.com/watch?v=2P829kk0-9M Passage to Freedom slide show & Khánh Ly's song 1954 cha bỏ quê, 1975 con bỏ nước

> *away from oppressors*
> *One day in 75, I am at the dead end,*
> *As the demons chase me into the ocean!*
>
> Translation by Dr. H. Tuong

In my opinion, the Vietnamese patriotic sentiment longing for Independence has been developed even deeper as the people start appreciating their first, short but sweet, Independence from China. They appreciate it even more when they have lost it to the French invasion. On the one hand, the fuzzy transition from the French to their major WW2 ally, the Americans, has indeed given Hồ Chí Minh an excellent strategic weapon. Hồ must have studied the American way during his stay in Harlem. Consequently, his encounter with the OSS during the anti-Japan campaign and the obvious U.S. allegiance to France[139] have given him ideas to formulate a logical rhetorical plan to instigate and set the naïve and illiterate Vietnamese peasants up against the U.S. involvement. It is also convenient for Hồ that the Vietnamese have been raised to be obedient, devoted and loyal to whom they perceive to be a good leader. Hồ has worked hard at liberating Việt-nam from the colonials and, consequently, he appears to the people as a patriotic hero and a reliable leader. On the other hand, it is very clear how much the Vietnamese have hated being dominated by foreign countries, whether they are eastern or western powers. Việt Nam long history of foreign domination has forged them to become merciless against foreign invasions. With that in mind, we can see now how the climate has been set up for a bloody war.

In the eyes of Hồ Chí Minh and Gen. Võ Nguyên Giáp, the U.S. had openly sided with Việt-nam's enemy, the French, when they refused to help them fight the latter. After the French had left Việt Nam, the U.S. had stayed behind and became visibly involved by giving aids to South Việt Nam. Both men had the perfect weapon for their psychological warfare. On the one hand, they proved to the naïve and submissive peasants that the U.S. was the new invaders and, consequently, the new enemy of the perpetual invader-fighter Vietnamese people. On the other hand, they also proved that the nationalists in the newly created RVN, since they had sided with the Americans,

---

[139] Ignoring his letters to Pres. Truman, refusal to assist him against the French, 1 Billion U.S. dollars in aid to colonial France to combat the Viet Minh in the early 1950's.

were no longer their own kind, but "Tay Sai của Giặc Mỹ" (servants of the American Invaders), just like those "Tay Sai của Giặc Pháp" who had collaborated with the French. "Tay sai" were always the enemy of the patriotic people! If this reasoning stood firm, of course, the Americans would inherit from the French the Vietnamese people's generational hatred that had cumulated during their history of over thousands years of domination. This hatred against foreign invaders had been engraved deeply, for some more than others[140], in the Vietnamese DNA[141]. What followed would help Hồ show to the pre-literate peasants that his logic had stood as firm as a mountain!

      Hồ Chí Minh and the People's Democratic Republic of Việt-nam People DRV knew exactly how to use this propaganda to manipulate and recruit naïve submissive members for their camp in their Eternal Revolution campaign: fight days and nights, as long as it takes, at any cost, until victory. In order to apply Sun Tzu "Know them, know ourselves…" strategies, anyone child, woman, elder, handicapped… could be used to retrieve intelligence, attack openly or underground, launch sabotage, propagate psychological warfare… as they were used in the past against other invaders. No one who perceived being oppressed by the establishment would refuse to become Hero of the Ancestral Land? Music had shown they would eternally be celebrated for their sacrifice.

      Lưu Hữu Phước did not compose just one historically patriotic song "Hội Nghị Diên Hồng" to instigate the uprising against the French. In 1939, he also created another master piece calling on students' patriotism and making them up rise by evoking Việt Nam thousand years of history of repelling invaders: "Sinh-viên Hành khúc" or "March of the Students."

Này sinh viên ơi ! Đứng lên đáp lời sông núi.
      Hear Ye, Students! Stand up and respond to Rivers and

---

140   Vietnam has always relied on agriculture, much more than industry (North), to boost its economy. People from the economically-poorer North (Tonkin) and Northern Central (An Nam) have always been struggling and working very hard in tiny parcels of field to make a living. On the contrary, those in the South (Cochinchina) have always been enjoying excellent crops in vast borderless fields throughout history, and have usually been the spoiled and easy-going ones as economy has been in their favor. This might be perhaps the reason why the French kept An Nam and Cochinchina when they gave away North to Hồ Chí Minh.

141   By "engraved in their DNA," I really meant it. It was as if they were born with the belief readily implanted in their brain…

Mountains.
Đồng lòng cùng đi, đi mở đường khai lối.
With one soul we go, go and clear our path
Vì non sông nước xưa, truyền muôn năm chớ quên.
For our old-time mountains and waters, which are passed on to us, for thousands years we will never forget.
Nào anh em Bắc Nam, cùng nhau ta kết đoàn.
We Northerners or Southerner, will band together
Hồn thanh xuân như gương trong sáng.
Our serene youth shines like a bright mirror,
Đừng tiếc máu nóng, tài xin ráng...
We shall not regret losing our hot blood, please do your best with your talents....
....
....

Này sinh viên ơi! Dấu xưa vết còn chưa xóa.
Hear Ye, Students! Our old path has not been erased yet.
Hùng cường trời Nam, ghi trên bản vàng bia đá.
Our Bravery in the South, has been engraved on gilded plaques and stone markers.
Lùa quân Chiêm nát tan, thành công Nam tiến luôn
Chasing and crushing Champa, success has pushed us Southbound
Bình bao phen Tống Nguyên, từng ca câu khải hoàn.
Repelling from Tang to Yuan[142], we have always sung Victory songs.
Hồ Tây tranh phong oai son phấn.
West Lake has fully been glorified.
Lừng tiếng sát Thoát Trần Quốc Tuấn.
Renown for killing (Yuan's Prince) Toghan[143]: Trần Quốc Tuấn
Mài kiếm cứu nước nhớ người núi Lam.
Sharpening our swords to save our waters, we will always remember Emperor Lam (Lam Sơn)
Trừ Thanh, Quang Trung giết hằng bao đám...
Rid of Qing's (Emperor Qianlong) Army, Emperor Quang Trung has taken out a great many...

Translation by Dr. Ha Tuong

---

142   China's imperial dynasties.
143   China's general.

This song was later adopted and revised[144] in 1956 to become the South Việt Nam National Anthem (the first RVN under Pres. Ngô Đình Diệm through the second one under Pres. Nguyễn văn Thiệu) "Tiếng Gọi Công Dân" or "Calling on the Citizens." New lyrics were inserted into the old tune of the original "Sinh-viên Hành khúc" above. My father did make a comment that the South Vietnam national anthem reminded him, and probably many other Vietnamese patriots, of the anti-French revolutionary song "March of the Students," especially when it included the word "giải phóng or liberation" in the first verse. This meant it would always remind patriotic listeners of the glorious Việt Minh led by Hồ Chí Minh against the French.

Này Công Dân ơi! Quốc gia đến ngày giải phóng.
Đồng lòng cùng đi hy sinh tiếc gì thân sống.
Vì tương lai Quốc Dân, cùng xông pha khói tên,
Làm sao cho núi sông từ nay luôn vững bền.
Dù cho thây phơi trên gươm giáo,
Thù nước, lấy máu đào đem báo.
Nòi giống lúc biến phải cần giải nguy,
Người Công Dân luôn vững bền tâm trí.
Hùng tráng quyết chiến đấu làm cho khắp nơi
Vang tiếng người nước Nam cho đến muôn đời!
Công Dân ơi! Mau hiến thân dưới cờ!
Công Dân ơi! Mau làm cho cõi bờ
Thoát cơn tàn phá, vẻ vang nòi giống
Xứng danh nghìn năm giòng giống Lạc Hồng!

Oh citizens! Our country has reached the day of liberation.
Of one heart we go forth, sacrificing ourselves with no regrets.
For the future of the people, advance into battle,
Let us make this land eternally strong.
Should our bodies be left on the battlefields,
The nation will be avenged with our crimson blood
In troublesome times, the Race will be rescued,
We the People remain resolute in our hearts and minds.

---

144   It was actually revised as Thanh-niên Hành khúc or The march of the Youths in 1948 for use as national anthem of the new Provisional Central Government of Vietnam, which later on became the State of Vietnam (later, Republic of Vietnam)

Courageously we will fight such that everywhere,
The Glory of the Vietnamese forever resounds!
Oh citizens! Hasten to offer yourselves under the flag!
Oh citizens! Hasten to defend this land,
Escape from destruction, and bask our Race in glory
Be forever worthy of the descendants of Lạc Hồng!

Translation from Wikipedia[145]

Besides, my father did also make a profound remark that most of the songs popular in the South were "so sad and self-defeating," as they were always mourning someone or grieving the disappointing destiny of separation from beloved RVN soldiers, lamenting war-torn country, or blaming losses on death... "Such songs would very much break the heart of the toughest heroes, instead of heating up their unbroken spirit and boiling up their patriotism... On the contrary, communist songs are more uplifting..." Psychologically speaking, what father said was true. His remarks made perfectly sense when comparing NVA (forbidden in the South during the war) to RVN songs... The antiwar movement in Việt-nam had also made good use of them to melt the hearts of the republic and lower the morale of its combatants...

As soon as the French withdrew their last troops from An Nam (Central) and Cochinchina (South) per Geneva Conference (Accords) of 1954, Emperor Bảo Đại started his exile in France and supported Ngô Đình Diệm, a Catholic anti-communist and anti-colonialist, to be the Prime Minister of the new State of Việt Nam. A Geneva Conference resolution had prescribed that Diệm, along with Hồ Chí Minh, were to implement a National Election within a few years to unify Việt Nam. In the meantime, the U.S., under Pres. Eisenhower, started pumping in humanitarian aids to help rebuild the country, evacuate and resettle approximately one million peace and freedom-seeking families from the North, and, at the same time, military aids to build up and strengthen a new armed forces for the State of Vietnam (South). This unstable period of transition became even more unstable because of religious conflicts between various Buddhist and Catholic groups, and the political and military chaos caused by the war lord "wannabe's" in the South. This socio-political strain caused by the needy newly resettled Northern refugees, added to serious

---
145   https://en.wikipedia.org/wiki/National_anthem_of_South_Vietnam-#Tiếng_Gọi_Công_Dân_-_Call_of_the_Citizens_(1956–1975)

disagreement between Hồ and Ngô, political divisions between the State of Vietnam factions[146], political "power-looting" had inevitably caused even more clashes between diverse groups.

Hồ Chí Minh's match: Ngô Đình Diệm – Another turning point

Having known Hồ from past experiences that had forced him to go into a short exile, Prime Minister Diệm deflected from the National Election prescribed by the 1954 Geneva Conferences and initiated the First Republic of Việt-nam a year later, in 1955, when he became its first President. He focused both national resources (remember that the South was fertile and rich) and incoming American aids on Agricultural and Political Stabilization ("Bình-định Nông Thôn, Bình-định Chính-trị") while reinforcing the budding military. He could see clearly what Hồ could do to take over the South: use the conspiracy tactic that the "US is going to invade and colonize Việt-nam" to instigate the hateful, anti-colonial, but mostly illiterate and politically naïve, nevertheless obedient, and devoted population. The sole purpose was to cause major uprising against the nationalists. For that reason, Diệm believed a strong army was needed as the future of the South looked bleak when Hồ had left his "5th Army[147]," the "un-uniformed" Việt Minh fighters, behind in the South to sabotage nationalists' efforts, while recruiting communist Southern sympathizers to move North for warfare training. "Tập-kết ra Bắc." "Uncle Hồ" was up to something "not good!" The U.S. was ready to help Diệm rebuild his new Republic.

In the meantime, Hồ Chí Minh launched a major campaign against the Northern landowners. The latter started being prosecuted and sanctioned, oftentimes executed by clubbing and stoning to death or by firing squad by the People Court "Toà Án Nhân Dân đấu tố" in public. Vindictive farmers could now take their "sinful" (for being rich) land owners out for purging using the 1953 "Cải Cách Điền Địa" Agricultural Reform law[148], so they could possibly take over their land.

146 https://en.wikipedia.org/wiki/Ngo_Dinh_Diem
147 This 5th Army was to become the National Liberation Front, nicknamed the Việt Cộng, organized by Commissar Lê Duẩn.
148 **Đấu tố địa chủ & cải cách ruộng đất phần 1 (Prosecuting landlords and Land Reform)** *YouTube documentary, re-enacted movie* Chúng Tôi Muốn Sống

As reported by an Indochina War correspondent and author Bernard Fall, between 50,000 to 200,000 land owners were gravely clubbed or even executed on location in big cities as well as in remote areas[149], as recounted as well by by-standers and the Southbound fleeing refugees. Việt Minh had also applied Land Reform in the South, but "đấu tố" was less harsh. Thousands of Huế landowners were executed by the People's Court, and a great many had to abandon their land and flee to Southern provinces.

Rule by Fear

The fear of communist retaliation against the "bourgeois" and the better-than-poor population (most of the Southerners belonged to this middle class, as the South was economically much more comfortable than the North and Central), as well as its harsh treatment of the freedom lovers had urged the State of Vietnam to quickly get organized and stand up against Communism. In a short order, Prime Minister Ngô Đình Diệm and the Southerners readily embraced the U.S. offer for help. The great fear of Communism had quickly outgrown any desire for unification with the North, and therefore deviated the South away from a general election, as per the Geneva Conference of 1954 (See section on Ngô Đình Diệm).

What happened in the incoming years reinforced the belief that Communism was aiming at abolishing the growing Freedom in the South, especially when the world could witness China Mao Tse-tung launch his bloody Chinese People's Revolution of 1949[150] and Cultural Revolution in May 1966[151]. These two upheavals had destroyed China thousands of years of civilization and its meaningful traditional economic and socio-cultural infrastructure. The horrendous massacre of many millions of lives in both processes had terrified the world, especially those countries in proximity. Living proofs

---

1956 (We Want To Live) YouTube, YouTube documentary photos and newspaper clips
https://www.youtube.com/watch?v=vvRZBT72Ov0_
149   https://en.wikipedia.org/wiki/Land_reform_in_Vietnam Land owners prosecuted and executed in North "đấu tố"
150   https://history.state.gov/milestones/1945 Mao's Chinese People's Revolution of 1949
151   https://en.wikipedia.org/wiki/Cultural_Revolution Mao's Cultural Revolution of 1966

were the Chiang Kai-Shek's groups that evaded mainland China to resettle in Taiwan, creating the new freedom-loving Nationalist Republic of China. These cataclysmic floods of refugees had proven how much fear Communism could inflict on nationalists' soul.

On the Western front, in WW1 "not too long ago," during Lenin's bloody Russian Revolution of 1917, the Bolshevik's rise to power[152] was recorded in history for having decimated millions of "bourgeois" and better-than-poor people (middle class). Fidel Castro's and "Che" Guevara's counter-cultural Marxist-influenced revolutions in Cuba and in many other South American countries, with their terrorizing firing squad, and communist-style land reforms, loved by the have-nots and feared by the have-some or above, had also shown to the South Vietnamese, most of whom belonging to the middle and better-than-poor classes, what could happen to them if Communism flooded the South.

Mao Tse-tung's, Lenin's, Castro/Guevara's revolutions did cause major impact on Southern people's mind and made them become weary of communist invasion. Nevertheless, in my opinion as I later found out during our 1973 Phoenix Campaign[153], they were more scared of their family members being harmed by the communists than their Ancestral Land being ruled by Communism. Except in some provinces that were controlled by the VC and the NVA, such as Thừa Thiên (Huế), Bình Định, et. al. where anti-communist feelings were extremely strong, in general, the peasants in other peaceful areas might be more egocentric: their political inclination was more humanitarian than political, and they thought more in terms of praying for Peace and minding their own business to make a simple living. Understandably, they would fear more the communists harming them (or even executing them) and their loved ones than fear being accused by or even tortured by the nationalist establishment. They would rather collaborate with the enemy than with the nationalist! Consequently, many were not assertively anti-communist and tended to be bilateral (sympathizing with or being submissive to), seeking

---

152   https://en.wikipedia.org/wiki/Bolsheviks Lenin's Bolsheviks
153   Before graduating from boot camps, officer cadets were assigned to political pacification teams in remote areas to train peasants on the RVN versus Communist ideologies. To my astonishment, I found out that the people of Quy Nhơn City, Bình Định District, where people were utmost either anti-Communist or Communist, were ignorant about Democracy and Freedom. This was not the same as the American counter-espionage Phoenix Program in the RVN.

status quo and leaning towards who pressured them the most. Naturally, communist threats would cause much more fear and submission (meaning they would chose to cooperate with the enemy), especially when they were out of nationalist protection zones. Such could be the case of Mỹ Lai villagers who were massacred by U.S. troops having moved out of their protective strategic hamlets built by the RVN back to VC-controlled zone (to be discussed further more).

Historically, being dominated by foreigners for a long time, the majority of the people had been trained to subjugate themselves to extreme authority figure. As a result, they would bite their lips and try to handle tyranny as their ancestors had under the Chinese, French or Japanese in a docile manner. They could withstand pain and mental degradation easily if they were to be tortured or tortured by the communists and would not complain or fight back to avoid being terminated. Nevertheless, what they feared the most would be their relatives or family members being tortured or killed as exemplary guinea pigs. This was why the VC would use the tactic of threatening to kill family members to pressure them into cooperating with them[154]. This was another reason the peasants would collaborate with the communists rather than with the nationalists, as the latter were limited by the Geneva Convention. Huế civilians suffered the most during the 1968 Tết Offensive because they had dared resist[155]!

...Moshe Dayan, Israeli Army Colonel observer, during his visit in Việt Nam to learn about counter-guerilla warfare, had witnessed a VC Việt-cộng (of the National Liberation Front NLF, Southerners converted to Communism) interrogation. He was stunned when he heard the interrogated person (VC) say, after all gruesome methods of torture had been applied to him: "I won't tell you anything! Go ahead and kill me. Many others will rise..." Dayan might not know the POW feared

---

[154] As rumored, the NVA or PAVN, probably ebbing trained for conventional war, might be more humane then the VC (National Liberation Front). The VC, who were locals, knew who were related to ARVN soldiers (especially the Regional Forces and Militia) and RVN government employees, so they tended to exploit corporal manipulations on these victims much more than the NVA would. Ironically, the VC always behaved as if they were blood related parts of direct family of the victims (as told by my mother-in-law who had stayed behind, but had later joined us in the U.S. via the INS family orderly departure program).

[155] Tết Offensive of 1968 with 7,800 innocent Huế civilians massacred by NVA and mainly the VC.

more his relatives or family being retaliated by the communists than him/herself being tortured to death in an interrogation. The ARVN had to stop using tortures (creative interrogation) to comply with the U.S. Government conditional mandate for military aid eligibility (Geneva Convention).

As reported by Phạm Tuấn, an ARVN recruiter of ethnic intelligence and Special Forces personnel, such as the MIKE Force Khmer-Krom (supported by MACV), whenever VC were captured, they were immediately transferred to MACV for intelligence extraction. ARVN intelligence had no jurisdiction over prisoners until after they were interrogated: jailing or re-habilitating for the Chương-trình Chiêu Hồi Open Arms Program[156]. This was facilitated by the MACV personnel attached to each ARVN platoon, who also coordinated strategic and political efforts between ARVN and U.S. Command. The ARVN regarded this arrangement as controlling, also resulting in budding reluctance from their part to cooperate whenever there was a conflict of policy (to be discussed later). Tuấn resettled later in Minnesota.

Diệm's dilemma

The political situation became more complicated when clashes occurred between two major religious groups: the Catholic, as perceived by the people as represented by an oppressive presidency, versus the Buddhist, which was the predominant religious group in the country. It became much more complicated when it turned into a more racial conflict as the majority of the Buddhists were Central and Southerner people, while the Catholic were more originally from the North. Among the million 1954 refugees, a great many were from the larger Catholic villages such as Bùi Chu and Phát Diệm[157]". The Buddhists strongly believed that the new Catholic immigrants were protected by "their Catholic president," via Pres. Diệm's older brothers, the oppressive Ngô Đình Nhu and the Catholic Archbishop Ngô Đình Thục.

In the meantime, like zombies raised from the dead,

---

156 "Chương trình Chiêu Hồi" welcomes the former enemy to the rank of informers or collaborators.
157 Even though not all refugees from the North were Catholic, whenever there was a conflict, opposing groups stood out because they all spoke different Vietnamese dialects and had a heavy accent.

Hồ's "left behind" 5th Army, remnants of the Việt Minh which had defeated the French, started regrouping and purging into a new communist National Liberation Front (Việt-cộng VC) insurgence against Pres. Diệm's regime. Confusion turned into major chaos. The situation was aggravated even more as the internal political infrastructure of the country was destabilized by rebelling insurgent groups. Many small local leaders believed they could do better than Pres. Diệm. They popped up all over the country...

While many nationalists loved Diệm as a patriotic anti-French and devoted anti-communist civil servant, many others hated him mainly because he had allowed his siblings and extended family share and abuse his power[158]," a presidency with an abusive "gia-đình trị" regime (family ruling by nepotism). Out of the five siblings, his two most renowned brothers, tyrant Mr. Advisor Ngô Đình Nhu, who was married to an also-renowned feminist activist Madame Nhu, and Huế's tyrant Mr. Ngô Đình Cẩn, had probably cost him the Vietnamese's respect. Both brothers had their own secret service, the undercover police "mật vụ" and, also, had allegedly CIA-trained special forces oppress (by imprisonment, torture) the Buddhist and intellectual protesters; many of the latter ended up dead or simply disappeared. They were suspected and accused for either uprising against the regime, or cooperating with the communists. My math tutor, a university student, was incarcerated and tortured, but, luckily, he was released for lack of evidence of belligerence. Back in the U.S., Kennedy became weary of the "nepotistic and despotic family."

Even though he was more permissive with his siblings, Diệm's rule was still unacceptable to the non-Catholic oppressed citizens. Nevertheless, many believed Diệm was more clairvoyant than people had thought. Probably having known Hồ Chí Minh and his multi-faceted strategies so well, and understanding the Vietnamese's painful obsession about being invaded for over ten centuries, Diệm might have foreseen that the U.S. getting directly involved in the war would give Hồ his perfect weapon against the new Republic and the Americans: "another foreign invasion by the U.S., this time aided by the nationalist traitors[159]!" As the U.S. was a great ally of France, Việt-nam's colonizer for nearly a century, its

---

[158]   http://alphahistory.com/vietnam/ngo-dinh-diem/ Ngo Dinh Diem as seen by a more impartial commentary.
[159]   http://www.vietnamgear.com/bio/5.aspx Pres. Diệm knew it...

fighting in a WW2- like war in Việt Nam would certainly imply it was invading Việt Nam with the intent of colonization. Consequently, he made all efforts to "localize" the war so that it would be just an internal conflict between the North and the South, while keeping the U.S. in the out-of-bound area. He wanted the U.S. to be just providing aids and advisors without fighting the war itself. If this could come true, the U.S. and, perhaps, China and the Soviet Union too, would have to stay out of Vietnamese politics (to avoid international conflicts) and let North and South Việt Nam resolve their "internal issues". Many politicians and analysts during the time had accused Diệm's plan leaning towards a civil war, a war between people in the same country like the American Civil War. Nevertheless, the Southerners had always considered the North to be a totally separate country. Even myself, a Southerner-born from Northerner parents, had bought into it. If there were a war, it would not be a conflict or civil war, but one between two countries with different ideologies, cultures, histories, directions...[160]

According to an RVN documentary[161], Lt. Gen. William Westmoreland, a hero of 1952 Korean War, and, later, Commander of XVIII Airborne Corps (82nd and 101st), had discussed with Diệm the U.S.'s strong intention to use the same conventional tactics as they more or less had successfully in WW2 and the Korean wars. This information concurred well with Darin Nellis's documentary film JFK: A President Betrayed as he had revealed that the Kennedy's Eisenhower-influenced Joint Chiefs of Staff JCS[162] were all pushing for a direct involvement in Vietnam War. They strongly believed it would favorably end the war fast. Nevertheless, the U.S. plan of direct participation might not have the same outcome as the one involving Korea, Europe and Japan during WW2, considering his weaker and under-equipped troops (at that time), as Hồ was deploying a different tactic learned from the American OSS and from the remnants of the Japanese army: the adapted guerrila warfare.

---

160   In this matter, the French "divide and conquer" policy had unequivocally succeeded in dividing the Northerners and Southerners from each other.
161   <u>Lich-sử Việt Nam 1945-1975 - Part 2 Vietnam History 1945-1975 documentary video by GMD Music Entertainment, DVD Center</u>
162   JFK Joint Chiefs of Staff were mostly generals of WW2 era.

Pivoting points for the Republic of Vietnam: Democracy

While objecting to the U.S. direct involvement in the war, Pres. Diệm had enthusiastically agreed on military and humanitarian aids. His siblings, especially brother Ngô Đình Nhu and his infamous spouse, "Madame Nhu," even though they had on the one hand abused power by building up their own empire, while on the other hand, had actually contributed by helping fight against the budding invasion from the North (in spite of Hồ absolutely denying his troops were invading), contain Hồ's 5th army (the embedded communist Việt Minh passed for local insurgent VC), as well as pacify the local Southerner insurgent lords (such as "Bảy Viễn's" Bình Xuyên), the pro-French factions (such as Gen. Nguyễn Văn Hinh and his army), and the non-Catholic Buddhist sects of Cao Đài, Hoà Hảo… Nhu had successfully resettled nearly a million Northern refugees (Passage to Freedom) in the South. Nevertheless, whether the brothers liked it or not, the U.S. started sending troops to Việt Nam. At the beginning, it was in an advisory capacity, with 1,000 troops in 1961, this pace gradually increasing to 17,000 in 1965. Many of these troops, mainly the Special Forces units, had decided to participate in combat duties "clandestinely" to assist the newly formed ARVN, out of compassion and comradery, in spite of central high command ordering them not to.

Even though the U.S. government considered Diệm as Việt-nam's Winston Churchill, it actually disliked him more and more because of serious nepotism and his defying U.S. policy. He survived the "coup d'état" Airborne siege[163] of November 1960 and the Presidential Palace bombing of February 1962[164], but not the one three years later on November 1, 1963, which followed a few months after he issued a decree forbidding Buddhists to display their religious flag while encouraging Catholics to raise Vatican's ones in celebration of his brother Archbishop Ngô Đình Thục's promotion. This move was gravely considered oppressive against the Buddhist congregation by the U.S. decision makers. It resulted in the Buddhist Crisis in Huế, followed by many monk hunger strikes and culminating with the self-immolation

---

163   https://en.wikipedia.org/wiki/1960_South_Vietnamese_coup_attempt Coup d'état by Airborne Lt. Col. Vương văn Đông and Col. Nguyễn Chánh Thi

164   https://en.wikipedia.org/wiki/1962_South_Vietnamese_Independence_Palace_bombing The (old) Presidential Palace Dinh Độc Lập was bombed by two RVN Air Force pilots flying Skyraiders fighters.

by Reverend Monk Thích Quảng Đức.

Real leaders or "wannabes?"

Both Pres. Diệm and fierce tyrant advisor, brother Nhu, were tragically executed by dissident high-ranking ARVN officers while being transported inside an M113 tank as prisoners by the coup plotters. Leading the coup against the Ngô's brothers was Gen. Dương văn Minh[165], or "Big Minh," who...*ironically, had helped Pres. Diệm at the beginning of the latter's rule pacify most post-Điện Biên Phủ rebels and independent armies (late 1950's). Eventually, he was also the same person who, as default interim President of the RVN, had unconditionally surrendered the RVN to the victorious NVA and VC forces at the end of the war (April 30, 1975)....*

"Big Minh" was supported broadly by the most powerful generals such as Generals Tôn Thất Đính, Đỗ Cao Trí, and Nguyễn Khánh, these being I and II Corps Commanders. The coup was, most importantly, "more or less blessed" by the U.S. government (via Ambassador Henry C. Lodge[166]) and acknowledged, if not blessed[167] by Kennedy himself, the latter having "closed its eyes and let the coup happen."

Declassified in 2003, a U.S. government file stated that, before being overthrown and executed, Pres. Diệm and his brothers had foreseen and protested against the U.S. that the latter had been undermining their government. The Ngô brothers alleged that the American government had staged

---

165 https://en.wikipedia.org/wiki/Dương_Văn_Minh

166 http://www.seasite.niu.edu/crossroads/russell/vntimeline.htm and National Security Archive on JFK and the Diem coup by John Prados http://nsarchive.gwu.edu/NSAEBB/NSAEBB101/#audio Document 20 and 21: Pres. Kennedy's plan for newly assigned ambassador Henry C. Lodge "in case" of "a coup d'état." Documents 7, 19, and 22 suggested that JFK had led discussions with CIA possibilities of coup, balancing between pros and cons, while Document 16, 17 showed JFK discussing Diem's successors and U.S. funding plotters $42,000 for executing coup. There is no evidence JFK has ordered the Ngo brothers' execution. Also, JFK Library has the official posted information at https://www.jfklibrary.org/JFK/JFK-in-History/Vietnam-Diem-and-the-Buddhist-Crisis.aspx. Also, 2011 Declassified The Kennedy Counter Insurgency Program – Pentagon Papers (page 10/197). Page 23/197: On Oct 18, 1961, Diem said he wanted no U.S. combat troop for any mission... and requested bilateral defense treaty (more support and combat equipment for the ARVN.

167 JFK's Cable 243 regarding the overthrow of the Ngo's Presidency https://en.wikipedia.org/wiki/Buddhist_crisis

and financed the anti-RVN rebellion organized by different South Vietnamese political and Buddhist factions. A very simple reason precipitating to this overthrowing plot could be that the brothers had made the biggest mistake of openly refusing to be the "American puppet government" that would comply with Kennedy's Counter Insurgency Program[168] CIP (Roger Thompson-Hilsman's proposal) and give the U.S. "carte blanche" to be directly involved in the war. The Central Intelligence Agency believed the Ngô brothers had their own plan to resolve differences with Hồ Chí Minh, not by a major war, but through negotiations with the North: the North-South Solution "Giải-pháp Bắc-Nam" (1963) that could probably be similar to nowadays North and South Korea one. Would North and South Vietnam be able to negotiate fruitfully a similar peace accord that would allow each country to have its own sovereignty Socialist Communist in the North and Republic in the South?

  This softening policy of the 1st republic might seem to be weak in the eyes of the U.S., especially with the Eisenhower's clan. Nevertheless, Diệm's plan might have led Việt Nam closer in the directions prescribed by the Geneva Conference (French Indochina) of 1954, not to unify the two Việt-nams, but, at least, to bring out Peace like in the two Koreas. If he were not overthrown or assassinated, would there be a democratic election that would at least facilitate a status-quo that would allow them to coexist in Peace, separated at the 17th Parallel? Nevertheless, with the Ngô brothers removed, there would be no one to bar the U.S. from bringing troops in and wage war.

  In my opinion, as confirmed by my brother-in-law Anh Bảy[169] later in the early 1970s, Diệm had served a good example of stubbornness leading to his own disaster. The next leaders of the RVN would definitely be "much easier to work with" and, consequently, were fully supported by the U.S.. They had seen what would happen, as it did to the two sibling leaders of the 1st Republic, if they dared contradict the American might. Words went around that the new RVN leaders might be receiving "favors" from the U.S. Government as rewards for cooperation. Corrupt leaders would give birth

---

168 In Vietnamese and partial English https:// www.scribd.com/document/322552500/Từ-Giải-Phap-Bắc-Nam-1963-Đến-Sự-Cố-Marigold and in English https://www.mtholyoke.edu/acad/intrel/pentagon2/doc119.htm a strategic concept for Việt Nam.

169 He was a member of the 2nd RVN Presidential Cabinet.

to many other echelons of corrupt little leaders… Corruptible or obedient easy-going leaders would not do much good for a country that needed good leaders; however, the U.S. leadership would have much more ease implementing its Việt-nam policy with them in power.

Source: Sài Gòn Echo - Chuyện Tình Ấp Chiến Lược
(Strategic Hamlet Love)
Saigonechoinfo

By the same token, as time went by and after subsequent coups d'état, the U.S. realized there were plenty of other wannabes who were eager to lead. There had been so many "coups d'état" Ambassador Henry C. Lodge had to summon in the main generals (under Gen. Nguyễn Khánh's Junta), potential "trouble-makers," to let them know "the U.S. had enough with the coups[170]." This was more evidence that the U.S. was really meddling with the RVN sovereignty. What else could the Americans do when the country they are supporting was seriously damaged by its own people, at the expense of the Americans' taxes and lives? Nevertheless, the lack of stability in the RVN did actually give Secretary of State Robert McNamara major difficulties in strategic planning: How could the U.S. help fight this war if the hosting country (RVN) kept switching non-stop from one revolution to another? This country was heavily divided and had no concrete selfless leadership!

---

170   https://en.wikipedia.org/wiki/Nguyễn_Khánh

Source: Phú Yên Online - baophuyen.com.vn
Phá Ấp Chiến Lược ở Hoà Đồng
(Let's Destroy the Strategic Hamlet at Hoà Đồng)

What caused it? The new taste of Democracy had given wannabe's the idea they could compete for power: they were ambitious and felt they had to do something to prove to their followers they were the real leaders. This might be what Kennedy had foreseen when he, as a senator, was reflecting about formerly-colonized countries that just recovered their independence (to be discussed later).

The brothers Diệm and Nhu left behind the legacy of a successful Strategic Hamlet Program "Ấp Chiến lược" (assisted by the U.S. Green Berets as depicted in the iconic John Wayne's movie The Green Berets), this program being developed at the same time as the formation of the indigenous militia "nghĩa quân" and the rural construction cadre "cán bộ xây dựng nông thôn." It had proven to be one of the most successful programs that would at the same time protect the civilians and keep them safe away from communist manipulations: just like in the European Medieval Era, when the VC attacked, the peasants were to retreat inside protection walls. When not working on their field, they ought to live inside these walls, too.

Unfortunately, after Pres. Diệm was killed, the naïve peasants, probably misled by the communists, started moving out of the strategic hamlets, returning to their land, which was indeed infested by VC, who were probably their relatives or even loved ones. Once resettled in VC-controlled zone, the peasants had no choice but collaborate with the communists, and, therefore became the enemy. Many had even destroyed the "Ấp Chiến lược[171]" that had been protecting them… That was the case of both massacres at Mỹ Lai (Mar 16, 1968 part of Tết Offensive) and Phong Nhất/ Phong Nhì[172] (Feb 12, 1968 Tết Offensive). Many massacres could have been avoided if the peasants had remained in their strategic hamlets!

Shocking news in the U.S.

While things were becoming more and more unstable in South Vietnam, interestingly and sadly, back in the U.S., a major world event occurred just a few weeks after Diệm's assassination in Sài-gòn. On Nov. 22, 1963, within the same month, the U.S. Pres. John F. Kennedy was also assassinated while officially visiting Dallas, Texas, for his 2nd term presidential campaign event. The assassin, civilian Lee Oswald, was also assassinated a few hours immediately after his successful attempt on Kennedy while in the secure custody by local law enforcement. Oswald's assassination occurred while being filmed in real time by American television. He was the only witness who would know who was behind the plot. Conspiracy theories popped up left and right meanwhile all investigations associated with the assassination of both Kennedy and Oswald went quiet. They ended abruptly when the Warren Commission concluded that the assassin had acted alone, and, therefore, no link could be made between various conspiracy theories. The materials most recently declassified under Pres. Trump's administration in November 2017 specified the attempt on Oswald's life was to prove to the public that Oswald was the only one responsible for Kennedy's death.

While the Warren Report ended the official government investigations, many of the older conspiracy theories remained alive while more new ones emerged as current issues in the form of news media documentaries. One that suspects a United

---

[171] Vietnam: A Television History – Episode 3 - YouTube
[172] https://www.youtube.com/watch?v=UPfXnlqlNZk and https://en.wikipedia.org/wiki/Phong_Nh%E1%BB%8B_and_Phong_Nh%E1%BA%A5t_massacre Massacre by Rep of Korea 2nd Marine Division

States Government cover-up remains popular today and is often a topic for discussion and speculation by conspiracy theorists. The big question still remains: How could the government investigations end so abruptly? By cross referencing information from various sources, including the ones behind the scene, the Warren Commission could have dragged out a great many leads.

Mysteries and Speculations about the death of President John F. Kennedy

In order to understand better what was going on during that time, we would need to go back into the past and examine the course of history prior to Kennedy's death. The United States was transitioning from a victorious WW2 era Republican Presidency into a new exciting but confusing Democratic agenda. This meant that the government was like shrimp in an ecological mix of salt sea and fresh river water environment (estuaries), as we usually say in Vietnamese "nước chè hai". The shrimp were confused not knowing where to swim because they had grown up in fresh water but had to face incoming salt water from the ocean[173].

Darin Nellis in his documentary JFK: A President Betrayed[174], as also confirmed by Cable News Network CNN series The Sixties (2015), pointed out that there was a major conflict between JFK's team policy making efforts to make Peace with the Communist Block via negotiations, and his Joint Chiefs of Staff JCS, most of whom were WW2 heroes, as well as some of his advisors' who were strongly opposed to peace initiatives. They strongly believed in conquering the communists by military supremacy the same way they did the Fascist's and Nazi's in WW2. For some rightists, JFK was a traitor ("JFK a Traitor" flyers were disseminated in Dallas on Nov. 22, 1963 the day he was assassinated), and, for others, "he was a patriotic who was betrayed" (Nellis's documentary). In the Smithsonian Oct. 3, 2013 documentary film The Day JFK died[175] on television, Jacqueline Kennedy's secret service

---

[173] For some reason, the Vietnamese believe these shrimp taste better than either salt water or fresh water.

[174] *JFK: A President Betrayed (2013) by Darin Nellis, Agora Productions based on* Something to Talk About documentary film archives. This film had won many Film Festival awards

[175] http://www.smithsonianchannel.com/shows/the-day-kennedy-

agent Clint Hill stated that the succeeding Pres. Lyndon B. Johnson's and Kennedy's clans *"were not close"* and oftentimes had conflicting ideas. Does this insinuate Johnson was in tune with the *"hawks' clan"* or that he was an island himself?

Darin Nellis concluded that Kennedy would do anything to keep the U.S. out of Việt Nam. He was probably the only U.S. President who had a historian, Robert Schlesinger, as a special advisor with whom he could consult and cross-refer. Perhaps he had realized that thorough knowledge about past history would help make successful plans for future action, and, consequently, Kennedy had changed his mind about Việt Nam to avoid a major war that might end up with a defeat as the French Indochinese war had. In spite of his military staff's wish, a month earlier, Pres. Kennedy had decided to remove all military personnel from Việt Nam, starting with 1,000 in 1963 as shown in his signed National Security Action Memorandum 263[176]. His confident John Galbraith, American Ambassador in India, whom he relied on for his "back channel" consultation, had always been advising him to do away with[177] military involvement in Việt-nam. Having himself witnessed the 1954 French defeat at Điện Biên Phủ, Galbraith, as JFK's "back channel advisor," might have discussed the feasibility of a major U.S. intervention in Việt-nam and its risks involved as a result.

This situation might reflect well what Kennedy, when a senator, had referred to as a new democracy in need of stabilizing. Concerning formerly-colonized countries, Kennedy had made a statement then that more or less explained his philosophy of U.S. engagement in newly liberated nations:

"Before the U.S. moved in at any degree, independence must be granted to the people. The people must support the struggle and any intervention by the U.S. will be bound to be futile." Việt-nam's Independence was not granted, but won at the battle field. Hatred would make people blurry and therefore, they would not be able to perceive and process clearly. Perhaps

---

died/0/3387792 is a newer released documentary

176   National Security Action Memorandum 263 (see addendum) signed by JFK ordering U.S. troops to be withdrawn from Vietnam as Stage 1 of disengaging the U.S. from the "Vietnam Conflict."

177   *Pentagon Papers – Overthrow of Ngo Dinh Diem (1963) Gravel Edition Vol. 2, Chapter 4, Released June 13, 2011.* https://www.mtholyoke.edu/acad/intrel/pentagon2/pent6.htm and https://www.scribd.com/doc/57888473/The-Overthrow-of-Ngo-Dinh-Diem

he had conceived that, once a country had recovered its democracy, it would need more time to grow and mature with its new status in order to achieve self- determination. Not until then would the U.S. be able to intervene successfully without getting entangled in a mess... Psychologically speaking, it would be difficult for new democracies to transition all by themselves out of their past structured co-dependency on their colonizer into a confusing unstructured independence.

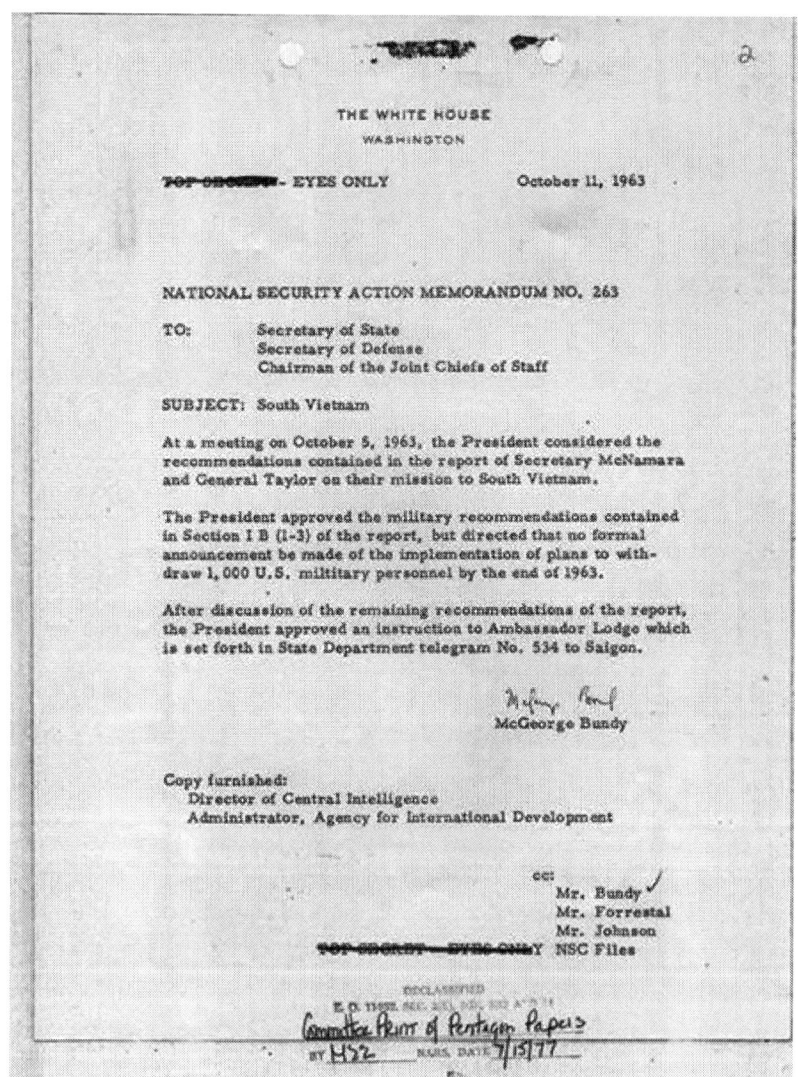

Source: National Security Action Memorandum 263
Commons.wikimedia.org

Renowned American Psychologist Jerome Maslow, famous for his landmark work in human development, developed a concept of "Hierarchy of Human Needs", which is a five tier model usually presented in a pyramid structure. According to the Maslow way, when applied to changing and emerging governments, people might have to restart from the bottom: reshaping their entire way of knowing and understanding of the world and how their new nation fits into it. In some cases, they might stumble at the "Belonging Level" (level 3 of the Hierarchy of Human Needs) if they could not overcome obstacles, and consequently, develop a resistance to or a combative attitude towards others. In a later chapter, I will discuss at length his theory and how it can be applied to people when adapting to a new environment or a new form of government.

Kennedy's original 1961 stance was to overpower Cuba and the Soviet Union, even if the U.S. had to resort to the use of nuclear weapons ("nuke them before they catch up!"). His Eisenhower-influenced Joint Chiefs of Staff JCS, who strongly believed the U.S. would win the war against Communism in Việt Nam, considered him to be "weak" (1962) when they saw him start leaning more towards privately negotiating for Peace with the Communist Block by negotiating with Soviet Union Nikita Premier Khrushchev without discussing with his Cabinet or going through his secretaries of State or Defense. This difference in policy between JFK and his generals had taken a divergent and abrupt turning point after the April 1961 Cuban Bay of Pigs incident[178], and especially after the major clashes between the Soviet Union and the U.S., from the erection of the Berlin Wall in August 1961 to the "October 1962 Cuban Missile Crisis[179]," during which the two countries were on the brink of nuclear war. One could never believe a disastrous near

---

178   As admitted later by JFK, the CIA had formed a secret commando group of Cubans immigrants in exile in the U.S. to invade Cuba and overthrow Fidel Castro - CNN *The Sixties documentary and History.com Bay of Pigs Invasion begins* http://www.history.com/this day in history/the-bay-of-pigs-invasion-begins . A big question mark stands out bright: Why did the CIA change its support and plan on the Cuban liberation forces?

179   http://www.history.com/topics/cold-war/cuban-missile-crisis, https://en.wikipedia.org/wiki/Leonid_Brezhnev and http://military.wikia.com/wiki/Premier_of_the_Soviet_Union from the Cuban Missile Crisis to the beginning of an nuclear reduction agreement

nuclear-WWIII Cold War could turn into a pro-active nuclear reduction treaty between the UK, the U.S. and the Soviet Union!!

It was obvious that the tensions that spiked up high, escalating from the CIA-backed attack of Cuba (Bay of Pigs) to the Soviet installation of medium range nuclear ballistic missiles in Cuba, had really caused major fear among Americans of an aggression by the communist bloc. Every family was geared up in case of a nuclear attack from Cuba and an invasion of countries in Europe and Asia (JFK's Domino Effects) by the Soviet (and China). This was like a wakeup call that had urged all citizens to prepare for war, thus opening their mind up to Vietnam War later. It was also a nerve tension created to prepare the people of the U.S.A for a war of attrition against Communism!

What followed was an incredible philosophical swing of both U.S. President Kennedy and USSR Premier Khrushchev to a more peaceful nuclear-containment talk that would have squished out the Cold War if it had been allowed to continue. Kennedy was concerned that his military JCS and the State Department were undermining his executive power. Many of the JCS, according to many political analysts, might even have tried to instigate Cuban-crisis entangling plots against Kennedy, such as Pentagon Chief Lyman Lemnitzer[180] or Air Force Gen. Curtis E. LeMay [181]. Among the military, Kennedy seemed to trust only Gen. Maxwell Taylor, whom he later made chair of the Joint Chiefs of Staff. Besides Taylor, he had to rely on his faithful brother Senator Robert Kennedy, and his other "back channel and third party connections." Among these were John Galbraith, the American Ambassador in India and the "JFK citizen representative," journalist Norman Cousins and his family. Pres. Kennedy had also relied on Galbraith as the back-channel contact with Hồ Chí Minh to explore possibilities of toning down the insurgence in Việt-nam in exchange for the

---

180   Fifty Years after JFK Murder, the Finger Finally Points to Pentagon Chief Lemnitzer by Author Richard Cottrell http://progressivepress.com/blog-entry/50-years-after-jfk-murder-finger-finally-points-pentagon-chief-lemnitzer

181   LeMay Chairs the Joint Chiefs of Staff by U.S. Col. Walter M. Higgins, Jr. and JFK Coup d'Etat – the Administrative Details by William E. Kelley (http://jfkcountercoup.blogspot.com/2012/04/lemay-chairs-joint-chiefs-of-staff_11.html. Kelley: "This historical administrative record shows that the assassination was not the work of a lone-nut nor a renegade CIA-Mafia-Cuban intelligence network, but a well-planned, coordinated, integrated and official program – an inside job – coup d'etat by a domestic, anti-Communist network active in the anti-Castro Cuban project." (2012)

U.S. not getting involved there. As mentioned earlier, Galbraith was in Việt Nam during the Điện Biên Phủ event. Norman was Kennedy's liaison with Khrushchev and had successfully facilitated the three-sided July 25, 1963 Soviet Union-United Kingdom.-United States Nuclear Arm Test Ban Treaty. If Kennedy and Diệm had not been assassinated, they both could have eased the U.S. and Việt-nam (as well as Cambodia and Laos) out of the bloody war! The question still lingered, "Who was behind JFK's assassination?"

David Myers, a retired mental health worker, stated in his blog in Quora[182] when discussing the reason why he was assassinated (2/11/15): "JFK had been deemed a 'commie sympathizer' by his own Joint Chiefs of Staff because he was negotiating with Nikita Khrushchev to end the cold war, end the nuclear arms race, and mutual disarmament of all nuclear weapons - threatening to end the war machine's way of live for all the vested elements of the military industrial congressional complex." Unfortunately, with Kennedy being assassinated in 1963 and Premier Khrushchev (also First Secretary of Central Committee) being replaced in October 1964 by pro-Cold War Premier Alexei Kosygin (with Soviet Central Committee General Secretary Leonit Brezhnev), the course of events had unfortunately redirected the Soviet and the U.S. back onto the Cold War path.

Fate must have been determined that Vietnam War be expanded in the next decade and millions of lives become expandable! This major turning point where two presidents were assassinated had opened up the RVN to a new era and led both the U.S. and the RVN into a new incredible journey to a near-WWIII arena! It was all about Destiny!

Interestingly, the most recent History Channel series on "<u>JFK Declassified: Tracking Lee Harvey Oswald (May 2017)</u>*" was developed by former CIA agent Robert Baer and a police officer (in charge of the 1963 investigation) using some two million declassified Warren Commission files on JFK's assassination. This "private" investigation was conducted by looking into the major conspiracy theories. Actual visits to the sites where Oswald and potential leads to his activities had led the investigation to some conclusions, such as that the Soviet was not involved in the plot, and that leads stopped at the last clue Sophia Duran, an employee of the Cuban Embassy in*

| 182 | https://www.quora.com/Why-was-Robert-Kennedy-assassinated |

Mexico with an obscure background, the last person Oswald had contacted when trying to apply for a visa to escape from the U.S. to Cuba. She was arrested by Mexican Police, probably upon a request by the U.S. the day after JFK was assassinated. Was Oswald actually seeking a way out from CIA pressure (theory of the CIA involvement in assassination) by trying to flee and hide in the Soviet Embassy in Cuba? No further record was found in the Warren Commission file... Sources of information having all disappeared have made the theory of CIA or U.S. government involvement in JFK assassination become even more plausible[183]!

My questions remain the same: Why would the Soviet want to take JFK out when he had successfully connected with USSR[184] Premier Khrushchev and successfully negotiated the July 25, 1963 three-sided mutual Soviet-U.K.-US Nuclear Arm Test Ban Treaty? Were Oswald trips to the Soviet, his marrying a Russian wife from there and his throwing Soviet "sickle and hammer" signs when escorted by Dallas Police between interrogations, staged to sidetrack people and make them think he had acted on behalf of the Soviets? How would Cuba, the Soviet "little brother," be strong and resourceful enough to organize such a plot thousand miles away? Nevertheless, if Castro held grudges enough against JFK's Bay of Pigs attack and Missile crisis, he would be bragging responsibility for the assassination, instead of silencing the only witness, Lee Harvey Oswald. This leads us to the next conspiracy theory: the U.S. military and CIA connection[185]. Only insiders would try to destroy all witnesses and evidence...

More behind the scene

The military viewpoints took the stance of JFK's JCS, of course, describing both Presidents Kennedy and Johnson as being uncooperative with them, and being responsible for leading the U.S. into the Vietnam War. This totally contradicts what had been revealed by Darin Nellis (JFK: A President

---

183   Especially if the CIA could plot Fidel Castro's assassination using James Bond's types of devices (CNN – Newly declassified documents under Pres. Trump Nov. 2017), or successfully supported Pres. Diem coup...
184   Union of Soviet Socialist Republics https://en.wikipedia.org/wiki/Soviet_Union
185   https://en.wikipedia.org/wiki/CIA_Kennedy_assassination_conspiracy_theory Did CIA assassinate JFK?

Betrayed), CNN series The Sixties, Jacqueline Kennedy's secret service agent Clint Hill, or JFK's civilian "unofficial" advisors Robert Kennedy and Ambassador John Galbraith as I had discussed earlier.

Gen. H.R. McMaster[186]'s book and Brian C. Darling's blog on Dereliction of Duty had both alluded to the fact that both Kennedy and Johnson had concurred on marginalizing their JCS in the matters of U.S. involvement in the war. McMaster blamed this administration for not listening to military advice. They also stated that the White House had played the chiefs against each other and ignored their advice. Furthermore, they stated that, even after Kennedy's death, Johnson was still maintaining that Việt Nam issues were "unwanted" and that resources should be directed to domestic programs.

This could mean, contrarily to what Jacqueline Kennedy's secret service Clint Hill had said about Kennedy and Johnson "not being close," that both were in agreement with Diệm's goal of keeping the U.S. forces at bay from direct involvement. If this were true, why would Johnson drastically increase troops in Việt-nam to 500,000 after he took over presidency? Was he under pressure by the JCS? When McMaster was researching for his book "Dereliction of Duty" in 1997, the declassified files used in both films "JFK: A President Betrayed" and "The Day Kennedy Died" were not released yet (2013); nevertheless, he did confirm there was a major disagreement between Kennedy's policy and his military advisors. This lead us back to the same question: Who was behind JFK's assassination?

Turning the history page

A few months after JFK's assassination, the new U.S. President Lyndon B. Johnson reversed all of JFK's orders that were going to disengage military involvement in RVN. Defense Secretary Robert McNamara, in spite of his statistical and concrete sequential report showing the RVN lacked political and military stability, supported the U.S. increased intervention

---

186   Dereliction of Duty (1997) by Gen. H.R. McMaster and Brian Christopher Darling's Feb. 21, 2017 guess's post Dereliction of Duty (guess) on http://blogsofwar.com/dereliction-of-duty/

in South Vietnam[187]. The two Tonkin Bay incidents[188] (8/2/1964 and 8/4/1964) gave the U.S. military a reason to move combat troops into Việt Nam and expand the war, even though it was an undeclared one. From there, he cooperated with the JCS and quickly escalated military involvement by sending more and more troops, transitioning from advisory to combating, probably the way JCS had always wanted. Did Johnson heartedly lean towards chiefs of staff's advices, or did he have no choice but give in and follow the JCS's war protocol? Was the military pressure the reason why he refused to run or be nominated for a second term?

Nevertheless, towards the end of his short term filling out for Kennedy, Johnson decided not to run for another term. Had anyone investigated to find out why he had decided to give up? Was he intimidated because of other public political figures such as Ngô Đình Diệm, John F. Kennedy, Dr. Martin Luther King and Democratic Senator Robert Kennedy had all been assassinated? What did he know about JFK's assassination? Was it because the U.S. economy and political agenda were getting dangerously out of hand? Was it because the highly revered CBS anchorman "Uncle Walter Cronkite" had determined, after covering the Tết Offensive of 1968, that the war was a "stalemate" and that the only way out for the U.S. would be to negotiate (Feb. 27, 1968)? Those questions remained muffled up and un-answered, and no major conspiracy theory was born… yet!

It was so convenient that Palestinian Sirhan Srhan was Robert Kennedy's assassin[189] (June 5, 1968), and that his execution verdict terminated further investigations! Like JFK's case, Robert Kennedy's assassination investigation was hushed shut, at the most strategic moment when the senator was soaring towards a presidency. Was it because Senator Robert Kennedy, like his brother JFK, was thinking about withdrawing the U.S. from Việt-nam? Before running his presidential campaign, Robert was given an advice by France President Charles DeGaule to "get out of Việt-nam." If this were true, his assassination might have something to do with the pro-war clan! It was so convenient that both brothers, who had the same ideologies that the rising Democrat movement

---

187   http://www.nytimes.com/1995/04/09/world/mcnamara-recalls-and-regrets-vietnam.html
188   https://en.wikipedia.org/wiki/Gulf_of_Tonkin_incident
189   http://www.history.com/this-day-in-history/bobby-kennedy-is-assassinated

would support, were eliminated. Many political commentators and analysts believed Sen. Kennedy would win the presidential Democratic Party Nomination and would be a significant opponent against Republican candidate Pres. Richard Nixon (more so than candidate Hubert Humphrey), if he were not murdered. Sirhan and his patriotism to Arab causes versus Sen. Kennedy's support for Israel (promising 50 fighter jets to Israel if elected President[190]), in my opinion, were the very simple reasons to cause Kennedy's death, but, as Israel was still far out of reach on the political agenda, the explanation did not really fit into the big picture. There had to be a more plausible political reason why Sirhan had assassinated Robert Kennedy...

One can see that I am gradually leading this conversation to a conspiracy theory that Robert Kennedy's death and his brothers are connected: they both have the same ideology. If we extrapolate this theme, we might be able to connect all of the Kennedys' death, including the most recent accidental plane crash that kills John F. Kennedy Junior (determined to be caused by "a spatial disorientation" while piloting the plane[191]) and his wife.

David Myers, a retired mental health worker, in his 2/11/15 blog on Quora Q/A <u>Why Was Robert Kennedy Assassinated</u>[192]*, had stated: "...Upon election as President, he (Sen. Robert Kennedy) would have turned the full force of the power of the U.S. Presidency against the powers (including the ones within the government) that engineered the assassination of his brother. Those forces included elements of the CIA (which JFK had promised to 'smash into a thousand pieces and scatter them to the winds' because of their unaccountability and their misrepresentation to JFK during the bay of pigs invasion), J. Edgar Hoover's absolute control over the FBI and his use of blackmail on politicians and government officials, including JFK himself (the Kennedys were going to let mandatory retirement finish J. Edgar's long run in 1964) - and Hoover also engineered the cover-up of the JFK assassination for his good buddy LBJ; the mafia, whom RFK had been trying to eliminate and who provided some of the manpower and money to assist with JFK's assassination."*

190	https://www.quora.com/Why-did-Sirhan-Sirhan-kill-Bobby-Kennedy
191	National Transportation Safety Board news release – 7/6/2000
192	Excerpt from David Meyers's Quora blog at https://www.quora.com/Why-was-Robert-Kennedy-assassinated

At this point, readers can draw their own conclusion... and the conversation continues "Would there be a Vietnam War if Pres. Kennedy had not been assassinated? Who wanted to escalate the war and for what reasons? This might be the beginning of the growing split between the conservative Republican's and the ever changing and expanding Democrat Party. The split turns into clashes, notably at the 1969 Democratic Convention in Chicago, then into bloodier and bloodier confrontations between the Nixon establishment and the antiwar movement. Who will benefit from these clashes? Like an onion, the more one peels it, the more complicated internal layers it will present!

From that point in history, pressure kept building up ("graduated pressure") fast, and Pres. Eisenhower's doctrine was delivered by both Pres. Johnson's JCS and Republican Defense Secretary McNamara pushing for a war of attrition. Hồ Chí Minh jumped into this occasion to declare a solemn patriotic cause, calling on Vietnamese patriotism against foreign invasion by the U.S. This allowed NVA Gen. Võ Nguyên Giáp carte blanche and full power to utilize docile but hateful followers to launch and expand the National Liberation Front in South Vietnam.

"Know our Foes, Know Ourselves, One hundred Battles, One hundred Victories." Knowing the enemy would not be enough: one had to know one's own people and ones allies well too, as the latter could be one's foe in disguise and living in our house, as coined by the Vietnamese phrase "raising stinging bees in our own sleeves," and getting stung to death by them!

In this jungle of conspiracy theories, I tend to lean more *towards the possibility that JFK's assassination is plotted by a complicated U.S. inside job, rather than by a Soviet or Cuban conspiracy. Remember the fliers "JFK is a traitor" found in Dallas on his visit there? Remember the serious conflict between the JCS and JFK? And* <u>JFK's National Security Action Memorandum</u> *263 ordering troop withdrawal from Việt-nam? And LBJ's not running for second term? As for a Soviet involvement in JFK's assassination, rationally speaking, it does not make sense to link it to the Soviet, for the very simple reason that, as discussed in an earlier paragraph, with the facilitation of Norman Cousins, JFK has recently succeeded in appeasing and establishing fruitful communication with* the Premier Nikita Khrushchev[193], as well as negotiating with him the three-way 1963 Soviet-U.K.-US Nuclear Arm Test Ban Treaty. It does not make sense either that Fidel Castro, the Soviet little brother that is militarily, and, probably economically dependent on the Soviet, would go against the man who just negotiated successfully with his big brother... Besides, Cuba is still recovering from its recent revolution and might not have enough might to organize such a complex covert mission far away from home base.

As a matter of fact, I believe that both assassinations of Presidents John F. Kennedy and Ngô Đình Diệm are linked together? In my opinion, both leaders happen to have similar policies that will incapacitate the U.S. military might from increasing direct involvement in the RVN! They have both attempted to stay away from the U.S. Cold War policy, and that might be why they are eliminated, so that the war can ravage in South East Asia for the next decade! Perhaps that same plot has also caused Sen. Robert Kennedy's death less than five years later, when he is running for Presidency. Maybe, the same political force has wanted to prevent JFK's pacifying political approach and to perpetuate it through his brother Robert Kennedy. This sounds like another conspiracy theory... The Cold War might be extinguished, and the Việt-nam Conflict localized and contained as a manageable skirmish between the "brothers and sisters," allowing many million lives to be spared so that the U.S. and South East Asia will experience a totally different Future! Unfortunately, that has not happened and...

The Unites States goes to war

On August 2nd, 1964, less than a year after both assassinations of "weak" Pres. Kennedy and "refusing-U.S.-direct-involvement" Pres. Diệm, the Tonkin Gulf incident[194] swiftly gave both Pres. L B Johnson the political reason to officially enter the war against the Vietnamese communists, and Hồ Chí Minh his pretext to summon the people against invaders. Many sources confirmed that the U.S. had never declared this war.

---

193   https://en.wikipedia.org/wiki/Nikita_Khrushchev
194   Even though proofs of the two incidents of U.S. Navy destroyers being attacked by North Vietnam boats were not substantiated

Once more, the song Hội Nghị Diên Hồng Conference was summoned again:

"Toàn dân! Nghe chăng? Sơn hà nguy biến!"
  To the people: Can you hear me? Mountains and Rivers (Ancestral land) are at risk!
"Hận thù đăng đăng! Biên thùy rung chuyển"
  Hatred is piling up! Borders are quaking.
"Tuông giày non sông rền vang tiếng vó câu"
  Stamping Mountains and Rivers (enemy), cavalry is thundering
"Gây oán nghìn thu"
  Causing Hatred throughout thousand years,
"Toàn dân Tiên Long! Sơn hà nguy biến!"
  All people, Children of Angel and Dragon! Mountains and Rivers are at risk!
"Hận thù đăng đăng! Nên hòa hay chiến?"
  Hatred piling up! Make Peace or War?
… …   Translation by Dr. H. Tuong

Eventually, the U.S. became more and more involved in the war with constantly and rapidly increasing military aids and troops, moving from advisory capabilities in 1960 (900 advisors) to actual combating at battle fields in 1964 (from 1961: 3,200 advisors to 1965: 184,300 combat troops[195]). Unfortunately, the White House foreign policy became more and more confused and gradually disoriented because of the American press-media-fueled antiwar reports. This contributed to an up-creeping of an emotionally angered and panicky population. A poor and aching economy and the civil rights movement also complicated the emotional burdens the country was feeling and experiencing. In the U.S. as well as in Việt Nam, the **climate**[196] "Thiên Thời" became more and more unfavorable to the war efforts, while the **environment (terrain)** "Địa Lợi" in both the U.S. and Việt Nam turned sour and harsh with protests (US) and strife at the battle fields (VN),. In both nations, the **people's heart (human harmony)** "Nhân Hoà" was definitely polarized. Those three

---

195   American War Library: Allied Troops Levels 1960-1973) - http://www.americanwarlibrary.com/vietnam/vwatl.htm
196   The elements for success: Celestial Timeliness, Favorable Environment and Mindful Harmony, "Thiên thời, Địa lợi, Nhân hoà."

success factors Thiên Thời, Địa Lợi, Nhân Hoà that we always needed for successful campaign were clashing more and more seriously, as the war became more and more intense with terrifying outcomes. The cultures also conflicted and abused, but no one paid attention to and could foresee the results. It was a bad Omen!

In the meantime, the Southerners (the RVN) grew more and more cognizant of communist calamity as history had shown millions of defenseless lives being meaninglessly wasted during both the Russian and Chinese revolutions. In Việt Nam, the horrendous local "đấu tố" (conviction by the People's Court[197]) stories in particular, as mentioned earlier in this chapter, told by witnesses since the days the communist Land Reform of 1953 took over land in the North and in some part of the South, had indeed wakened up and seized many naïve souls with fear.

Politicizing the Dissident North/Central versus the Docile South

The fact of the matter was that a great many Southerners, who had been status-quo type of farmers, and submissive citizens under prior generational foreign domination, were not educated enough in more contemporary general knowledge and politics. As a result, they were not prepared for the new Freedom and Democracy ideologies. Perhaps the only thing they really cared about at that time was Peace so they could make an easier (as they were hoping) living: cultivate their land, raise farm animals, sell crops, and keep on farming. They might have been more concerned about the weather than politics, and would close their mind to and obey whatever authorities might have control over their land. As a matter of fact, they were still behaving like vassal servants

---

197   People's Court in Vietnam, probably based on the Chinese People's Revolution model, was nothing like the ones on TV in the U.S.. It was set up by the Communist Party so that the court was run by political cadres in conjunction with the Communist Party's ruling. People who had grudges could bring anyone they had problems with in the past out to court for strict punishment without any due process or legal support. The people's court of the mid-1940s to the 1960s had seen a great many land owners, alleged-traitors, collaborators of old regimes... executed or gravely beaten on the spot, as recounted by survivors as a portrayed in the 1956 iconic movie We want to live (Chúng tôi muốn sống). Video in Vietnamese (closed captioned in English): http://minhduc7.blogspot.com/2014/09/cuon-phim-co-canh-au-to-thoi-cai-cach.html

obedient to their new government, or their local lords, as they did under the French colonization. This was "Pire que l'eau qui dort," as they said it in French: the worst thing that could be was sleeping water. Dormant and politically uncommitted citizens could turn around and bite back at you, especially when the instigators imposed pressure and fear on them. This explained why one minute peasants could be your friends and, the next minute, your most fierce foe!

When students, myself included, could not understand or avoid discussing politics, or when neighbors just minded their own business and bent to whichever directions the wind was blowing, I could see, on the one hand, the difficulty a new RVN would have when trying to convince the Southerners it would be worth it to fight against Communism. On the other hand, many politically uninterested or neutral people still nurtured a sentiment that the communists were heroes as they had just gloriously repelled the French and Japanese invaders. They would not believe the brave and patriotic VC would grab their land (Land Reform Law) as refugees from the North claimed? How could they fight against these Independence heroes?

On the contrary, the peasants (most Vietnamese in pacifying areas during the early 1960's were peasants) could see new foreigners bringing cannons and tanks and airplanes to attack them and their heroes. Was that for Freedom and Democracy? What were those strange words? For hundreds of years, the (docile) Southerners had been more or less satisfied with the past governments, even under the French. They were always allowed to work on their fertile fields and bring rice home. Wasn't that Freedom? Good docile Confucians would not need Democracy? They had been happy (as docile peasants) whether they were ruled by a monarchy, Colonials or the Republic.

It was totally different in the North and part of Central. These areas had historically been dissident (see map on the next page) and had always stood up and rebelled against invaders since the Chinese domination throughout the French and Japanese occupations. While Northerners had continued this tradition fighting foreigners, Southerners, with some exceptions, had been less dissident as their life was much fulfilled. The people in North and Central Vietnam were poor, while those in the South always rich. As a result, instead of fighting as the former had been doing, the latter had been maintaining political status quo, if not cooperating with the colonial establishment, during the French colonization[198], thus had benefitted from it.

Not even among my own direct family and family friends (except those working in the government and the ARVN, and the Passage for Freedom refugees from the North) could I detect any strong anti-communist conviction or sentiment, but, instead, I could feel some shreds of respect for the Việt Minh's efforts of gaining Independence from the French. As predicted by Pres. Diệm, Hồ Chí Minh had made extremely good use of the

Source: alphahistory.com
Vietnam War 1950 Dissidents Map
From CIA Map Supplement 1/5/1951

---

[198] Note: the South was not a part of Vietnam during Chinese occupation, not until the 1300s – as discussed earlier

patriotic anti-foreign-invader sentiment and redirected it against Americans. This explained very well how fast most people I encountered while trying to find my way out of Sài-gòn at the end of the war (see <u>Vietnam: Peace or Freedom: Part 1 – Two Minnows</u>) had quickly switched to communist side: they changed camps and sided with the VC and NVA in a blink of an eye, when the latter entered Sài-gòn.

  Did the South support the U.S. intervention in Việt-nam? Silly question! The truth of the matter was that the Northern refugees of 1954 and those Southerners who had first hand experiences with Americans were enthusiastically embracing American intervention, as they had really understood the reason why the Americans and allies were present in South Việt Nam. They full-heartedly appreciated the U.S. involvement and efforts helping the South recover from the French war and rebuild it. Those who were collaborators and sympathizers with the communists were obviously anti Americans, while those mind-my-own-business ones were vacillating between avoid or cooperate: this was a large group of submissive people that politicization (psychological warfare) would be badly needed to make them true nationalist allies.

  Could it have been a big mistake on the RVN part not to have politics or psychological warfare as part of its school curriculum? It was obvious that Hồ Chí Minh had brainwashed his youth by forcing them to learn from his "red book[199]." Their rewards was to be called "con cháu Bác Hồ" (uncle Hồ's children) or to wear a red scarf around their neck[200]. The essential politicization to show why nationalists had to fight Communism was covered neither in Lycée Jean Jacques Rousseau which I attended, nor in the Vietnamese Baccalaureate program which I was also in. At the college level, the dose of anti-communist propaganda which I had received during my freshman through junior years in our summer military training was not enough. In my opinion, besides the optional coverage in the news (newspaper and public television), the RVN had not done enough in politicizing

---

[199] Hồ Chí Minh's book was similar to Mao Tse-tung's red book that was used as the principal guidelines under the Cultural Revolution. When a child had achieved an acceptable level of Communist teaching and proven good "uncle Hồ's" citizenship (still true up to current time 2017), he/she would be allowed to wear a red scarf around the neck.

[200] This red scarf is still the symbol of alliance to "Bác Hồ" ("uncle Hồ" nowadays).

its youth, as well as with its adult citizens. This was detrimental to the cause of the RVN. No matter how full of nutrients (armed and equipped) the body was, if the soul, in this case political-warfare, was missing or lacking, there would be no way one could be useful to the Party and to its national cause!

CHAPTER 5
The Psychology Playbook

Disclaimer

As resettlement conditions are different depending on socio-geographic locations, commentaries in this book about the Vietnamese veterans and immigrants, these being possibly applicable to most, psychologically-speaking, pertain more to those resettled in the United States, and, in particular in the state of Minnesota with which I am the most familiar and whose community I have worked the most closely. By extension, as the cultures of the countries in Europe, Australia, Asia and the United States differ much from each other, sociological adaptation of these immigrants to the environment will be, as a result, very different. Nevertheless, the general comments that are based on the internet information, documentary films and written literature are current world-wide.

    After they have received Independence from the French, the Vietnamese still had a lot of mental processing to do, as they just came out of their submissive subjugation to the French. They were starting to evolve psycho-politically out of their co-dependency on a foreign power, and into a new era of socio-political accommodation that would eventually lead to self-realization: the moral and mental deliverance from subjugation. They needed time to learn. Such were the self-realization and self-determination phases they had to go through. Maslow's theory could explain well this stage of mind as the people were trying to fit into a new environment. Unfortunately, time was moving fast, whether they had time to mature politically or not... The war would not wait! In order to dissect this mental complication, we need to use the magical magnifying glass of psychology to look into the culture.

Psychology: Maslow and the Hierarchy of needs

What has puzzled a lot of Vietnamese nationalists nowadays is why their own people are so divided. They know they are divided and wowed they will prevent it from propagating, and yet, they continue to sink themselves even more into the ocean of polarization! I am not sure whether or not it is true in other states in the U.S., but for sure, in Minnesota, the older immigrants from Việt Nam, as they themselves have admitted, are "totally divided," so divided they cannot do much together in the community as a team. Consequently, the Vietnamese communities operate their businesses and organizations at very small scale, e.g. family operations or social self-help clubs, if compared to the other three SE Asian immigrant groups Hmong, Laotian and Cambodian. Many who have been exposed to Western psychology believe it is the "cái Tôi" (self or, in French, "le moi"), the egocentric narcissistic impulse that fuels their eagerness to stump others down, in order to elevate themselves or, at least, give them the impression they are superior to others. This attitude has certainly caused them not to trust each other.

Maslow's "Hierarchy of Needs," in my opinion, as a retired career educator who has worked with some 30,000 Minnesotans of all ethnicities, genders, walks of life and cultures, can clearly explain this strangely harmful mentality. By expansion, his explanation will clarify mental issues from frustration, anger, maladaptation, rejection, mental health... to mutual sabotage, vindication, bullying, oppression... These behaviors are manifested when people experience problems adjusting self to a new environment, such as new job, new culture, new government, new relocation… or new Freedom and Democracy. The most interesting part is that, in my opinion, this mental crisis has been developing through generations mainly since the French colonization. Maslow's theory can be applied to also explain the mental struggles of Vietnam Veterans[201] (or any veterans) trying to cope with

---

201   This appellation refers to the U.S. army personnel who had served in Vietnam during the War. South Vietnamese army veterans are not considered by the U.S. government as Vietnam Veterans (referred to in Vietnamese as "cựu chiến binh Hoa Kỳ"), but just called former ARVN soldiers ("cựu chiến sĩ QLVNCH").

being estranged by their own compatriots and loved ones when returning home.

According to Abraham Maslow[202] (Maslow's Theory of Motivation), people are activated by a Hierarchy of Needs. Before they can fully move to the next higher level in the Pyramid of needs, they have to fulfill the current one by completely becoming aware of how they fit in such environment in terms of those needs (see following diagram). During the time a child grows up from infancy to adulthood, s/he goes through many stages of needs, one at a time as s/he tries to figure out her/his place in the community or peer group at that specific need level.

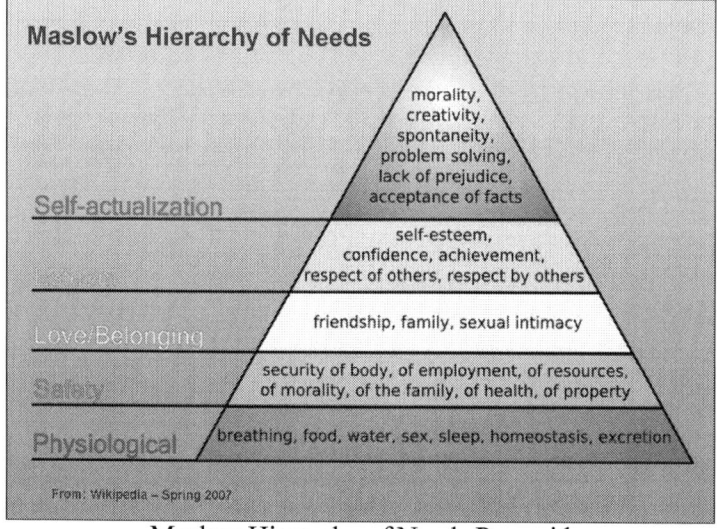

Maslow Hierarchy of Needs Pyramid
Source: Wikipedia 2007

---

202   https://en.wikipedia.org/wiki/Abraham_Maslow American psychologist of the 1960s, psychology professor at many American universities, and one of the 10th most cited psychologist in the XXth Century (Review of General Psychology, 2002). His Hierarchy of Needs has been one of the major components in most Colleges of Education and psychology health classes. http://www.edpsycinteractive.org/topics/conation/maslow.html Education Psychology Interactive has even dissected this hierarchy of needs into more detailed sub stages.

As an educator, I have always paid much attention to children's needs, and do not realize until later that adults who have been through major changes in their life (migration, new job, new relationship, new promotion or demotion…) also evolve though the same hierarchy. The only difference is that the adults might try to handle more than one level of needs at a time. My experiences while being a political refugee, teaming with diverse self-help groups, such as community volunteers, professional teachers or principals' associations… or working with diverse cultural groups have helped me understand what individuals have to go through when examined through the lens of Maslow's needs theory. I believe it is probably applicable in all cultures, with some "micro-cultural adjustments". Each individual grows and matures in the environment at each level of needs, and reacts to it as s/he is trying to realize what her/his position and role she or he has in it. Babies have to develop in a lower level fully before expanding to the next level. When they encounter difficulties adjusting that force them into refusal, they will feel rejected and will take a fetal position, resulting in crying (fright) or having a temper tantrum (fight back).

The Hierarchy of Needs has been applied in K-12 school settings to accommodate children; nevertheless, it can also be applied by extension to adults who have already gone through these five levels of needs when growing up, but who are now facing a new situation or environment that might be a detriment to their growth.

When repeating the cycle, they might try to speed up the process by sometimes combining more than one level at a time. They oftentimes make attempts to apply the old experiences they have learned and mastered in their original culture with the hope it will conveniently work, which is not the case most of the time. If there are conflicts of values (the old system they are familiar with versus the unfamiliar new one), they will experience difficulties and react negatively against the new environment (misfit). That situation might be disappointing and will discourage them, leading them to rebelling or developing mental health issues.

In order to avoid this counter-reaction, they seriously need coaching, guidance or counseling, to prevent them from reverting to their root culture and going into "hiding" in a cultural ghetto where they feel safe and comfortable (refusal to mainstream or "acculturate[203]"). Unfortunately, people from the old culture rarely admit they need counseling, as it's mentally degrading for the family or personal honor. This leave them only one way out: go into hiding…

Below are the five levels of needs Maslow has described in his "Hierarchy of Needs" theory, starting with the lowest one on the bottom of the pyramid (for beginners), and culminating to the highest one on the summit (where the individual has well adapted to the environment).

1. Physiological needs are the lowest level. At this level, the child (or individual adult in the process of adapting to new environments) is more concerned about eating, sleeping, while the adults have additional needs such as sex, maintaining equilibrium of the mind.

2. Safety is the next level. At this level, the individual is concerned about his/her own and family security, health care, owning resources and possession, self-preservation (balancing between the host and the root culture as well as language allegiance).

    The two levels above make the individual focus more on self (child) or family or clique (adult), rather than on the larger community. S/he is more self-absorbed for her/his own protection. This is where adult immigrants try to hang on to their cultural/ linguistic group, and still have fear of acculturation and mainstreaming. The individual may have little interest in learning the new language and culture, and still refers to "my homeland" as "my country." For many, level 2 is filled with mixed emotions and feelings (dual allegiance), as the individual starts grieving some losses and abandonment, and developing a fear of the unknown. At these stages, the individuals are the most fragile.

3. Belonging is the first level where s/he adapts to others and adopts the new environment. The individual cares

---

203 Become more atoned to the new environment

about her/his relationship with others, either of the same or different culture or ideology, and tries to fit into a larger group (a child will relate to the family and some in the community of proximity). Learning about the new system becomes an honor, and allegiance to it starts growing. Mixed emotions still exist as the individual sets off for a new journey. Any negative feedback from the host environment will cause catastrophic reaction.

4. Self-esteem grows when Belonging stage is successful. In my opinion, this is the most difficult stage so far to accomplish, as self-esteem and self-confidence depend on how the individual fits in while going through the previous Belonging step. Encountering difficulty, hostility or failure there will result in the development of feelings of despair, not being accepted and respected; this failure to belong will lead to mental health issues (such as in the case of "Mentality of the Colonized" (as discussed in Vietnam: Peace or Freedom - Part 1 Two Minnows), Vietnamese-on-Vietnamese shaming/ blaming/ crime/ anti-social behavior, or even rebellion, suicide...).

A great many adult immigrants might be stuck at this level longer, or indefinitely, if they fail to open themselves up and learn new things and adapt in the lower levels, overcome their fear of the unknown or outgrow their grieving of losses (survivor's guilt). They might never develop high self-esteem, and end up hiding behind their mono-cultural community, where they can feel much more self-confident and think they can be respected. The only way to outgrow fears is to be flexible and open our mind in order to learn and adapt for one's and everyone's good. Failure to fit in definitely pushes them into hiding in cultural sanctuaries. For the Vietnamese immigrants who feel misfit, communities where there is a presence of a large and strong monolingual and monocultural population, such as in big cities with a "Little Sài-gòn" or California Bolsa type of conglomeration, would certainly be a sanctuary that makes them feel the safest and, consequently, helps them survive. The negative side of sanctuaries is that it is so convenient and might psychologically prevent the immigrants from making efforts to learn English and mainstream with the community at large.

5. Self-actualization is the top level of needs. One can reach this level if and only if one is able to acculturate and achieve compassion and respect from others, and has consequently achieved good self-esteem from the previous level. At this level, the individual will start looking into higher critical thinking matters such as reaching out into current issues that matter to, not the immediate ethnic community, but the one at large (e.g. interacting with social media or politics, having concerns about social justice, accommodating diversity and facilitating mutual support...), coping with it, resolving issues that affect everyone, and, most importantly, coming to peace with unresolved matters in the root cultural world... The individual can activate an understanding of one's position in the system and is willingly ready to be a constructive part of it. The person can actually see her/his position in the greater common good. Typically, s/he is well-educated, fluent in the majority language and has a secure and better job. S/he is interested in current issues and is concerned about the well-being of the community at large.

Those stuck in levels 3 or 4 might likely call "sold-out[204]" those people who are successful in the community at large (achieve self-actualization). The former, being unsuccessful in understanding higher level of thinking and mental processing skills in a rational way, and being under the control of emotions (fight or fright mode as described by Daniel Goleman in Emotional Intelligence), oftentimes have ideas and ideologies that mismatch to the current ones in the mainstream. One might feel their rationales and attitude about socio-political matters, which might be naturally acceptable in their root culture, may actually be irrelevant, incompatible, or even awkward and oftentimes ridiculous, thus causing racial clashes and/or intergenerational conflicts.

---

204 Based on my experiences working with a great many African Americans (my doctoral research on African American high school students), Hispanics and Asians (new and old timer immigrants or American born), as well as American Indians and Caucasians of age groups 11-65..

How does PTSD fit into this?

The forthcoming chapters will oftentimes refer to Maslow's Hierarchy of Needs when discussing the status of mind of the new Westerner grunts or even more experienced combatants (military or support personnel from Australia, the U.S....) when they first arrive "in country" in Việt-nam as well as when they return home a few years later at the end of the "tours." This psychology may also be applied to the "Boat People" and former ARVN soldiers and their family, as well as to other refugees and immigrants from the Republic of Khmer, the Kingdom of Laos (including the Miao tribes) or, more recently, from Middle Eastern and Eastern European countries, when they are resettled in new countries at the end of the war. It explains well the causes of post traumatic syndrome when applied in conjunction with the Snapshot and Emotional Intelligence Flight or Fight theories...

People newly resettled in an unfamiliar location always worry about physiological needs first (is edible food available, where am I going to sleep?...), then the environment (who am I going to work with or against, who and where is the potential opposition?). In the next step, they will start assessing how they fit into the environment, and react depending on how they adjust to the environment and how the latter reacts to their presence. This is when the returning GIs experience when they hear their own calling them "baby-killers, hut burners..." and start questioning "why do my own people whom I have served[205] hate me so much, why...?"

When Hierarchy of Needs is combined with Daniel Goleman's Emotional Intelligence, which I will explain next, the former relating to change processes while the latter expanding into effects of changes, some of the mysteries in human beings coping with the environment can be revealed and, hopefully, will lead to solutions that help the individuals become more understood and, therefore, honored. As for the ARVN and the Boat People, we need to explore more the Asian psychology and fuse it with Maslow's and Goleman's psychologies in order to understand the process they have to undergo in order to cope with the new environment. Goleman will explain why most go into hiding, while some others fight back by rebelling or seeking revenge.

---

205   Oh yes, they have served! They have done what some others have not, or have refused to...for their country!

Psychology: Daniel Goleman's Reptilian versus Neocortex Brains

Most of us "normal people" who think killing others is unacceptable might be wondering how some others can be desensitized so that death does not matter any longer. Psychologically-speaking, as explained by Psychologist Educator Daniel Goleman and Neuroscientist Joseph LeDoux[206] when researching Emotional Intelligence and Brain, human perception of happenings, the stimulus, as received via our senses, is processed via many parts of our brain, such as the Thalamus (reception and perception of stimuli) in the first stage, then transferred to the "Reptilian Brain" (unprocessed emotions, causing Flight/ Freeze or Fight[207]) and Amygdala (for a quick reference screening via the Hippocampus that initially processes emotions and feelings – Limbic Brain), then finally to the

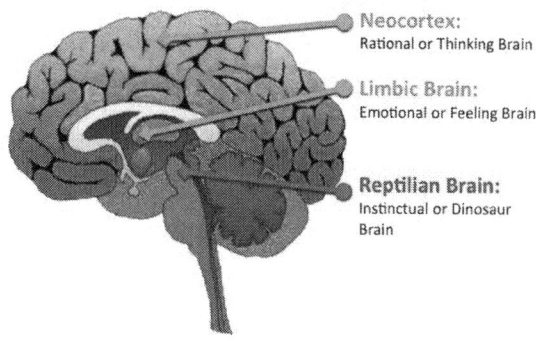

Psychological theories to help you communicate better with anyone.
Source: Blog.bufferapp.com

---

206   Emotional Intelligence: Why It Can Matter More Than IQ by Daniel Goleman (1997) and https://en.wikipedia.org/wiki/Amygdala_hijack and https://en.wikipedia.org/wiki/Joseph_E._LeDoux
207   The Reptilian Brain cannot actually process emotions rationally or reference past experiences.

Neocortex for analysis (the rational thinking brain) for decision making. In the meantime, the Hippocampus gland will spontaneously determine *from past experiences whether or not the stimuli can be a danger factor.* It will send out an emotionally-controlled (by hormones) alert urging the individual to fright/ flight or fight[208] as immediate human reaction that can be unexpectedly irrational and destructive. Ledoux strongly believes that Emotional Learning is possible so that Amygdala impulses can be controlled and reactions be guided in a predetermined way, as if the Hippocampus can create a pattern of trouble-shooting leading to a possible prescribed course of rationally "thought-up" action.

The VC who have been "brainwashed" to kill rather than make a mistake releasing the wrong person (this will be discussed later) have probably been pre-programmed at the Hippocampus level on what to do: kill! Brainwashing is a form of predetermining hormonal brain function that simulates "made up past experiences" that will trigger desired automatic responses. For instance: to massacre the over-run U.S. or ARVN remnant troops who should have the chance to surrender (per Geneva Convention – to be discussed in post-Paris accords POW exchange) will intimidate the RVN and give a strict warning to the civilians never to cooperate with the ARVN.

Death, for some combatants, can be redefined by the Hippocampus as an acceptable or even preferred consequence as some situations required, thus making the reactive Amygdala to signal to the "rational thinking brain" Neocortex to kill or to be ready to sacrifice oneself for a "greater cause."

Obviously, when facing the enemy, only the smart "reptilians" can manage their emotions in order to not kill right away, but weigh pro-and-con choices, Life and Death, and urge them to take alternative courses of action. "Smart" reptilians are able to "hold still" and let stimuli bypass and go right into the neocortex phase of processing and making rational appropriate actions. Unfortunately, the process has too

---

208   In most terrorist attacks or mass massacres (during VN War or most recently in the U.S.), the victims, no matter how numerous they were, seemed to flee for their life or freeze and let themselves be killed, rather than fight back. In those cases, stimulus received by the Limbic system was overwhelmingly interpreted as imminent threat and might have incapacitated the management, understanding and usage of emotions in conjunction with rational thinking via Frontal Brain system.

many choices that result in slow decision-making. The ARVN protocol is determined by a list

The Emotional Brain

- Prefrontal Cortex — Emotional Manager
- Eye
- Amygdala
- Fight or Flight
- **Limbic System** — Instant, automatic response
- Visual Cortex

Emotional Brain chart which closely resembles Daniel Goleman's. likesuccess.com

of rules, "strict implementation of superior's order of the day[209]," humanities, nationalist norms, Geneva Convention and U.S. military aid eligibility, not as simple as the communist rules "kill-them-all, and take no prisoner" which is the opposite. The American forces might be limited by much more rules besides the ones on the ARVN list: human rights, due process, American socially accepted solutions, culturally-appropriateness, press media,... This is where the U.S. and the RVN conflict each other! This is also the reason why **it is extremely difficult, maybe impossible, to fight a war humanely against an inhumane and extremely cruel enemy!**

The not-so-smart "reptilians," who can become

---

[209] Diverting from superior's order might have serious consequences, therefore, it is very difficult for the low ranking ARVN officers to follow U.S. advisors alternative action plan when their commanders are incapacitated. This was interpreted by the U.S. as incompetence or cowardliness when the ARVN hesitate (Ken Burns The Vietnam War). American advisors are not their superior and therefore cannot override the chain of command.

unstructured as they fail to redirect emotions to the rational brain, will need to go through a structured automatic decision-making mechanism called the desensitized "brainwash process," such as the Communist Party indoctrination or the routine emergency response taught in boot camps, to give the individual the expected conditioned response in cases of crisis, e.g. assault or counter-attack, kill others or sacrifice oneself. These conditioned reflexes have to be kept simple so the individual soldier does not have to think and can avoid getting confused, thus failing the mission. The simplest reaction used by terrorists, in this case the VC (not the NVA which is more conventional and disciplined), can be "kill them all" or self-sacrifice for the sake of the mission.

The worst case scenario is the one with "out-of-whack" reptilian: it will be definitely flee and abandon, as consequently, reveal positions to the enemy… For the latter, there are always the superior officers who will shoot the execution "coup de grace[210]" shot with their Colt 45. Deserting VC are usually executed, and so are their family members as collaterals.

Contrarily to the "reptilians," when growing up, some rational-thinking people might have educated their brain to by-pass the Hippocampus-Amygdala stage and refer stimuli to Neocortex to actually perform critical thinking. The latter must balance out decisions between Life and Death, Humaneness and Sanction, or take a magnanimous course of action before committing the last resort of killing. Of course, since the nationalist (ARVN and allies) brain has been exposed to so many choices, between various, and sometimes contradicting cultural and socio-political values that they are torn into many directions, causing them to not function as fast as a single-minded one (the VC) that has been desensitized to the idea of Killing or Self-sacrificing. Hesitating and slow decision-making can be viewed as weak by the enemy and incompetent by the allies.

---

210   Meaning "execute." I remember having heard our boot camp company commanders saying that officers' side arm 45 caliber 1911 pistols (usually a Colt 45) is for executing soldiers who are deserting at the front line, besides close-quarter combat. I always carry both, a Colt 45 and an M16 rifle.

Psychology: Snapshots and the haunting

Socio-political truth differs from one version to others depending on what sources of truth or the foundation on which critical thinking skills were based on (whether it was a political, sociological, psychological, economical... or anthropological approach, whether it was a quantitative or qualitative research method) to determine a process so researchers can triangulate[211] and elaborate the various facts and pieces of information that surround the perceived image (the context). All vary in "the Eye of the Beholder." Each interested person lives in her/his own world, and acts based on his/her own paradigm. As a result, people will see things differently and don't really know which one reflects the real truth, as their personal viewpoints (opinion) and their choice of background information (from a vast ocean of sources) can definitely affect their perception.

"Snapshots" are psychologically defined as the very first picture or glimpse at an image we perceive of an event that strikes our emotions so much it will remain encrusted into our memories as a heart-breaking or disgusting experience. These images, sounds or even thoughts can be particularly deceptive and dangerously corruptive as they might and will cause our rational and analytic mind to combine them with pre-existing and pre-conceived thoughts, thus leading us far away from the Truth, if we fail to look for more informative images or data that accompany the original image. The more supplemental "snapshots" we can get the better idea we can form into a bigger picture, which is actual a "video or film" of the happening.

Imagine we are looking at a strikingly horrendous photograph that keeps haunting us: what we see can be interpreted in many possible different ways; however, it is insufficient and we do need to get another photo of what has happened before and after that same event as a follow up, then we might be able to see more clearly. If we can secure a series of photos, the story can be better visualized, and, of course, the best big picture will be a film or video of the event... Failure to expand the perception will result in drawing falsified pictures

---

211 Analyze and compare information gathered from various, oftentimes contradicting sources. The common denominator usually reveals the better and more accurate truth.

that will dupe our reasoning process; therefore, our judgment will turn into fake news.

If the snapshot is not processed carefully, a falsified judgment based on insufficient supporting information will eventually affect our reptilian brain (as discussed earlier in Daniel Goleman's <u>Emotional Intelligence</u> - Reptilian Brain) as the image engraved in it will create an emotional response that will cause us to jump into a quick conclusion and react automatically at the sight of the image or thought about it. This is how people who are ill-informed about the Vietnam War have developed antiwar sentiments based on snapshots of battle field inhumanity (selective[212] pictures of Mỹ lai Massacre in 1969, of the VC captain prisoner being executed in 1968…). The news media have actually successfully used the snapshot effects on people reptilian brain to manipulate their feelings.

In the same way, at the end of their service, the Vietnam Veterans who return home with internalized snapshots of their brothers in arms taking their last breath at the battle field, or images of the peasants or enemy they might not even see killed will always be haunted by those images when experiencing Post Traumatic Syndrome (PTSD). Moreover, the estranged feelings of being labeled as "baby killers" or "hut burners" will affect their self-esteem sense belonging making a great many feel home is not really home, but where they just come back from, Việt-nam, is.

A typical "snapshot" example is the major trauma caused by the iconic photo of a VC being executed in public on the street of Sài-gòn by the RVN National Police Chief Col. Nguyễn Ngọc Loan. VC and NVA prisoners were most of the time treated as per Geneva Convention guidelines. Unfortunately, a few exceptional cases, as described earlier, were "very effectively" used by both the U.S. and international press media and probably by antiwar movement to instigate and cause even more protests in the United States.

The photographer's (Eddie Adams) photo that shows Col. Loan executing a handcuffed VC prisoner, Lieutenant (or Captain?) Nguyễn Văn Lém, on the 3rd day of Tết Offensive of 1968, had infuriated the whole world. Eddie later wrote an apology letter to Gen. Loan, as the former failed to give a

---

[212] The news media had selected to show these photos on the news, but not the atrocities committed by the communists such as 1968 Tet Offensive massacre of Huế, or the 1972 Easter Offensive carnage on the Highway of Terror.

great many people around the whole world more explanation to accompany his photograph. Lt. Lém, head of a VC death squad, prior to being shot in the head by Gen. Loan, had just committed a killing spree executing 34 innocent civilians (three generations), one of them was an 80 year old mother of a Lt. Col. Nguyễn Tuấn who refused to cooperate with Lém's order (to show him how to operate tanks?)... The 34 deaths (using the old Chinese "chu di tam tộc," or execute three consecutive generations, to terrorize the survivors) had consequently triggered the animal instinct and the Reptilian Brain in Col. Loan and made him ignore the 1949 Geneva Convention [213].

    I could assertively say the RVN was not the only army among our allies losing it and executing prisoners. U.S. personnel had done it out of anger seeing their comrades killed, like the machine gunner in Charlie Company who had admitted on public television[214] that he had given in to his Reptilian Brain and shot a VC with his Colt 45, instead of having the helicopter take the latter back for intelligence extraction. The fierce South Korean army was well known during the war for "punishing" civilians suspected of cooperating with VC or being part of the communist insurgence. Furthermore, before the Paris Accords Agreement was signed, NVA/VC troops rarely took prisoners (prisoners were usually executed during and post war)... and did not hesitate to execute innocent civilians... Civilians who had witnessed massacres during the war all have horrifying snapshots; nevertheless, the television viewers in their family living rooms in the U.S. had even more significant snapshots as they lived remotely from the war (which I used to sarcastically, and jokingly, referred to in French as "la guerre de salon") and had much less background information about the war than the civilians in Việt-nam.

    The best snapshot used by the North Việt Nam Army

---

[213] Interestingly enough, some commentators are still trying to defend, on many occasions during the 1968 Tết Offensive, as perhaps they think Lém has massacred "only a few dozen" innocent lives compared to the American G.I.s' 504. All of his victims were defenseless innocent elders and children of many families ...Of course, pro-Communists would argue to their teeth to show Lém was a patriotic hero who had dared killed enemy indiscriminately, alleging these civilians were traitors. The savage are always savage! http://consilientinductions.blogspot.com/2011/06/nguyen-van-lem-hes-just-plain-vanilla.html . The 1929 Geneva Convention, later upgraded to the 1949 WW1 and WW2 Geneva Convention III, as agreed on by most countries in the world, calls for the humane treatment of war prisoners. https://ihl-databases.icrc.org/ihl/INTRO/305?OpenDocument

[214] National Geographic documentary on Charlie Company in Corps IV Mekong Delta Brothers in War.

was the flag they flew in battles: It was always the NLF colors of blue/red with a red star in the center. NVA troops rarely fought under their own People DRV flag, which was Red with a yellow star in the center (this is also their current 2019 Democratic Republic of Vietnam national flag). On April 30, 1975, when entering Sài-gòn and forcing through the gate of the Independence Palace, NVA tanks, motorized vehicles and their main army "Bộ Đội" also flew the NLF flags. NLF flags had flooded the streets to "snapshot" the world the picture that it was a Civil War, with people from the South liberating their own country, that the North did not invade and that the U.S. had no business being there meddling with Việt Nam sovereignty. If we looked at documentary photos of antiwar protesters in the U.S. and European countries, we would most of the time see the NLF flags, this making people think it was truly a civil war. GIs fighting in Nam had always looked forward to finding an NVA flag; they would understand now why this red DRV flag trophy was so rare during the war!

The best thing truth researchers can do is to rely upon information that is provided from the "horses' mouth," authenticated documents (including declassified ones) or real-life action visuals (audio tapes, film and videos, such as declassified materials) and use socio-psychological-based critical thinking skills and scientific cross-reference methods to come up with possible justification of political behavior. Still, this does not guarantee that the truth has been unveiled, unless the information gathered from various, sometimes contradictory, sources can be secured. This is when investigators need to triangulate and come up with possible plausible conclusions The approach described hereby is called the socio-political methods of research-based analysis. It can be either quantitative (the traditional research based on statistical data and events) or qualitative (the newer approach, also based on existing data, but not necessarily statistics[215], that digs into special exceptions that have been neglected or unaccounted for by quantitative research). The analysis in this book is more qualitative than quantitative.

This is the reason why a great many commentators do not agree with each other's conclusions or speculations. As a matter of fact, many, including especially the People DRV or the RVN Vietnamese politicians and activists in the free world, might disagree with my version of the story (or

---

215   Statistics are referred to, but are not used as foundation for a conclusion.

part of it) if they only use their own intuition or "gut feeling" about events without any formal proven and accepted research methodologies that are better suited to more accurately detail past events [216]…

---

[216] This means that they have to be perfectly bicultural/bilingual and highly trained in socio-political science. In most of the cases of unfounded argumentation, the contesters rely on peer controversial theories or gossiping rather than on impartial and validated sources.

CHAPTER 6
Communist versus Nationalist Brain

The Vietnam War is an unfair war between the Humanitarian Democracy and the Totalitarian Conformism. It is Humane against the Inhumane! The Vietnam War is not a black and white one like WW1 and II, where all sides have to respect more or less strictly a more humane agreed on pre-Geneva Convention[217] or some sort of civility. During the Vietnam War, the nationalists (including the allies) fight an honorable conventional war, while the communists an unconventional one. The latter have chosen to do whatever work[218] for them, whether it is ethical or not, as long as they can execute "the intended plan" and achieve their ultimate goal!

As I have brought up in an earlier chapter, Northerners and Southerners do not think and act the same way. The more I think in terms of comparative warfare, the more I can see the differences in thinking, acting and reacting styles between the Northerners[219], who have been generational rebel historically fighting invaders overtly or covertly (Sun Tzu's[220] way has been their way of life), and the Southerners, who, except those who have been politicized or Sun Tzu story fans, tend to be peaceful mind-my-own-business pacifiers, therefore are more collaborators, sympathizers or just peace lovers. I know for sure that my family, which is originally Northerner[221], has always preferred story books that usually reflect Confucian compliance and loyalty (patriotism) and that also promote using Chinese Sun Tzu-influenced strategies to outwit Chinese invaders.

---

217   This is not the 1954 Geneva Conference on Indochina and Korea, but an agreement (1929 & 1949 Geneva Convention) on many WW1 and II issues, including the humane treatment of prisoners of war.

218   As they were considered by the free world as rebels in a civil war, they had ignored the international 1949 WW2 Geneva Convention and 1954 Indochina Geneva Conference until near the end of Paris Peace Talk.

219   Including refugees and migrants from North Vietnam, now living in the South.

220   As discussed in earlier chapter, Chinese strategist Sun Tzu has featured many popular war stories read by Vietnamese.

221   Northerners might be more rebels as they had a much longer history of fighting invaders.

The Communist brain

Kill or not kill? Attack in ocean waves "tấn công biển người" even if we will lose lots of lives? Or rig booby traps[222] on the fairground to cause harm and terrorize civilians? Launch missiles into civilian crowds or residence? Throw a grenade into city bus? Explode a train? Or massacre to terrorize or revenge? Bury them alive? Shoot down or behead husbands in front of wife and children, or both parents in front of children[223]? The communist ultimate goal is to intimidate and subjugate the survivors to submission by using fear and terror. Extreme fear will tackle and lead the victims into "fright" (Goleman). After all, the victims are all either "rebels" who choose to cooperate with the RVN and, therefore, "traitors against the People of Việt-nam." There is no reason to have mercy!

"Use our body to block a downhill run-away cannon? Use our body to carpet a mine field so comrades can cross it? Heroes sacrifice, don't they? We won't take prisoners or spare witnesses' life… as they are burden or might turn against us! We can use children and women to spy for us, or sell GI's opium or weed to corrupt their mind… They can fight, too, for their Ancestral Land… Worse for worse, they can be our shields against the ARVN and American invaders! How do we take prisoners? VC have no way to keep prisoners… **Kill them all. Who will know whether or not we have killed innocent civilians or the enemy after they surrender, if they are all dead and there is no witness[224]? What the heck with Geneva Conventions[225]!** The whole world thinks we, rebels,

---

222 VC's booby traps were usually undetectable by mine locators as they were made out of non-metallic natural components, such as undetectable (by metal detectors) gravel and bamboo, especially the triggering mechanism. They were both explosives and silent spring-loaded piercing traps, and were usually hidden in mud or bushes. Conventional explosives were used only under corpses or other metal pieces of military equipment.

223 During the last years of the war, one of my close friends and her children had to witness a VC squad shoot her husband at point blank. She could never forget seeing his blood splash all over the walls, and had vowed to never return to Vietnam, not even for a short visit.

224 That was the reason why a substantial number of allies and ARVN were shot down when they were sleeping during attacks, and not many, if any, were taken prisoners. Only when it was too obvious (such as American pilots shot down in DRV), then prisoners were taken.

225 NVA seemed to be more concerned about the 1949 Geneva Convention than the VC and they can be construed as invaders. Their strategy was to pass for

are fighting a Civil War anyway! We abide to no rule!"

"Our choice is very clear and simple, as we have only one choice, for 'Uncle Hồ and the Party:' we will just do what the Party is expecting. Our comrades are watching me, and I am watching them... Whatever happens, I will go into Martyrdom and my family will be honored!"

Such might be what a simple-minded illiterate VC/NVA soldier might think... or maybe they are not allowed to think as the Party has already done the thinking for them. It has actually become their combat instinct. Series of similar processes have been implanted in the VC/NVA's brain, and, when in crisis, their reptilian brain will take over and, instead of fight or flight panic taking over, corresponding automatic responses will order them to act accordingly to the "pre-programmed plan." At the culminating time period of 1975, the Party has taken some extra measures when it realizes its combatants are overwhelmed with fear and defeatist intention. Executing defecting individuals is not enough especially with their sniper units that are highly dispersed. Hormones have been used to overcome defective brain.

"Just a few months before the end of the war, young children (VC) were found dead, hanged at the ankle from their sniping position on top of coconut trees, with trace of mind-control drugs in their system. The RVN believed the drugs were used to help them 'fight courageously and push them to the extreme limit and achieve the undoable,' as I vividly remember having read about it in the Sài-gòn newspapers."

The bottom line is that they have one single Belief that directs them to act consciously or unconsciously in order to achieve superiority over the nationalists and total control of the population, even if it requires annihilation of the innocent.

Mass massacres and executions are committed as the VC[226] have gagged victims with telephone and electric cords, chain or barbed wire, buried them alive or clubbed them to death, e.g. the case of civilians of the Imperial City

---

VC to avoid the World knowing they were invading. As a result, they could do anything as VC were fighting an unconventional war and disregarded Geneva Convention rules of engagement. Evidences were so clear during the Tết Offensive of 1968 and the Ester Tide Offensive of 1972, as well as the final offensive of 1973-1975, with indiscriminate massacres.

226   Not the NVA, as verified by Huế survivors and NVA Gen. Bùi Tín

of Huế during the Tết Mậu Thân Offensive of 1968[227], when approximately 7,800 have been mercilessly executed and buried in mass graves. Mass executions of civilians by the VC and NVA have happened frequently since the 1949-1957 (Hồ Chí Minh's Land Reform Law[228] prosecution and executions by the People's Court, modeled after the same Chinese law[229] by Mao 1947-53) throughout the Vietnam War in North, Central and South Vietnam. Allies and ARVN soldiers have also been executed instead of being allowed to surrender as prisoners of war. Here again, the purpose is, not just to punish, but mainly to intimidate the survivors. As History has proven, dictatorships (Lenin, Stalin, Mao, Hitler, Hussein, Kim…) always use rule by fear and terror in their way to power and total control.

Not like the communists who rarely take ARVN prisoners[230] or who usually get rid of civilian witnesses, the RVN always gives a second chance for the VC to repent in redemption and to convert to nationalists' side[231]. It has allowed a great many VC to surrender, either collaborate with the RVN or just return to their family. Some of those who have switched side have rendered great counter-espionage and counter-guerilla services to the nationalists.

The Nationalist brain

To launch an attack against a VC hamlet, the nationalists have to abide to a long list of items determined by American warfare rules of engagement, besides the actual strategic plan of engagement. Who is the enemy? Where are they? When and how do we shoot? What about civilian casualties? Where and how should we keep the prisoners? What to do with sympathizers, aspiring VC, or prisoners?

---

227   Vietnamese documentary film *Lich-sử Chiến-tranh Việt Nam 1945-1975 - Part 2*

228   YouTube in Vietnamese: https://www.youtube.com/watch?v=jn5octvGq-yM

229   https://en.wikipedia.org/wiki/Land_reforms_by_country#China

230   The Paris Accords negotiation could have been a strong motivation for the VC and NVA to take more prisoners, especially when RVN complained about mass graves and massacres the Communists had committed during the 1968 Offensive.

231   For that reason, the ARVN had launched a major campaign inviting the VC to switch side to the RVN in the "Chiêu Hồi" or "Open Arms" program, giving a chance to the VC, the South Vietnamese converts, to join the RVN side and return to their home.

Do we have enough food and medical supplies for civilian casualties, and prisoners? How do we take prisoners? We have to make sure we take care of their injury? Do we have food for them? Do they need cigarettes? Make sure we avoid destroying our historical sites? Am I violating the Geneva Convention if I do this and that? What to do to maintain good communication with allied forces? Alternative plans to reduce civilian casualties? What would the press media think about this and that? Actually, as there are two distinct modes of conducting a campaign, the traditional Vietnamese one as opposed to the American one, the commanders might have to pick and choose. This unresolved dilemma causes confusion, not just to the high ARVN command, but mainly to the lower hierarchy combatants when the higher command is incapacitated or killed. "Whose plan should I continue to follow? My direct commander's (who is no longer available), my own improvised plan (which might also be Sun Tzu influenced Vietnamese) or the American advisor's one?" It's all in the eye of the beholder's sight and mind...

In the case of U.S. forces, even more questions arise: Are we sure those are not innocent civilians? Everybody looks alike...Civilians are shooting at us? How will the press media in the U.S. interpret this? What will the American people and the world think about this? How is soldiers' morale? Defense Attaché Office DAO[232] opinion? Is it against Washington D.C. politics? What are ally positions? What do our allies think about this? What should I do when ARVN commanders refuse to follow our plan? Is my family at home going to see this? And so on... There is a great amount of weighing and balancing between the "should's and should-not's;" this brain process always slows us down or even inhibits reaction.

When there are so many variances that interfere in the decision process, there is always a myriad of confusion, hesitation, questions, and, consequently, Doubts[233]. It is always difficult to decide when one is not the only decision maker, particularly when the remote protesters back home (US) or in Sài-gòn all want to have a piece in this process. For the nationalists and allies, they have to strictly observe that long list of objections and restrictions and go through a series of algorithms that require a great deal of "gray matter" and time

---

232   The U.S. Defense Attaché Office was the liaison between the U.S. military operation in Sài-gòn and the U.S. White House and Pentagon

233   In opposition to Belief.

to process, and multiple levels of bureaucracy and popular expectations to deal with. Definitely, it is much easier to fight during WW1 and WW2 in this matter when no one at home can see and make a judgment of what U.S. troops are doing. The world is watching Vietnam War and our own people are protesting and stabbing us in our back!

This explains the major differences between the nationalists and their Allies' and the communist camp. This has also, in my opinion as a cross-cultural socio- psychological educator and community organizer, distinguished the real heroes, who are able to make a conscious choice to act between matters of life and death, from the hero look-alikes, who act as if they are real heroes, but in reality out of simple monovalent and inhumane robotic reaction. Nevertheless, what cause major problems are the differences between East and West. These differences have led to miscommunication between the United States and the RVN forces, thus affecting the efficiency in coordination and efforts. It might even have caused the two dear allies not to fight side by side as they had at the beginning of the war, as shown in documentary films and written accounts. We rarely see them fight the enemy side by side!

CHAPTER 7
East-West Incompatibility

While the communists practiced merciless terrorism, the nationalist side, including the United States and Allies (except the army of the Republic of Korea[234]), fought the war with much more humaneness and compassion towards both the enemy and the suspected civilians. The Americans in general believed all people had the right to live and survive, and should be treated with fairness and magnanimity in the name of Humanity. Unfortunately, in real life, when enemies and innocent people could not be distinguished from each other, their practices were usually distorted or misinterpreted and used by the media to fuel anger in the United States and in Europe, when, in the meantime, enemy's acts of war crime and genocide seemed to be ignored, or rather seldom or never hit the news. Nevertheless, Humaneness and Compassion were well taken advantage of by enemy psychological warfare. As discussed in a later chapter, **the enemy really appreciated that the "Conscience of America" had actually helped them fight America's own forces[235]**. Unfortunately, the more humanely we fought, the less effective our efforts had become. The more humane we were, the more the enemy saw we were weak: by making us weak, our own "rear support" had actually uplifted the enemy's morale.

Israeli Col. Moshe Dayan was shocked to see how tough VC were during interrogation by the ARVN counter-intelligence. The truth of the matter was that the prisoner of war POW appeared to be tough or courageous, but, in reality, they might be more terrified that the commanders would retaliate against their family if they revealed VC military secret. On the contrary, the ROK troops' method was cruel but effective: "lấy độc trị độc" was the Asian way anyway – use venom to cure venom. As Gen Westmoreland observed when visiting troops in the earlier years, Asian lives were really cheap. That was why the ROK way worked the best: just push the first POW off the flying helicopter and the next one would give up and spill out all secrets. They had been atoned to the

---

[234] The ROK fought the VN War the Asian way, as they are not affected by the U.S. warfare demands of treating suspected civilians as innocent. Torture and even execution were allowed if the suspects show signs they have collaborated with the enemy, this being usually the case because the peasants had no choice but to collaborate: refusal will result the whole family being executed by the communists.

[235] NVA Gen. Bùi Tín

Chinese way since their more than three thousand years of domination: they had to weigh which terror would be scarier, to be killed or to lose their relatives to massacre. VC terrorists would be successfully handled only by terror, particularly those who believed they would be rewarded with a better life after death.

Modern strategies were combined with ancient Sun Tzu's military wisdom, this making it much more complex as the allies (except the Republic of Korea, which perhaps used Sun Tzu's) were not familiar with; thus making Vietnam War even more complicated for the allies as they tended to launch campaigns that were mostly independent from those led by the ARVN. There seemed to be a major conflict between the upfront and direct-hit modern Western warfare and the subtle twisty and undermining Eastern one. This conflict might have caused a warfare clash between the U.S. and the RVN commanders, besides differences in their political agendas and priorities, thus making team work difficult…

Sun Tzu's teaching has not been applied just in war. As I have said earlier when discussing Chinese Invasion and the Annamite resistance, the Vietnamese have learned to "live" what Sun Tzu has taught in all aspects of their life. The question is: "did the Army of the Republic of Việt-nam ARVN use at some point in time during the war the same Sun Tzu-inspired tactics as their ancestors had used in the far past? If it did, how different was it compared to the VC/ NVA's tactics"

Whether the ARVN has used its "own warfare" or not is the question that should be asked! As confirmed by many major American Vietnam War documentaries, the U.S. has forcefully demanded that the RVN adopt the exact format imposed by the American military. Its army is to be shaped, equipped, trained in the same way the U.S. army (NewsMax[236]) is, this being part of the military aids "deal!" The ARVN is expected to fight the war the same way the American troops do, using the same strategies and tactics.

These warfare differences might have, starting the late 1960s, caused a great divide between the allies and the RVN, the latter being torn between the imposed American warfare and the traditional Vietnamese one. It has initially popped up under Pres. Diệm (oppressive policy against the opposition and

---

[236] http://www.newsmax.com/Newsmax-Tv/documentary-vietnam-war-controversy-military/2015/10/14/id/696212/ on Lam Son 719 Laos incursion of 1970.

the suspected communist), and cannot be resolved under Pres. Thiệu (problems with infiltration into the Buddhist factions). The conflict has become even more serious when the Nixon and Kissinger team negotiates in secret with China and the People DRV at the expense of the RVN. According to some very subtle remarks made by my in-laws, **the U.S. oftentimes disregards what the RVN has to say as if it has no self-determination and sovereignty of the RVN...** The attitude of the U.S. towards the RVN in the Paris Accords negotiations, by having Kissinger (with Nixon present in the background) monopolizing the negotiations of peace conditions with his counterpart DRV envoyé Lê Đức Thọ, has deliberately left the RVN out of crucial direct participation in the process which will eventually decide its own life and death. It is a flabbergasting evidence of, not just the U.S. White House's despicable disrespect to the RVN and its people, but also an unacceptable betrayal towards its major ally.

For a while, after he had found out that the Paris Accords clause, as confirmed earlier by Ambassador Bunker (June 3, 1971) in Kissinger's presence, "that NVA would withdraw troops from the South at the same time as the allies would" had been overturned by Kissinger[237] in favor of the DRV, Thiệu withdrew his negotiation team from Paris negotiations. He even ignored Nixon's call from Honolulu. He strongly refused to recognize the legitimacy of the NLF. Recognizing NLF as a government would justify that the war was a civil war[238], and, by the same token, accuse the U.S. for illegally intervening in Việt Nam internal affairs, and, at the same time, minimize the fact that the North was invading the South. It would also justify HCM and the North had been fighting for a good cause: to defend their country against American invaders, as the communists had made their people to believe... Thiệu's uncooperative reaction was actually to avoid the term "civil war" and to defend the rational that the U.S. intervention in Việt-nam had actually been to legitimately assist the South fight an invasion from the North, instead of politically and militarily meddling with Việt-nam sovereignty,

---

237   The Socialist Republic of Vietnam, in a press conference in Hà-nội on Aug. 1st, 1972, had requested Kissinger to let NVA troops stay in the South and pressure Pres. Thiệu to resign. The second request coincided with the Buddhist protests in Viet Nam demanding the same. Source: Blog Việt từ Xóm Cồn, Hạ Uy Di, Sep. 2009 by Mường Giang, posted by Strategic Technical Directorate.

238   As the anti-war movement in the U.S. and Pres. Nixon/ Kissinger had been claiming, for the sake of an easier withdrawal from a war that had been escalated for almost a decade...

this being internationally unacceptable and illegal.

Predictability in warfare

Even though there were many excellent generals in the ARVN, the world seems to pay more attention to two: the U.S. Army Gen. William Child Westmoreland and the NVA Gen. Võ Nguyên Giáp. **It was as if the war was just between the U.S. and the North,** this reinforcing very well what I had commented earlier in the previous section about how the participation of the RVN in the war was disregarded.

Gen. Westmoreland, an avid chess player, started his career in WW2 and Korean War, and was considered to be a very effective "hard nose" officer. Gen. Giáp was the Việt Minh commander who had masterminded strategies to defeat the French army in 1954; he had also led the Vietnamese garrison against the Japanese until the end of the American-Japanese war. Gen. Giáp was ranked the 5th among the top military generals of all time[239].

If we examined how the war was fought in WW1 and WW2, then, transcending those findings into the Vietnam War, we would have to admit that Gen. Võ Nguyên Giáp was right about the American and the ARVN strategies (copied from the American's) used in Việt Nam. Their tactical sequence of military assaults in Việt Nam had always been the same since the World Wars, meaning predictable: heavy bombardment, either by air and/or artillery, followed by helicopter transported ground troop deployment to attack an invisible and unpredictable enemy. The same "appear-and-disappear" tactics that the Việt Minh (ghost army) had used against the French[240] were applied and improved by the communists, this time using tunnels, civilian-provided intelligence. Innocent people had also been used as screens to shield their soldiers when attacked by American and ARVN forces (e.g. the village with the "Napalm girl.").

---

239   https://www.thetoptens.com/top-military-generals/page11.asp based on a 30,000 sample survey of The Top Ten
240   It seems to me that the Americans had not consulted with the French for strategies... The Điện Biên Phủ mistake of encampment in a valley, if learned, could have prevented the Khe Sanh experience.

Westmoreland was chosen to head the joint U.S. and allied forces in Việt Nam and implement a war policy where "we will win every battle." The military analysts, who were knowledgeable about Sun Tzu's tactics being used in the war by his counterpart Gen. Võ Nguyên Giáp, believed that only part of Westmoreland's experiences in WW2 could be applicable in Việt Nam. As inspired by Chinese strategist Sun Tzu, Gen. Giáp had focused his efforts on out-thinking and out-witting, rather than outfighting Westmoreland: he always studied every American (as well as Allies and ARVN) tactical moves and figured out effective counter-moves... As an example, American routine tactics of carpet bombing the area before launching the ground troops had been counteracted by the ghost-like VC and NVA strategy that, when they heard U.S. bombing, urged them to hide in well protected undergrounds[241] and prepare to receive American ground troops[242] with ambushes, infiltration assaults or mass attack (ocean wave of combatants "tấn công biển người").

When engaging Americans, Giáp's troops always tried to flood the opposition fast and mingle in (mix their soldiers into American positions) so that no air or artillery bombardment would dare provide support for fear of "friendly fire." They usually launched direct attacks to engage and divert, and another indirect one to destroy (as in the war Đại Việt against the Yuan's dynasty army), or to lure the tiger away from the den ("điệu hổ ly sơn") in order to attack the den itself... Military and political turmoil created in Laos (to help the Pathet Lao Army, Communist Lao) before the Offensive were also planned to distract the Americans from the trail. Attacking Laos helped **lure the U.S. away from its focus on Hồ Chí Minh's trail,** according to Gen. Bùi Tín, was effectively executed in 1968, Tết Offensive of the Year of the Monkey Mậu Thân, when the communist forces launched an overall attack on most of the larger cities in all four Army Corps sectors. The 1972 general Easter Offensive was also a preemptive probe for the final one in 1975 to help them evaluate the effects of the American and allies' withdrawal program on the RVN stability

---

241   When stationing in Bình Định where the 3-star NVA division was operating, I learned that the NVA was hiding in underground shelters that were protected by a shock absorbing bamboo roof that was covered with indestructible rocks. These hiding places could withstand artillery and air bombardment.

242   Mark McNeilly, author of Sun Tzu and the Art of Modern Warfare, and Dr. Richard G. Gabriel, Distinguished Professor, War Studies Dept, Royal Military College of Canada: Sun Tzu Tactics used in Vietnam War. History.com (liveleak.com) and YouTube (Sun Tzu Tactics used in Vietnam 1, 2 and 3). Also, http://artofwarsuntzu.com/america_experiences_sun_tzu.htm

and, especially, the reaction of the American public back home: would they change their mind about withdrawing troops if they saw their ally, the RVN, being subdued? **During the whole war time of the 1960s and early 1970s, instigating antiwar movement in the U.S. was the NVA strategy of "nội công, ngoại kích:" attack the American government "rear support" ("hậu cần") on its own turf, while assaulting its forces in Việt-nam. Use Sun Tzu's tactics to defeat the Chinese. Use Sun Tzu's again against the Americans.**

Bùi Tín more than once had described the important role of the antiwar movement and the "Conscience of America" in the U.S. in helping "attack them at home to win the war at the battle." Outwitting them was nothing but to use their own forces against themselves.

Eventually, with further American involvement in the war, the ARVN was bound by American tactical protocols, as part of the U.S. military requirements and conditions for continued military support. As pointed out by NewsMax television[243] in its documentary series "Vietnam War," and as reflected in Star and Stripes article cited earlier[244], the army of the RVN was not just equipped the way the U.S. was, but also had to fight the war the way the U.S. did. It was as if the ARVN was an extension of American forces.

This also concurred with what my brother in law Anh Bảy's[245] comments about the U.S. policy makers' tendency of supporting and promoting easily corruptible civilian leaders[246] and military decision makers; thus making the RVN dependent on American resources. Commanders who showed signs of resistance or stubbornness (or uncooperating civilian leaders like Anh Bảy[247]) would soon be replaced or sent into deadly dangerous zones… On one occasion, Anh Bảy was seriously reprimanded by his "buddy" Pres. Thiệu for refusing to cooperate with the U.S. Air Force authorities when the latter

---

243 http://www.newsmax.com/Newsmax-Tv/documentary-vietnam-war-controversy-military/2015/10/14/id/696212/
244 Star and Stripes article *The Strength and Weakness of the ARVN of June 18, 1972, written by Jack Foisie for the Los Angeles Times*
245 He was one of Pres. Thiệu's Cabinet members.
246 It would be easier to have the RVN leaders agree with U.S. dictated policy.
247 Brother in law Anh Bảy was severely reprimanded by Pres. Thiệu for turning down American contractors' request to expand a civilian airport into one that would allow B52 bombers to land and take off. The reason for his decision was that the ground was too soft to support the weight of B52 planes.

requested him permission to reinforce and enlarge a runway to accommodate bombers B52 to take off and land at one of the air bases. The reason of the refusal: the plan was not feasible because the ground structure where the Americans wanted the airfield for B52 was too soft (sand), not suitable for heavy planes.

Incompatibility between U.S. and RVN policy

In one word, in this teamwork efforts, the RVN full compliance, if not submission to the American plans, was expected; otherwise, not fighting the war the way the U.S. wanted would be defined as failure on the RVN part, as judged by American warfare standards. This had probably limited the creativity of ARVN commanders, as Sun Tzu's wisdom would not be accepted by U.S. advisors. This, consequently, could be the reason why cooperation between ARVN and the U.S. high command had suffered from cultural incompatibility. This might also explain why, if it was true, ARVN and ally troops seemed not to be fighting together (side by side) in joint efforts (e.g. U.S. and ARVN Marines CAP combined action platoons) in large unit campaigns during the U.S. troop influx between 1964 and 1971.

This explained why joint efforts of troops in side by side action were not shown in most archival documentaries. American documentaries showed only American troops fighting and dying, while Vietnamese[248] news reels showed just ARVN in action! Of course, nothing showed both fight side by side in most visual records. We never saw our allies fight together with the ARVN in Việt-nam! This definitely supported Hồ Chí Minh and Lê Duẩn's claim that the Americans were invaders: U.S. flags on planes, PT-boats and tanks, U.S. flags flying from territories retaken by U.S. troops from VC, U.S. troops fighting by themselves... It was so obvious!

On the contrary, war history films usually portrayed American forces coordinating and teaming up well with the French and British in WW1, or French, British, Korean and Pilipino in WW2[249]. Nevertheless, at the operation level,

---

248   American advisors being embedded in the RVN do not count as "side by side" joint campaign. In my opinion, side by side campaign occurs only when an American unit fights together with an ARVN one.
249   One can only rely on cinematographic accounts of WW1 and WW2 battle

American troops and the ARVN had shown great enthusiasm working together whenever they had a chance to. It was a sad situation for both sides! Joint campaigns between U.S. and ARVN in special operations seemed to work well, though (e.g. commando[250], Special Forces).

The Star and Stripes' article "The Strength and Weakness of the ARVN" of June 18, 1972, written by Jack Foisie for the Los Angeles Times, quoted a pro-American ARVN regiment commander's statement, probably responding to a U.S. military follow up or feed-back on **"the first big test upon Vietnamization graduation[251] to see if the ARVN was capable of fighting the war without U.S. support"** *This statement was a grave insult at the ARVN who had been fighting Communism since the 1940s!*

"You gave us your weapons and you tried to teach us your military ways. But did you really expect that we would fight the war as you did? If you did, you haven't learned very much about me and my countrymen."

Taking another look at Competence (or lack of) would make us become more aware that this term would be defined differently "in the eye of the beholder." Let's consider this scenario: the standardized tests administered in public schools. Could a standardized test effectively and fairly classify an individual student as smart or otherwise? How could we be sure it would be academically and culturally unbiased so that every student would feel at ease taking it and being judged by it? Bias had always been the concerns. How could we rate the ARVN tactical competence based on American warfare standards? It would be like rating Macaroni and Cheese using Vietnamese culinary standards!

"A military operation involves deception. Even though you are competent, appear to be incompetent. Though effective, appear to be ineffective."

Sun-Tzu, *The Art of War. Strategic Assessments*

---

fields as real life film reports are rare, as the news media are not allowed to be embedded.
250   Strategic Technical Directorate "Nha Kỹ Thuật."
251   …as if the ARVN had never fought in this war before…

Whether the statement above made by Jack Foisie in the Los Angeles Times would later be used to further denigrate the performance of the ARVN commanders or their men or not, it definitely disregarded the importance of the cultural aspect of Sun Tzu's-infused warfare that the ARVN could have used in the war, the very same warfare being used by the Vietnamese communists to defeat the U.S. and the allies. Nevertheless, imagine how confusing it would be for the ARVN that grew up with Tzu's culture to have to regurgitate American-style strategies so they could adapt the two conflicting modes of operation and "effectively" fight the war as expected... This certainly reminded me of my friend's ARVN Navy captain being mad at his U.S. Navy advisor when the latter wanted to change course and strategic plan so he could take a swim in a beautiful bay...

This cultural incompatibility had no doubt caused major differences in expectations from both sides and resulted in misjudgment of anticipated outcomes. Did this also happen to the relationship between China, the Soviet Union and the DRV? Did the NVA have to operate under a strict set of rules? Most documentaries showed the NVA and NLF had been operating independently from Peking and the Kremlin, and had their own political cadre with Lê Duẩn's leadership behind Hồ Chí Minh.

Massacres as tool of terror or secret of winning a war?

When reflecting about Sun Tzu's wisdom, we would need to ponder how the South Korean expeditionary forces had used Tzu's strategies in their campaigns (separate from the U.S. and other allies). Technically, the ROK would not have to abide by the American rules of engagement, as long as they coordinate their campaign in order to avoid friendly fire. They were not receiving American aids. As reported by the Vietnamese peasants, the Korean divisions had caused terror among VC, NVA and, unfortunately, civilians. Their harsh handling of suspects ranged from village massacres, e.g. the villages of Phong Nhất and Phong Nhị, during the aftermath of 1968 Tết Offensive[252], to throwing individual VC suspects out of a flying helicopter during Intel collecting. These incidents were unreported by the press perhaps because they did not

---

252   https://en.wikipedia.org/wiki/Phong_Nhị_and_Phong_Nhất_massacre

allow press embedment, resulting in news about how the ROK forces had handled suspected VC collaborators and prisoners of war had not been released.

Nevertheless, the ARVN, who were expected to fight the war the American way, and the Western allies would not be able to adopt the Korean way because they would face immediate reaction from the world, especially from the American people and the Vietnamese protestors, all of whom taking advantage of the new "democratic wave" and the "Conscience of America" to ban the war. A typical example was the case of March 1968 Mỹ Lai Massacre which was widely reported by the media and had served to make people upset at the "cruel" nationalists to the present time, while keeping an extreme low profile on massacres by the communist massacre (e.g. the 7,800 innocent civilians in Huế less than two months before Mỹ Lai incident which was also a part of the Tết Offensive of January 1968 Phase 1)!

What was so special about the ROK way that was cruel but did not cause any major political wave? To deal effectively with a population that had generationally learned that they could resist, overtly or covertly, and succeed at it, the enemy had used rule by terror techniques and would keep pushing until they were killed. The Chinese had executed such fighters by beheading, having elephants stump or tearing victims in four directions by horses… sometimes up to three dissident's generations at a time. The French had executed by firing squad, beheading or hanging in public, and as well as the Japanese and HCM's Việt Minh. The enemy had been used to the sacrificial way of war, so sacrificing others or themselves meant nothing as long as the goals were achieved. Nevertheless, only terror to them as individuals or to their family can melt their resistance. As a matter of fact, Asian-style terror worked better than Western methods (single execution, extracting intelligence by torture, or just interrogating). The ROK troops knew that as they themselves had ancestors experience Chinese and Japanese rule-by-terror before.

Terror was used by the communists to tame the peasants and make the latter bend and collaborate… Terror was also used by the ROK forces during their military campaigns… "Use poison or venom to cure poison and venom" ("lấy độc trị độc" use enemy's harsh methods to cure harsh behavior) was cruel; however, it would help those who used it obtain total submission from the villagers. The RVN probably wanted to do the same, themselves believing in the same venom way, but the U.S. policy would strongly object to it. Torture, the least cruel measure in intelligence collecting, faded out since Pres. Diệm's ruling…

  The army of the Republic of Korea was considered by the peasants, and even by the VC, to be the most successful (feared by VC and NVA) for a very simple reason: they had used the Asian way to deal with Asians suspected for collaborating with the VC. American and RVN troops failed because they were compassionate and humane!

  RVN Gen. Lê Minh Đảo[253], commander of the 18th division that had defended Xuân Lộc city, Long Khánh during the last weeks before the Fall of Sài-gòn, had reported what he was told by his jailers during his 17 year reeducation camp incarceration: **"Do you know why you lost the war? You lost because you dared not use your guns to shoot at your people. As for us, if our job called for it, we would shoot them anyway…"** *"Các anh có biết tại sao các anh thua? Là tại các anh không dám cầm cây súng bắn vô dân anh; còn chúng tôi, có 'việc' thì vẫn phải bắn" (Tạp Chí Mị Dân). Gen. Đảo, in a get together to commemorate the fall of Sài-gòn in 2016, had added: "The RVN lost the war because of its humaneness – it had avoided to harm its own population and it was not cruel as the Communist North Việt Nam was… How could cruelty always win over kind honesty and the common good? It is impossible! I believe there is cause and effect in this life… I believe there should be Justice from above…"* This was the reason why I had stated earlier that **HồChíMinh-ism had made the Vietnam War a spiritual war: the war of the fanatics.**

  One of the reasons why JFK and the CIA had decided to support dissident ARVN generals' overthrowing the first RVN Pres. Ngô Đình Diệm was that the latter had systematically implemented a harsh and oppressive policy against his opposition; this being actually the traditional Asian way of ruling by fear and terror. Protesters were usually rounded up and tortured for information. However, as most of the interrogated victims were Buddhist, it appeared as if it was the Catholic oppressing the Buddhist. It was all about the harsh Asian way of controlling the opposition by intimidation.

---

253 https://en.wikipedia.org/wiki/Lê_Minh_Đảo

Perhaps up until now the Americans were still wondering mostly why the rebel generals had chosen to execute the Diệm brothers, as overthrowing was harsh enough in the American way!

The enemy had won the Vietnam War because the nationalists fought it with Compassion and Humaneness, as imposed by American warfare, while they used terror to dominate. It was still incomprehensible why protesters in Việt-nam, especially the Buddhist groups, still believed in communist preaching, which was very compatible with Buddhist humanity, against inhuman treatment whenever the RVN government took some action to ensure national security. This was the reason why I believed the RVN was weak at psychological warfare. Interestingly, a few decades after the war ended, both Cao Đài and Hoà Hảo Buddhist and Christian groups up rose and protested again, this time against the established communist system because of its authoritarian oppression, while a brown-robed Theravada Buddhist monk group[254] (state-sponsored?) was still seen rallying in support of the communist DRV government ...

Democracy in the Eye of the Beholder

Admittedly, the Asian way was crueler (e.g. the RVN torturing or even making exemplary sacrifices of suspected communists or sympathizers), but it might be the most effective way to sort out the innocent from the enemy. As a result, the ROK units were considered by analysts as the most effective forces in repelling the communist grass root because they "dared do it.". They had applied the same methods the VC and NVA had been using: terrorism[255]. The peasants, who would give in and collaborate with the enemy, would do the same to the ROK under the latter harsh pressure. A VC prisoner caught by the ROK who refused to cooperate and give up intelligence information was thrown overboard from a flying helicopter. That was why the ROK was successful when fighting the enemy and sympathizers feared them: they did not hesitate to

---

254   Theravada Buddhist monks (saffron/brown robe) were also protesting against the RVN during the war.

255   https://en.wikipedia.org/wiki/List_of_massacres_in_Vietnam List of massacres committed by U.S. and ROK, and incomplete list by VC/NVA/Khmer Rouge, and YouTube Vietnam's massacre by South Korea Army. Also, https://en.wikipedia.org/wiki/Phong_Nhị_and_Phong_Nhất_massacre by ROK.

use terror against terror. Nevertheless, they were still less harsh as they would not go to the extreme as the VC by torturing and killing their prisoners' family members to apply pressure when interrogating them and extracting Intel.

As I have heard, the Malaysian government has a very effective way of preventing Communism from spreading. If officials hear of any communist activity in a village, they will send the army to force all inhabitants out of their house overnight and burn the whole village to the ground. Use venom to cure venom is indeed the way in Asia.

The VC and NVA has no hesitation massacring to ensure that innocent civilians would not cooperate with the RVN and allies, or to make the peasants collaborate with them (supply medicine, food and intelligence). On my most recent visit to the Mekong Delta in Summer 2018, Vĩnh Long people, many of whom used to be more or less VC sympathizers in the 1970's (see "Vietnam: Peace or Freedom – Two Minnows"), have told me that a current head of hamlet, who used to be a VC combatant, has proclaimed publicly that his army "thà giết lầm người vô tội còn hơn thả lầm người phản động" would rather kill innocent people by mistake rather than release a dissident traitor." When an ARVN base is invaded during the war, the VC usually massacre all soldiers instead of taking them prisoners, as reported by Vĩnh Long residents who used to be sympathizers.

It was clear that the enemy had taken advantage of the Western military protocols abiding to the WW1 & 2 Geneva Convention restrictions. Given that there was a very fine line between innocent civilians ("innocent until proven guilty" as imposed by Western laws) and sympathizers (e.g. Mỹ Lai village?), the nationalists could not apply national security laws against them unless there was humanely-extracted evidence they were communist activists. Not even when the suspect was caught helping the enemy (this being apparently an evidence of guilt), he/she could not be considered collaborator because he/she could still say "I was helping my cousin." When they destroyed their own strategic hamlet the RVN had built to protect them and moved back into war zones controlled by the enemy, they could still say "I don't like to live far from my rice paddies" and get released. Since the nationalist government had no evidence the suspects had collaborated with the enemy, they would be safe. The Geneva Convention would definitely protect the hard-core guilty ones as the latter would never admit working for the enemy or reveal enemy secrets, unless they were subjected to near death torture. The Asian (VC, NVA, ROK) ways worked in Asian wars, where Western methods would fail out of cultural "weakness."

The Power of Fear

Ironically, any sign of lack of humaneness on the nationalist part was nitpickingly pointed out by the press and vehemently attacked by the Vietnamese civilian protesters, and, of course, exploited by Hà-nội[256] to fuel even more protests and political lobbying in the U.S., the "rear support," as well as in the world. "Hậu cần" was the strategic term both NVA Generals Võ Nguyên Giáp and Bùi Tín had used when referring to their own "rear" support, which for them was very strong, during their interview by the American press. In the meantime, the enemy was doing the opposite.

No matter how ferocious they could be, nobody cared about what the VC did because they were assumed to be "evil" already! As a result, they just frivolously committed genocidal acts, because the more inhumane they were, the more feared and submissive the peasants would become. If the VC shamed and threatened the peasants because of "lack of patriotism" when not cooperating with them, they would turn around and submit to enemy demand right away. Many press media reporters agreed that peasants were always caught in between: they had no choice but supply the enemy with food and medicine, stockpile ammos and arsenal for them, have their own elders, children and women spy on the nationalists, because questioning or resisting would certainly mean death to self or relatives. It was better and safer to be considered rebels by the RVN (they would stay alive if caught) than by the VC (the family might be executed if they refused to collaborate). This behavior gave the enemy the power of fear.

Informed and more educated[257] Southerners are

---

256   Anh Bảy, my in-law who was on Pres. Thiệu's cabinet, discussed in some family gathering alleged antiwar activities in the U.S. by South Vietnamese (students, monks...?), whether or not they were aware about it or not.

257   I stress on this word "informed," because a great many Southerners were then gossip-informed or misinformed, if informed at all, therefore easily misled because of their generational distrust of the establishment

convinced they better avoid Communism as much as they can. I also believe that the rest, the majority who are politically-uninterested Vietnamese, are more interested in minding their own business and passively making a living, than in committing themselves to the cause of the RVN. They are taking the little freedom they have for granted, and pray that Peace will come so they can run an even better business. I am grateful my Vietnamese in-law family had educated me on this topic!

Instigation and Fear

As a matter of fact, the enemy's psychological warfare alone did not just excel during the Vietnam War. It had actually received even more indirect support from the U.S. media when the latter were trying to prove to the compassionate Americans at home, who were supposed to be the backbone of U.S. support for the war, that their country was wrong fighting in the war. These voting and tax-paying antiwar activists would do their best to change government course in its involvement in the war by pressuring it to meet their demands. And indeed, the press media and the relentless Americans against the war had succeeded in incapacitating the U.S. government as the war went on! Consequently, not just South Việt Nam would suffer, but the very young men and women the U.S. had sent to war, the war heroes, were also betrayed and condemned on the way back! Sun Tzu's strategy of "nội công, ngoại kích" (striking inside-out while attacking out-side-in) as applied by Hồ Chí Minh had succeeded in directly or indirectly causing major turmoil against the U.S. at the front as well as at home (to be discussed later in infiltration)..

CHAPTER 8
Let the Game Begin… The Paris Accords

Just like in any game, whether it is football, baseball… or even a martial arts match or a Ballroom Dance competition, there are always two major sides and there are winners versus losers. The main players participate directly in the game as they play with the sole purpose of scoring goals and win, while the sideliners are sitting on the sideline to "sub" (team reserve) or support (cheerleaders). Cheering for whom they support or booing those they don't are the people who sit in the bleachers and watch the game, with their banners, noise makers, signs, snacks and beers… They either applaud, or make unbearable noise.

The only big difference is that millions of more lives are at stake and many millions more will be suffering mentally or physically as a result of the Paris Peace negotiations. Another particular characteristic is that the outcome turns out to be in the hands of just a few individuals, the key negotiators.

Introducing the Players and the Sideliners[258]

The four sides that fought in the war, the Provisional Revolutionary Government[259] in South Việt Nam (supposedly the government of the NLF National Liberation Front, VC or Việt Cộng), the People's Democratic Republic of Việt-nam (DRV or North Vietnam), the United States of America and the Republic of Việt-nam (RVN or South Vietnam), had started these negotiations in 1968 in Paris, just four years after the U.S. officially entered the war in 1964 (year of controversial Tonkin Golf incident[260]). As reported by a few of my professors in my Socio-Political Science Graduate Program (Political Development Master's Program "Chương-trình Cao Học, Phát-triển Chính-trị", University of Vạn Hạnh, Sài-gòn) who had actually attended some of the Paris Accord sessions, the Peace

---

258   Merriam Webster dictionary: "the one that remains on the sidelines during an activity is the one that does not participate."
259   This group was not recognized by the RVN Government as there was not such a government.
260   A U.S. Navy ship, U.S.S Maddox, was attacked by a North Vietnam in the Gulf of Tonkin.
https://en.wikipedia.org/wiki/Gulf_of_Tonkin_incident

negotiators who seemed to be leading the discussions and had control of the directions were Politburo Lê Đức Thọ (DRV) and the U.S. Secretary of National Security Dr. Henry Kissinger[261]. During their Peace Talk, these two negotiators started their own "secret" negotiations, leaving the other two negotiating "tokens" out of the conversation, while Pres. Nixon conducted his own "secret" negotiations with China and the Soviet Union.

Many documents, especially the most recently declassified ones like the Nixon's video and his October 1972 taped conversation between Pres. Nixon and Sec. Kissinger, had unveiled mysteries on how their intricate and rather tricky moves had reinforced their power in their positions as well as warranted their success in their future, at the cost of millions of Southeast Asian, American and allies' lives as well as the forty some millions of survivors' Freedom and Democracy. These moves were flexibly designed to please both the angry and confused People of the U.S. and the supposedly-enemy, mainly the DRV.

Minister Trần văn Lắm (1913-2001)
Republic of Vietnam Foreign Affairs
"...the muffled Sideliner of South Việt Nam"

Born from a well-to-do ethnic Chinese real estate family in Chợ Lớn (Sài-gòn Chinatown), Lắm was educated at Hà-nội University and trained as a pharmacist. A soft-spoken diplomat fluent in French and English, he gradually rose to power in the Republic of Vietnam Parliament and started his foreign affairs career when Pres. Ngô Đình Diệm appointed him as ambassador of Australia and New Zealand[262]. He returned to non-governmental life of a banker just before he became Minister of Foreign Affairs in 1969, when he started leading the Republic of Việt-nam negotiating team in the Paris Peace Accords Conferences.

Lắm signed the Accords Agreement for the RVN after it was "tweaked" a little[263] to please Pres. Thiệu. The latter had

---

261 They both received Nobel prizes afterwards for having "successfully" facilitated negotiations in Paris to bring Peace back to Việt Nam.
262 https://en.wikipedia.org/wiki/Charles_Tran_Van_Lam Biography of RVN Foreign Affairs Min. Trần văn Lắm
263 The current RVN government could remain in power, the U.S. will continue to give aid to the RVN, Nixon promised to resume bombarding North Vietnam

made many attempts to reject it. After the Paris Peace Accords Agreement was signed, Lăm came back to Việt Nam as the new RVN Senate president. Towards the end of the war in 1975, he successfully persuaded Interim Pres. Trần văn Hương (Pres. Thiệu's replacement when the latter resigned just before the fall of Sài-gòn, April 1975) to relinquish power to RVN Gen. Dương văn Minh (Big Minh), as demanded by the DRV for a power transition. Big Minh surrendered the RVN to the communists and Minister Lăm resettled in Australia.

Mr. Lê Đức Thọ[264] (1911-1990) Democratic Republic of Vietnam primary negotiator
"...an active Player with the ruse of a fox"

One of the founders of the Indochinese Communist Party (1930), Thọ was imprisoned by the French colonials many times because of his dissident activities. Upon his release in 1944, he helped lead and started purging the Việt Minh of its nationalist members, then oversaw the communist insurgence in South Việt Nam (1956- on), while continuing to purge the Party (1963)[265]. While representing North Việt Nam during the Paris Accords, Thọ had actually engaged in secret talks with his American counterpart Henry Kissinger (starting Feb. 1970), making oppressive deals at the expense of the fate of the people of the RVN. Besides agreements on ceasefire arrangements, democratic election in the South, recognition of the Communist Provisional Revolutionary Government PRG (VC, in South Vietnam), release of prisoners of war, and rebuilding foreign aids, **Thọ was able to lure Kissinger into agreeing to cut off all aids to the RVN and withdraw all allied troops, while letting all NVA troops remain in the South and to continue occupying territories they had taken as well as resupplying.** These last clauses had constituted an actual death sentence to America's closest ally in the war, the Republic of Việt Nam. Consequently, the Hồ Chí Minh Trail, that used to be a dirt path, became the HCM Highway with traffic controllers with convoys of resupplies.

---

if the Communists violated the agreement...

264   http://www.nobelprize.org/nobel_prizes/peace/laureates/1973/tho-facts.html Lê Đức Thọ, Nobel Laureate

265   https://en.wikipedia.org/wiki/Lê_Đức_Thọ Biography

After the Treaty was signed, Thọ was elected to be one of the two Peace Nobel Prize recipients in 1973 (the other one being Kissinger), but he declined it, claiming that there was no Peace achieved as the DRV accused the "Republic of Vietnam had chosen to continue the war." Rationally speaking, as history would prove later, it was actually North Việt Nam that had violated the Paris Accords when it had taken advantage of the "keep NVA in the South in occupied territories" and "let NVA continue resupplying" clauses that Kissinger had foolishly ratified allowing the North to continue waging war in the South. This war had turned from an insurgent guerrilla into a conventional invasion, but still **in the name of the National Liberation Front** as if it was a total uprising of the Southerners VC (Việt Cộng).

Did Kissinger realize what he had done to the South? He went to accept the Nobel Prize anyway while Thọ declined: the latter was honest enough to know he did not deserve the prize! He might even know his party was going to violate the treaty when he signed it! Later on, he was sent to Kampuchea and named by the Hà-nội regime Chief Advisor of the People's Republic of Kampuchea (Cambodia) to keep Khmer Nationalism (the Republic of Khmer) in check and support the Khmer Rouge government.

Minister Nguyễn thị Bình (1927-)
South Việt Nam Communist Provisional Revolutionary Government (PRG of NLF or Việt-cộng) Foreign Affairs
"...another Sideliner"

Born South Vietnamese in Sa Đéc in 1927, she was anti-French nationalist Phan Chu Trinh's daughter. Like her father, she had promoted non-violent resistance against French occupation by participating in some intellectual movements, such as the Women Advancement Group "Yến Sa[266]" and the political movement to protect intellectuals' Peace. Her activities resulted in her being incarcerated by the French in Chí Hoà Prison in Sài-gòn, until 1953 when the Geneva Conference was initiated. Later on, she was promoted to Minister of Foreign Affairs in the VC Provisional

---

266   Mme Nguyễn thị Bình http://www.haugiang.gov.vn/Portal/DATA/sites/10/chuyende/phunu/phan3/nguyenthibinh.html

Revolutionary Government, thus making her the representative of the National Liberation Front at the Paris Accords.

Nevertheless, the course of history proved that neither the NLF (VC in South Vietnam, officially recognized by the DRV the PRG in South Vietnam) nor the RVN representatives would be actively negotiating Peace. The U.S. and North Việt Nam (DRV) ones had done all the meaningful talking and maneuvering…but, interestingly enough, both were expected to also ratify the final agreement as if they were active participants… Mme Bình was later elected Vice President twice by the National Assembly of the Socialist Republic of Việt-nam (SRV used interchangeably with DRV) in the late 1990s.

U.S. Secretary of State Dr. Henry Alfred Kissinger (1923-)
Also U.S. National Security Advisor, U.S. primary negotiator
"…an active Player with the ruse of a different kind of fox"

Little Henry fled Nazi Germany in 1938 to London, and then he moved to New York with his affluent Jewish family. An intelligent student, he studied and worked at the same time until he was drafted into the U.S. Army. He was naturalized in 1943. His service in military intelligence had introduced him into the strategies of the "underworld." He later studied at Harvard University and achieved a PhD degree with focus on "Peace, Legitimacy, and Equilibrium." Like an eagle, he continued soaring into the field of National Security, Nuclear Weapon Policy, Foreign Relations, Defense Studies... His advisory affiliations with the "almighty rich" Nelson Rockefeller helped him get appointed National Security Advisor by Nixon in 1968[267].

His ability to "bend to whatever direction the wind blows" helped him accommodate both China and the Soviet Union policies, thus allowing him to relax the tension between these two powers and the U.S. ("Policy of Détente"), and, consequently, making him Nixon's (who interestingly "did not like Jews") most trusted advisor. Many political analysts believed Kissinger was the actual person who "ran everything

---
[267] https://en.wikipedia.org/wiki/Henry_Kissinger#Early_life_and_education

in the U.S. government at that time," especially Vietnam War. Many discovered he had been very discreetly involved "behind the scene" in the 1973 Yom Kippur war, and, thereafter, Israel-Arab complications.

Mark Hertsgaard in *The Nation* (Oct. 29, 1990): "... the former Secretary of State and National Security Adviser's record of achievement in that part of the world (Persian Gulf Crisis) has been checkered at best. Not only did he help to make the Middle East war of 1973 inevitable by 'his need to dominate then-Secretary of State William Rogers and his willful misunderstanding of the limits of Soviet influence inside Egypt," as Seymour Hersh argued in the *Price of Power*... **"Kissinger had admitted to wiretapping reporters and his own aides during his years as Pres. Nixon's National Security Adviser, pre-empting the Watergate scandals...** Kissinger's obsession with keeping everyone -- the citizenry, the Congress, even his own Administration colleagues -- in the dark about his actions is displayed in all its banal iniquity in the State Department document printed on page 492. So are his casual disdain for law and constitutional procedures, his disregard for the humane consequences of his policies, his bizarre personal paranoia and his petulant sense of self-importance."*(The Secret life of Henry Kissinger:* Minutes of a 1975 meeting with Lawrence Eagleburger[268]).

PBS, *American Experience*[269]: "With Nixon's approval, Kissinger concentrated foreign policy-making power within the White House under the National Security Council, circumventing the established foreign affairs bureaucracy and effectively curtailing the authority of Secretary of State William Rogers. **Nixon and Kissinger both favored 'back-channel' communications and used secret negotiations** to lay the groundwork for détente with the Soviet Union and open a new dialogue with Communist China. Similarly, Kissinger began secret talks with North Việt Nam in 1969 in the hopes of reaching a settlement to the Vietnam War. At the same time, though, he counseled Nixon to increase bombing of North Việt Nam and to expand the war into Kampuchea and Laos... With the July 1971 announcement of his secret meetings with Chou En-lai, Kissinger emerged into the limelight, achieving unprecedented international celebrity. The formerly obscure

---

268 http://www.etan.org/news/kissinger/secret.htm Secret Life of Henry Kissinger Minutes of his meeting with L. Eagleburger.
269 http://www.pbs.org/wgbh/amex/china/peopleevents/pande02.html Henry A. Kissinger

presidential adviser was now everywhere: on the covers of Time and Newsweek, profiled on the network news shows, and featured on the front pages of newspapers across the country. '[A]t the height of a brilliant career,' wrote Time, 'he enjoys a global spotlight and an influence that most professors only read about in their libraries.' When Nixon resigned in August 1974 and Gerald Ford took office, Kissinger retained his position and his unprecedented influence on foreign affairs. While he continued to pursue the 'détente policy' with Russia, it grew increasingly unpopular..."

    Even though he was one of the leaders from the Freedom combatants' camp, I believe Kissinger had **played a non-allegiance game** so well his comrades as well as enemies did not really understand and know what was going on. One minute, he arranged carpet bombing North Việt Nam with strategic B52 super fortresses, the next minute he was "chilling" with adversary counterpart, special People's Democratic Republic of Việt-nam delegate Mr. Lê Đức Thọ, whose country was being bombarded as his planned. He had masqueraded a B52 North Việt Nam bombing mission by deviating targets in the middle of deployment: The B52's sent to North Việt Nam were secretly redirected to fly to Kampuchea instead to carpet bomb an area where intelligence reported presence of NVA stockpiles. This constituted a blatant violation of international laws which both Nixon and he had tried to hide. It was an untraceable killing mission that the people of the U.S. (and the world) would never be told about. Even until the 2010s, this undeclared war was still kept in the dark... Perhaps he had transitioned this Sun Tzu's intriguing tactic of "play it difficult," "apply pressure then release," or "use one move to mask another" His game was played so successfully and unnoticeably he was able to boost up his power by the time he closed down the Southeast Asia Theater and smoothly switched to the Middle East. He was and is still the American version of Sun Tzu.

    In my opinion, Kissinger had chosen to finish off Vietnam War so he could secretly focus on Middle East and support Israel, either because he was Jewish, or that he had to make discreet moves because he knew Nixon did not like Jews, or that he wanted to expand his power over and beyond Asia ... or probably all of the above? Did this sound like a conspiracy theory? Maybe, maybe not, just do a triangulation reasoning for yourself...!

Knowing his boss Nixon hated Jews, Kissinger played the game so well during the Vietnam War even pro-Israel commentators like Dr. Gerhard Falk[270], and other commentators (in response to Kissinger's statement about "the U.S. had saved Israel in the Yom Kippur War[271]"), strongly believed he was a Jew who had betrayed Israel, if he was not tried his best to annihilate it. If no one could tell anything about his directions, perhaps **his allegiance was just to himself!** Nevertheless, definitely, he wanted to make sure that Israel knew that "the U.S." had saved Israel in the Yom Kippur War when it was on the brink of defeat: the situation was getting so disastrous that Gen. Moshe Dayan had pressured Prime Minister Golda Meier to use Israel nuclear stockpile to annihilate the overpowering military of Egypt.

In the meantime, Kissinger removed himself from the Sutheast Asian conflict theater and began to focus immediately on networking for his next assignment. In spite of his big mess caused by his phone line tapping (many sources pointed out that Kissinger was the one requesting phone taps) that had cost Nixon his presidency; Kissinger was appointed Secretary of State under the new U.S. Pres. Gerald Ford. He knew how to sell himself using his expertise in "détente approach," an asset that would help him continue to soar career-wise. For some unknown reason, he was able to strategically distant himself from the 1973 Yom Kippur war which started only 8 months after the Paris Accords were signed…

Could U.S. aids to Việt Nam have been cut off so abruptly because Gen. Dayan had manipulated or used Israel Prime Minister Golda Meir to threaten to use eminent nuclear warheads against Egypt if Israel would not receive American back up[272]? Could this political pressure have come through Kissinger from Dayan or Meir? It was strange to see in website

---

[270] Dr. Gerhard Falk stated, in his article for jbuff.com, Jewish Buffalo on the Web blog, that "few Jews have had the opportunity to defend the lives of the Jewish people against our enemies as did Kissinger when he was the Secretary of State of the United States in 1973. Yet, Kissinger, during his tenure in that position, did everything possible to bring about the destruction of Israel and the slaughter of yet another 5 million Jews. In that he did not succeed because Pres. Nixon, a man who ranted anti-Jewish epithets, saved the people of Israel from mass murder against the wishes and machinations of Kissinger." http://www.jbuff.com/c081210.htm

[271] http://www.haaretz.com/Israel-news/.premium-1.555704 Kissinger said "the U.S. saved Israel in the Yum Kippur War."

[272] *CIA Report on Yom Kippur War: Israel Had Nuclear Arsenal, but it downplayed it, for fear the U.S. would push for Peace-talking. Harretz Israel News, Feb. 03, 2016* http://www.haaretz.com/israel-news/cia-report-on-yom-kippur-war-israel-had-nuclear-arsenal.premium-1.501101

photos that amphibious M113 tanks due for Việt Nam (there were more amphibious M113 than M48 tanks in Việt Nam because of swampy land, rice paddies and rivers) were rerouted to the Yom Kippur desert war? Did Thiệu's Cabinet have any suspicion about Nixon or Kissinger's changing of camp?

Whether the American people knew it or not, stopping the war involvement did not really mean Peace in the U.S.! Military aids allocated to Việt Nam but frozen by U.S. Congress might have started being discretely diverted to Israel just in time for Yom Kippur. Archival photos of American amphibious tank M113 deployed in the desert of Yum Kippur War really jumped out of the internet... *Yes, indeed, I am not kidding... amphibious tank in the desert...* This bait (aid promises were made to Pres. Thiệu after Paris Accords) and switch (switch to Israel) move, according to many commentators, had perhaps indirectly and unnoticeably committed the U.S. into a long term support for this new major ally[273] for all the years to come[274], Israel. The trick was that not many people, especially in the U.S., realized Kissinger had successfully switched the Peace-loving Americans from one major war, Việt-nam, to another, this time even more major, at a much larger scale and level of seriousness: the Arab-Israeli conflict[275]. **This new frontline could have escalated into a world-wide pandemonium!**

...*US Department of State – Office of the Historian: Milestone: 1969-1976, The 1973 Arab-Israel War* reported: *"By October 9, following a failed Israeli IDF[276] counter-attack against Egypt's forces, the Israelis requested that America do*

---

273   Kissinger: 1973, The Crucial Year (July 2009) by Alistair Horne, Simon & Schuster or British-published Kissinger's Year (2009) by Weindenfeld and Nicolson; Amazon. "Sir Alistair's authorized version, while not uncritical, is certainly a partisan. Spectacular as Kissinger's diplomatic triumphs were, it can be argued that many of their long-term consequences were not always benign. His handling of the Yom Kippur war has left America with what may prove to be an unsustainable commitment in the Middle East. It is not that Kissinger, as a Jew, was too partial to Israel; on the contrary, as Sir Alistair shows, he leant over backwards to avoid that." http://www.economist.com/node/13983256

274   As Kissinger stated in a rare interview for an Israeli television documentary on the 40th Anniversary of the Yum Kippur War, as reported by Amir Oren on Nov 02, 2013 on the Israel News Haaretz http://www.haaretz.com/israel-news/.premium-1.555704#

275   http://www.nytimes.com/2003/10/06/opinion/the-last-nuclear-moment.html and http://www.haaretz.com/dayan-sought-show-of-israel-s-nuclear-capabilities-in-1973-war-meir-opposed-1.317786 M. Dayan suggesting use of nuclear arsenal in 1973 Yum Kippur war, Nixon and Kissinger's secret "back channel" negotiations with the Soviet and China, Kissinger's switching to Arab-Israel conflict.

276   Israel Defense Forces.

the same for them. Not wanting to see Israel defeated, Nixon agreed, and American planes carrying weapons began arriving in Israel on October 14. With the American airlift underway, the fighting turned against the Arabs. On October 16, IDF units crossed the Suez Canal. Sadat[277] began to show interest in a ceasefire, leading the Soviet Premier Brezhnev to invite Kissinger to Moscow to negotiate an agreement. A U.S.-Soviet proposal for a ceasefire followed by peace talks was adopted by the UN Security Council as Resolution 338 on October 22. Afterward, however, Kissinger flew to Tel Aviv, where he told the Israelis that the United States would not object if the IDF continued to advance… 'If Nixon chose not to do so,' Brezhnev threatened, 'We should be faced with the necessity urgently to consider the question of taking appropriate steps unilaterally.' **The United States responded by putting its nuclear forces on worldwide alert on October 25 (1973)[278]**… *The 1973 war thus ended in an Israeli victory, but at great cost to the United States. Though the war did not scuttle détente,* **it nevertheless brought the United States to a nuclear confrontation with the Soviet Union, even closer than at any point since the Cuban missile crisis.** *The American military airlift to Israel, moreover, had led Arab oil producers to embargo oil shipments to the United States, as well as some Western European countries, causing international economic upheaval.* **The stage was set for Kissinger to make a major effort at Arab-Israeli peacemaking. (Oct. 31, 2013)[279]"**

Despite who the American president was, Kissinger seemed to have all the presidential authorities to deal directly with opposing sovereignties: Nixon had always given him full power[280] even though he disliked Jews! Kissinger always seemed to excel in his art of sniffing out possibilities that would move him up the echelons of power and secure better positions …In an interview shown in Trials of Henry Kissinger[281] documentary film, **Kissinger bragged he always looked for better advancement opportunities** whenever he felt there was any shred of possibility. We should admit he was at his best in this field of opportunism. Kissinger definitely did not want the public world to see his "secret stash of documentary files" until after he passed away. The future was promising substantial

---

277 Egyptian President Anwar Sadat
278 Barely eight months after the U.S. totally left the Vietnam War.
279 https://history.state.gov/milestones/1969-1976/arab-israeli-war-1973
280 https://en.wikipedia.org/wiki/Henry_Kissinger#1973_Yom_Kippur_War
281 *Trials of Henry Kissinger (2001) by Christopher Hitchens*

revelations of what other harm he had committed against the welfare of the world. Prior to and through the 1990s and 2000s, he still sought to give advice to politicians, including Hillary Clinton, one of the 2016 Democratic Party runners (CNN, FOX news media)…

The Paris Agreement of 1973 officially ended U.S. involvement in Việt Nam and, supposedly, the so-called "civil war" between the North and the South. As recalled earlier, Kissinger was a co-recipient of the Nobel Peace Prize. He presented himself [282] at the Norwegian Nobel Institute in Oslo to accept this highly honored prize as one of "the two chief negotiators who succeeded in arranging the ceasefire." On the contrary, the North Vietnamese negotiator, Lê Đức Thọ, the other Nobel Prize winner, obviously had understood the sham that the Paris Peace Accords were. He fittingly declined to accept the award. Kissinger's move appeared to be without shame or remorse. It was a total lack of dignity on his part, which were seemingly all too common for at least some self-serving politicians and diplomats! So many more lives had been wasted because of his tricky inhumane war games. He had actually sold out his friends for his own benefits, and yet, there was no Peace!

In his Prologue to <u>No Peace, No Honor: Nixon, Kissinger, and Betrayal in Việt Nam</u>, Larry Berman (2001) summed up so well: "But Pres. Ford had already accepted the political reality that U.S. Congress (with a Democratic majority) would not fund another supplemental budget request and that America's involvement in Việt Nam would soon be over. Reviewing the first draft of his address to a joint session of Congress, the president read his speechwriter's proposed words: 'And after years of effort, we negotiated a settlement which made it possible for us to remove our forces with honor and bring home our prisoners.' **Ford crossed out the words with honor. Henry Kissinger also knew that American honor was in danger.**

In the cabinet room on April 16, the secretary read aloud a **letter from Kampuchea Prince Sirik Matak**[283]**, one of the Kampuchea leaders who had refused the Kampuchea American ambassador's invitation to evacuate Phnom Penh.** The letter was written just hours before Matak was executed

---

282   http://www.nobelprize.org/nobel_prizes/peace/laureates/1973/press.html Kissinger accepted Peace Nobel Prize 1973. Award Ceremony speech.
283   http://vnafmamn.com/black_april.html

by Pol Pot: 'Dear Excellency and Friend, I thank you very sincerely for your letter and your offer to transport me towards freedom. I cannot, alas, leave in such a cowardly fashion. As for you, and in particular for your great country, I never believed for a moment that you would have this sentiment of abandoning a people, which has chosen liberty. You have refused us your protection, and we can do nothing about it. You leave, and my wish is that you and your country will find happiness under this sky. But, mark it well, that if I shall die here on the spot and in my country that I love, it is too bad, because we are all born and must die one day. I have committed this mistake of believing in you, the Americans.' Most of Pres. Lon Nol's cabinet members and leaders declined U.S. Ambassador John Gunther Dean's offer to exile in the U.S.: They were all executed by the Pol Pot regime as soon as the latter took over the country..."

United States 37th President Richard Milhous Nixon – Republican (1913-1994)
"...the manipulative String-Puller"

Nixon grew up in a poor family. In spite of his busy life working to help support his family, he got involved in high school, then in college debate (many championships) and sports (football and basketball), these activities having perhaps kindled his interests in leadership in law school extra-curricular activities. His law career led him through many years of routine life in California, where he started his family. His moving to Washington, D.C. during WW2 opened up a new door to the U.S. Navy. After a few years serving in this branch, he retired in 1946 as a Navy lieutenant commander. In 1966, he retired from the Naval Reserve as a commander. Earlier after retiring from naval active duty, Nixon started campaigning in politics, leading to his successfully running for U.S. Senate in 1949 when he received the nickname of "Tricky Dick." Later on, he was chosen by Pres. Eisenhower as Vice President. Nixon visited Sài-gòn and Hà-nội Indochina in 1953 when the French was still occupying Indochina. He decided to increase devotion to foreign relations: he was the first "modern vice president" to have ever been much involved in foreign and domestic policies.

Nixon was defeated by John F. Kennedy during this

first presidential run. This allowed him time off to return to California. His failure in governor's campaign, in concurrence with other failures with the Republican Party in many areas made him reflect on his internal issues. A turning point presented itself with new hope for a better Destiny[284]: Vietnam War issues had dragged the ruling Democrat party down, this being followed by the 1968 Tết Offensive in Việt Nam, then the withdrawal of Pres. Lyndon Johnson's candidacy: these events had given Nixon the idea and momentum to run for 1968 presidential election. He worked hard at gaining popularity from the people, and learned he could win only if he had to bend, and bend even more to the demand of the people to withdraw troops out of Việt-nam, in order to get the populist support. He won! His being re-elected for his second term four years later told him he had done the right thing: **bend to people's demand and get out of Việt-nam at any cost!**

Even though Nixon disliked the Jews, he noticed that Kissinger's approach matched well with his style when the latter was serving as a part-time foreign policy adviser to both the Kennedy and Johnson administrations: a more pragmatic approach to foreign affairs while deploying a "flexible response" strategy[285]. Nixon immediately appointed Kissinger as his national security adviser after he was elected president. Working closely together, they might have realized they were much more compatible in a sense they were both ego-centric and would not hesitate to make changes to fit their major needs: attain personal power and strengthen their positions.

As more declassified materials became unveiled, they revealed that Nixon and Kissinger worked to achieve their goals regardless of the cost. They used any means, including misleading the American people, and resorting even to "dirty" strategies (e.g. betraying the allies), to secure their power. As an example, towards the end of his presidency, Nixon had worked out with Kissinger a negotiation plan on their own terms on how they had decided the fate of the RVN, and consequently, the fate of Kampuchea and Laos without even informing them, including fellow-counterparts RVN Pres. Thiệu, and probably Khmer Pres. Lon Nol and Laos King Vatthana. They had even made preemptive decision to go ahead with their final Paris negotiations with the DRV representative

---

284 Nixon believed in destiny as he often referred to it in his speeches. https://www.hbo.com/documentaries/nixon-by-nixon-in-his-own-words Nixon by Nixon

285 http://www.pbs.org/wgbh/amex/china/peopleevents/pande02.html PBS Henry Kissinger's biography

Lê Đức Thọ in case Thiệu refused to sign the final agreements. Nixon's shameful executive action, followed by U.S. Congress decision under the "weakest U.S. president[286]" Gerald Ford to completely ignore when the RVN, Kampuchea and Laos were agonizing (as NVA Gen. Bùi Tín had acknowledged) had definitely helped the Vietnamese Communist Party win the final battle in South Việt Nam, and the Pathet Lao Army PLA in Laos, as well as the Khmer Rouge in Kampuchea. The cancellation of strategic B52 carpet bombing Operation Linebacker II in Hà-nội [287] when the NVA VC violated Paris Accords signaled the latter they had the green light to launch their final victorious Spring Offensive of 1975. The RVN was left with no support against an enemy that continued to receive it, this time legitimately per Paris agreement by the White House.

Both Kampuchea and Laos[288], which were left out of the Paris Accords, might not even have any idea what was happening and the wrath that was going to fall onto their people in the aftermath of the negotiations. **The "Domino Effect[289]" that both Presidents Eisenhower and Kennedy had foreseen had truly come true thanks to Nixon and his complicit Kissinger.**

Second Republic of Việt-nam President Nguyễn văn Thiệu (1923-2001)
"...the muffled, angry and left out Sideliner"

Born South Vietnamese in Phan Rang (II Corps Central VN) and French-educated, like many other nationalists, Thiệu joined the Hồ Chí Minh led Việt Minh when WW2 ended in 1945 and was trained in jungle fighting with sticks as they had no modern weapons. Realizing the Việt Minh was

---

286   http://www.viet-myths.net/buitin.htm NVA Gen. Bui Tin on _How North Vietnam Won the War._

287   https://en.wikipedia.org/wiki/Operation_Linebacker_II B52 bombers in North Vietnam

288   Laos had a separate Vientiane Treaty signed on February 21, 1973, leading to the cease-fire between the Kingdom of Laos and the Pathet Lao in their "Civil War." This had nothing to do with the United States involvement in the secret war that also involved the DRV as the Paris Accords between the "four parties:" RVN, NLF, USA and DRV. https://en.wikipedia.org/wiki/Vientiane_Treaty and Minnesota PBS TPT _The Secret War with Dr. Dao Yang_

289   https://en.wikipedia.org/wiki/Domino_theory Domino Theory or Effect – Eisenhower and Kennedy

turning communist, he defected to the French United Army, attended the renowned Saint Cyr Military Academy in France (comparable to the U.S. West Point Military Academy in New York), but quit the United Army as he realized there was discrimination against Vietnamese officers (difference in pay). He then switched to the French-backed Vietnamese National Army of the State of Việt Nam, at the time Hồ Chí Minh proclaimed Tonkin (North) independent.

  Like the other prestigious South Vietnamese officers (e.g. high-ranking Generals Đỗ Mậu, Cao văn Viên, Trần Thiện Khiêm), he had fought in many battles against the Việt Minh in the North on the same side with the French United Army. In 1954, the year the French were defeated at Điện Biên Phủ, Maj. Thiệu led a battalion repelling his former comrades, the Việt Minh, from his native village of Thanh Hải. Converted to Catholicism by marriage, he later joined Cần Lao Party to support the first RVN Catholic Pres. Ngô Đình Diệm[290]. Interestingly, Maj. Gen. Thiệu later joined the "coup d'état" that overthrew his protector Pres. Diệm on November 1st, 1963, in the midst of the Buddhist Crisis caused by the Ngô's presidency. He was promoted to Lieutenant General in 1965. After being "figurehead" of state, he finally was elected president of the second Republic in 1967 with running mate Gen. Nguyễn Cao Kỳ. The latter was replaced by a civilian VP for Thiệu's second term because of increasing rivalry between the two of them.

  Pres. Nguyễn văn Thiệu and VP Nguyễn Cao Kỳ had done what Pres. Ngô Đình Diệm as well as the following state figure heads (mostly powerful generals) could not do when they were in power. They were able to stop the vicious cycle of constant, almost yearly, "coups d'état" launched by the other French trained generals: by de-polarizing the factions and the rivalry[291] between the men who thought "they could become better leaders."

  Unfortunately, the generals' self-absorbed ego-centric pride had set themselves against each other, causing unstoppable "coups" from as early as 1954 until Thiệu's 2nd presidency in 1967. There had been so many "coups" that the

---

290 https://en.wikipedia.org/wiki/Nguyễn_Văn_Thiệu in English and https://vi.wikipedia.org/wiki/Nguyễn_Văn_Thiệu in Vietnamese

291 All the generals who had been rebelling against each other are covered in these two websites: https://en.wikipedia.org/wiki/Dương_Văn_Minh , https://en.wikipedia.org/wiki/Nguyễn_Khánh

U.S. was getting turned off from the war, to the extent that Pres. Kennedy had considered, with council of Sec. Robert McNamara, withdrawing troops from Việt-nam. Ambassador Maxwell Taylor had to summon them in to "scold" them as if they were children... Nevertheless, in spite of his success in more or less realigning the military commanders, Thiệu's handling of the Buddhist politics had caused him quite a bit of turmoil[292], especially when the secret police started discovering VC's infiltration in Buddhist temples, leading him to believe that the Buddhists were collaborating with the enemy, **in the name of Peace and Humanity.** It helped that VP Kỳ was Buddhist to counter balance Pres.Thiệu being Catholic, otherwise it could give the protestors the idea that the presidency's motive for repression was religious discrimination (as it had happened under Pres. Diệm), instead of political and national security matters.

Many analysts and bloggers alleged that Monk Thích Trí Quang [293] took part in helping the VC NVA in the Tết Offensive of 1968 and that Ấn Quang Buddhist temple had harbored enemy agents. Protests persisted until the last weeks before the Fall of Sài-gòn.

Many major events occurred during Thiệu's two terms, such as the Tết Mậu Thân Offensive of 1968, the 1968-73 tricky and disempowering Paris Accords negotiations, the 1971 Campaign Lam Sơn 719 into Tchepone in Laos, the 1972 major Easter Tide Offensive (bloodier than the 1968 one), the 1973-74 abandonment of Central Việt Nam due to drastic cuts in military aids, and finally the 1975 collapse of Sài-gòn. Similar to Mr. Ngô Đình Nhu[294]'s Tactical Hamlet Program, Thiệu's "Người cày có ruộng" (Farmers have their own land) land reform program was a legacy, a model program for third world countries at that time[295]. He and his family resettled in the U.S. after the war ended.

The French educated cabinet members might think Thiệu was pro-American; however, the Vietnamese who were pro-American believed otherwise, that he was not pro-

---

292   http://www.nytimes.com/1971/09/03/archives/buddhist-monks-quietly-lead-campaign-against-thicu.html
293   https://thong-bao-tin-cho-nhau-vn.blogspot.com/2016/03/chua-quang-hang-o-ran-oc-cua-viet-cong.html in English, https://ditmecodosaovang.wordpress.com/tag/phat-giao-an-quang/ in Vietnamese
294   Pres. Ngô Đình Diệm's elder brother, presidential advisor.
295   Washington Evening Stars and New York Times.

American enough when he had different opinion and policy disagreements with his Vice President Kỳ's who was actually a true U.S. "baby." Thiệu had boldly banned the 1972 Paris Accords arrangements by Kissinger and Nixon as it allowed the North to maintain NVA existing troops in the South (that would be 370,000), to continue occupying territories taken from the RVN ("the flag campaign[296]"), with continued resupplying of troops and ammunition and replacing worn-out weapons. The RVN had to recognize the VC's PRG as a legitimate government of the NLF (VC) and participant in the Paris Accords negotiations[297]. In order to bring Thiệu's team back to the negotiation table, Nixon had to promise Thiệu the U.S. would resume strategic bombing of Hà-nội if the latter violated the treaty. Would he be able to keep this promise?

Second Republic of Việt-nam Vice President Nguyễn Cao Kỳ (1930-2011)
"...the other RVN distant Sideliner"

Unlike Pres. Thiệu, Kỳ was born in the North (Sơn Tây). He attended a semi French-Vietnamese high school (Lycée du Protectorat) and later joined the French-backed State of Vietnam National Army[298]. Afterwards, he was trained into Air Force career at a French Air Force Academy in Marrakech, Morocco. Kỳ became the first pilot officer of the new ARVN. As Thiệu, he participated in the 1963 coup overthrowing Diệm and thereafter was promoted many times, up to Major General and member of a few Military Councils, living through many upheaval political events and coups. 1965 was the year he had to prove tough against his rival Lt. Gen. Nguyễn Chánh Thi, a true French-military career officer (commander of the 1st Company of the French 6e BCP or "6ème Bataillon Commando- Parachutistes"). Kỳ believed his former Central Vietnamese friend Thi (they both were flamboyant in their impeccable uniforms with a similar moustache, so similar I used to have them mixed up when growing up) was a left-wing provocateur who had instigated the major uprising of the Buddhist opposition. As a precautious measure, Thiệu and Kỳ

---

296 Corresponding flags (DRV versus RVN) were displayed according to which side had control over that territory.
297 http://www.history.com/this-day-in-history/paris-peace-accords-signed
298 https://en.wikipedia.org/wiki/Nguyễn_Cao_Kỳ and https://vi.wikipedia.org/wiki/Nguyễn_Cao_Kỳ RVN VP Nguyễn Cao Kỳ's biography in English and Vietnamese (very limited information)

deported Gen. Thi to the U.S.

In 1967, Kỳ joined Thiệu's presidency for one term and was replaced by civilian VP Trần văn Hương during the second term. VP Kỳ was the one who strictly cautioned Thiệu not to recognize the communist PRG (representing the Việt-cộng National Liberation Front in the South) in Paris Accords documents. He was also, perhaps in revenge for Thiệu's efforts to expel him from a 2nd term presidency partnership, the whistleblower on Thiệu's family underground illegal luxury merchandise trade and even drug trafficking, these being probably one of the reasons the U.S. had decided to cease support for the RVN. This incident made Kỳ a primary rival adversary. Thiệu had considered Kỳ to be a pro-American who undermined his authority. After the war ended, Kỳ and his family resettled in the U.S., but decided to move permanently back to Việt Nam in 2004. He allegedly reverted to pro-communist sentiments and started promoting foreign investments in Việt Nam. Naturally, his decision was not viewed well by the anticommunist Vietnamese abroad.

Like many other RVN generals, Thiệu and Kỳ, both French-trained from the early days of French United Army, might be an odd couple. They looked like a good team from the outside, but, behind the scene, they were actually rivals. How could I say it? They were mutually antagonistic? Thiệu, born in Central Phan Rang, seemed to work well with Vietnamese and French-trained and educated advisors and Cabinet (remember that my brother-in-law Anh Bảy was a member of), with the exception of two close American-trained advisors. One of them was young U.S.-degreed Hoàng Đức Nhã[299], Thiệu's Central-born (Phan Rang) cousin. He was Thiệu's Special Advisor and, later, Director of "Tổng-cục Dân Vận và Chiêu Hồi" People Mass Mobilization and Open Arms. He became Minister of Information towards the end of the war. The other one was Northern-born (Thanh Hoá) U.S.-degreed Dr. Nguyễn Tiến Hưng[300], Minister of Economic Development and post-

---

[299] http://www.zoominfo.com/p/Hoang-Nha/143385004 in English and https://vi.wikipedia.org/wiki/Hoàng_Đức_Nhã in Vietnamese. Mr. Nhã was educated in French Lycée Yersin, Dalat, before he attended Universities in the U.S... Mr. Nhã, along with Dr. Hưng, were close advisors to the President on Paris Accords. My brother in law Anh Bảy kept nicknaming him as "the American guy" as he was the littlest one trained in the U.S...

[300] Dr. Nguyễn Tiến Hưng received his Math Bachelors' Degree from the University of Sài-gòn, then he went to the U.S. for his Economics Masters' and PhD degrees at the University of Virginia. He and Pres. Thiệu had made efforts to curb Pres. Nixon to extend support to the RVN even though U.S. Congress had made

war historian and political analyst of the Vietnam War. Some of his work was used as references to support this book (see Bibliography).

      Surprisingly, VP Kỳ was considered to belong to the American "clan," perhaps because people assumed that he, being an American-trained jet pilot, would receive more benefits from and would be more trusted by the U.S. At the end of his presidential term 1967-71, some commentators believed Thiệu "had had enough" of Kỳ, and therefore had Trần văn Hương replace him as Vice President in his second term in 1971-75. Other sources described Kỳ as having more support from ARVN generals than Thiệu did, while, interestingly, the latter was believed to be more supported by the U.S. Nevertheless, Pres. Thiệu had received from Kỳ, and other political advisors, political and strategic advices that (as I could evaluate by cross-referencing with newly-declassified materials) were useful: they were more or less accurate or well-founded. His refusal to participate in the concluding phase of the Paris Accords was not a strategic game card he wanted to play against Nixon and Kissinger's moves. He did see clearly that the negotiations were turning extremely sour for the RVN when they required participants to recognize the PRG (of the NLF National Liberation Front as a set up with the DRV, the real "string-puller") and when it oppressed the RVN into a fetal submissive position as described earlier.

---

up its mind about cutting off all aid and withdraw from Việt Nam. Dr. Hưng had included important declassified-based documentaries in his books, such as 1986 The Palace Files revealing correspondence between Pres. Nixon and Pres. Thiệu, 2005 Khi Đồng Minh Tháo Chạy (When Your Ally Cuts out and Runs) and, most recently 2016, Khi Đồng Minh Nhảy Vào (When Your Ally Jumps In) Vietnamese May 2016 release https://en.wikipedia.org/wiki/Nguyễn_Tiến_Hưng and YouTube interviews of Mr. Nguyễn Xuân Nghĩa, Dr. Hưng's comrade.

CHAPTER 9

Disclaimer

As resettlement conditions are different depending on socio-geographic locations, commentaries in this book about the Vietnamese as well as non-Vietnamese veterans, psychologically-speaking, pertain more to those resettled in the state of Minnesota with which I am the most familiar and whose communities I have worked the most often, in education, community service and, later on, in political lobbying. Nevertheless, the general comments that are based on the internet information, documentary films and written literature are current world-wide.

Kampuchea and Laos were left out of the Picture
(see footnote below regarding terminology usage[301], e.g. Miao vs. Hmong)

Not like the Indochina War of the early 1950s, the Geneva Conference had included Cambodia and Laos in its negotiations and resolutions for maintaining peace and neutrality of the **Five[302] French Indochinas**. Even though the Vietnam War had seriously affected the security of Kampuchea and Laos, and, in many ways, caused the loss of millions of lives, these unnecessary losses were neither fairly recognized nor compensated but by a meager United States House of Representatives resolutions to accept refugees from these nations into the U.S. For fear of being condemned by the world for having meddled with Kampuchea and Laos sovereignty, Nixon and Kissinger had sneakily used clandestine and covert approaches, thus instigating civil wars in both countries. As for Laos, the turmoil had actually started "secretly" long before people had known about it, as early as the end of French colonization...

---

[301] As I had mentioned at the beginning of this book, names, terminology and appellations are used as it has been according to the historical timeline. For that reason, Miao or Meo are referred to in events that occurred before the 1970s, and Hmong as of 1975.

[302] Cambodia, Laos, Tonkin, Annam and Cochinchina.

Republic of Kampuchea or Khmer (Cambodia, as renamed by President Lon Nol)

During his long but intermittent rule of Cambodia (1941-2004), Prince Norodom Sihanouk had been maintaining a peaceful status-quo policy with the French transitioning into its newly-acquired Independence, and a more or less adequate diplomatic relations with the Socialist Democratic Republic of Vietnam (SDRV, also Democratic Republic of Vietnam DRV). Unfortunately, this status-quo had allowed Hồ Chí Minh to take advantage of his common national cause with the Prince, to bring about Independence from France, so that the NVA could "use" Cambodia as a stepping stone into the RVN via the Sihanouk Trail, the actual extension of the HCM trail from Laos. Prince Sihanouk's "deal" with Hồ had precipitated Nixon and Kissinger to alter the politics of Cambodia (changing from Kingdom to Republic of Kampuchea or Khmer[303]) by "replacing" the old leadership and installing pro-American Pres. Lon Nol (March 18, 1970 coup d'état[304]). This new regime accommodated U.S. strategic bombing (both covert and overt), intelligence gathering, as well as incursion campaigns on and across the border. It also caused political factions, such as Sơn Ngọc Thành's anti-communist group, to rebel.

Lon Nol was a magistrate in French colonial Cambodia (1937) and, later on, the commander of Cambodian anti Vietnamese communist army (1952), then army chief of staff (1955) and finally gravitating to Premier (1969) under Prince Sihanouk. As the latter had allowed the NVA to move troops in via the Hồ Chí Minh and Sihanouk trails and to "settle down" in semi-permanent military bases in Cambodia, Lon Nol, with U.S. support, overthrew the last prince of Cambodia, Prince Norodom Sihanouk, and abandoned the national neutrality from the South East Asia war[305]. With the subtle support of the U.S., he pushed his country into a war that was kept in the "dark." In spite of his controversial policies related to Việt-nam, he promoted on-going massacres of some 800 Vietnamese

---

303   The French called this country Cambodge, which in English was Cambodia. Prince Sihanouk kept it as Cambodia, until he was overthrown by American-backed Lon Nol, when the Republic of Kampuchea or Khmer was used to replace until the 2010s.

304   https://en.wikipedia.org/wiki/Cambodian_Civil_War

305   https://www.britannica.com/biography/Lon-Nol Lon Nol's uprising into power and joining the SE Asian War.

immigrants "cáp duồn[306]" in Kampuchea[307] that urged the RVN, under the command of Lt. Gen. Đỗ Cao Trí, to launch the incursion in 1970 to, on the one hand, protect and liberate Vietnamese descendants in Kampuchea, and, on the other hand, cooperate with the U.S. Army and destroy their caches of ammunition and supplies along the Sihanouk Trail and the RVN border.

As a new college student, I remember major demonstrations by Sài-gòn students and people in front of the Embassy of Kampuchea not too far from my house and in front of the House of Parliament. It was not easy to forget when tear gas used by the RVN National Police to protect the embassy flew right into your bedroom. Eventually, I joined in a peaceful sit-down protest in front of the House of Parliament on Lê Lợi Blvd (now Opera House) downtown Sài-gòn, also against Kampuchea for its genocidal act of Cáp Duon against the Vietnamese population in that mysterious country. Many times I had to run away as tear gas had sieged us from all corners, and used lime juice to flush my eyes. The Lon Nol government had twisted the history between Đại Việt and Champa and blamed the Annamites for invading and taking land, thus instigating the Khmer people to behead Vietnamese, instead of accepting that it was an annexation after two of their Kings had married Đại Việt Princesses (as explained earlier).

Lon Nol's presidency was highly supported by the White House and the Pentagon. Consequently and inevitably, the participation of Kampuchea in the war had escalated to major B52 carpet bombing (see earlier section on Kissinger) and across-border incursion… a completely full-fledged conventional war that had piggy-backed on the Vietnam War but which was kept secret from the American public and never recognized[308] until decades later. Prior to the

---

306   http://tranthachsalc.blogspot.com/2014/06/cap-duon.html in Vietnamese: ongoing "cáp duồn" massacres of Vietnamese settlers in Cambodia in revenge of Đại Việt invasion and annexing of Champa and Khmer (as early as 1769 AD as recorded by French missionary writer Louis-Eugene Louvet). Contemporary "cáp duồn "victims were gathered by family, and beheaded: their heads were thrown into the Mekong River and floated back to the Mekong Delta in Vietnam. One of the two-prong plans of the 1970 ARVN incursion into Cambodia (led by Gen. Đỗ Cao Trí) was a show of force and the rescue of potential "cáp duồn" victims https://en.wikipedia.org/wiki/Cambodian_Civil_War#Massacre_of_the_Vietnamese

307   Cambodia was the name the French gave to Kampuchea as part of Indochina. Lon Nol reverted to the Khmer's pronunciation as Kampuchea, and then Pol Pot (Khmer Rouge leader) changed it to Khmer.

308   https://en.wikipedia.org/wiki/Cambodian_Campaign Cambodian incursion and https://en.wikipedia.org/wiki/Do_Cao_Tri RVN Gen. Đỗ Cao Trí

bombing "incident," the U.S. and RVN had started clandestine operations to gather intelligence about the NVA bases on the Sihanouk Trail (extension of HCM Trail). These were the bases that, as allowed by Prince Sihanouk, had sheltered incoming NVA divisions entering from the DRV (North Vietnam) via the two trails, and served as weapons, equipment and food supply depots, both feeding into the war in South Vietnam.

The "Secret Khmer Special Forces" that were the most "secret"

B52 bombings and the Kampuchea incursion would not be made possible without the Intel provided by the "Forces Spéciales Khmères FSK" under Brig. Gen. Thach Reng[309] and other clandestine groups such as the Khmer-Krom, the South Vietnam Khmer ethnic groups MIKE Forces, sometimes nicknamed "American Commandos" by the ARVN, Cpt. Steve Yedinak's 200-member team[310], the Khmer Serei militia led by Sơn Ngọc

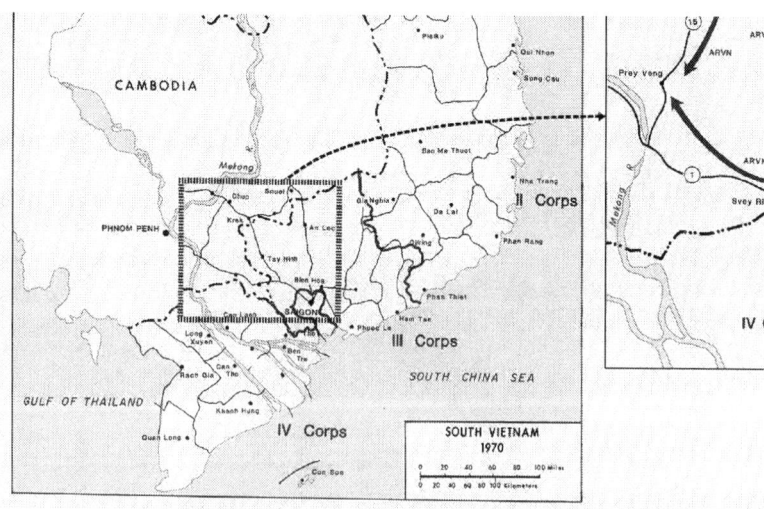

Kampuchea Campaign of 1970
Source: https://upload.wikimedia.org/wikipedia/commons/2/2f/
Map_Cambodian_Incursion_May_70_from_USMA.jpg

---

309   Forces Spéciales Khmères FSK https://www.revolvy.com/topic/Khmer%20Special%20Forces and https://en.wikipedia.org/wiki/Khmer_Special_Forces
310   https://www.voanews.com/a/effort-to-honor-khmer-krom-fighters-allied-with-us-in-vietnam-war/4031914.html

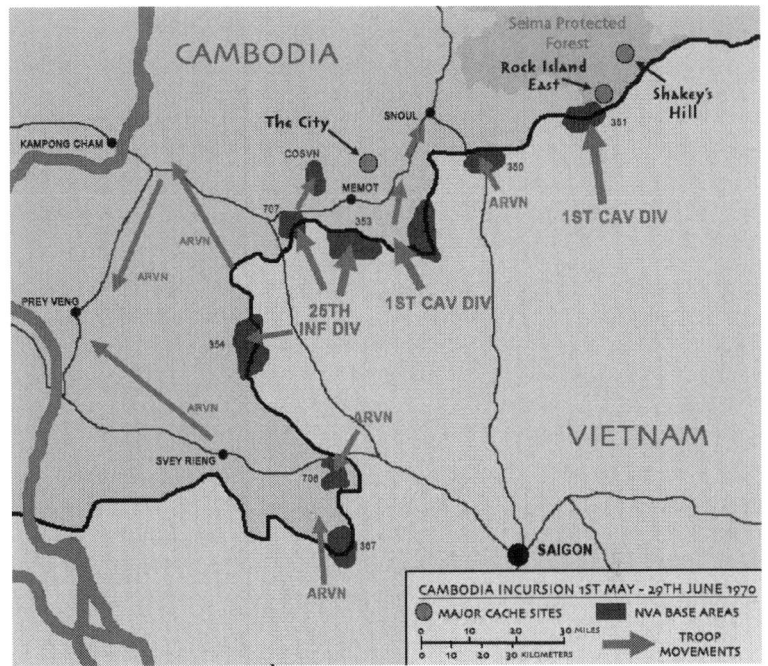

Kampuchea: the Hồ Chí Minh Trail, that branched out into Sihanouk Trail, lead to the Kampuchea Incursion of 1970 involving both RVN and the U.S.
Source: adst.org

Thành[311]... They had been providing tactical support to the "regular" National Republic of Khmer Army against the Khmer Rouge (1971-75). Had anyone in the U.S. known about them? I doubt it!

The Cambodian Campaign started in Mid-April 1970 by the ARVN armored and Ranger, Airborne units under Lt. Gen. Đỗ Cao Trí's[312] command, was joined a week later by the U.S. armored and Infantry units under Brig. Gen. Robert L. Shoemaker in Suay Rieng Province, cutting into Parrot's Beak

and Fishhook. Operation Toàn Thắng was a success with the ground work done by the clandestine groups. These quiet brave men ought to be recognized and honored as the Miao SGU had

---

311   https://en.wikipedia.org/wiki/Khmer_Serei Khmer Serei militia
312   Gen. Trí was Mười's cousin on her mother's side in Biên Hoà. This part of the family had members in both sides VC and RVN high ranking military.

received for having assisted the U.S. and the RVN access and identified the NVA bases along the Cambodia-Việt-nam border at the end of Sihanouk Trail (extension of HCM Trail from Lowland Laos)[313].

Unfortunately, in spite of its serious involvement in the war against the NVA and the major loss of nearly 3 million lives, Kampuchea never had a chance to voice its wishes at the Paris Accords negotiations. The U.S. Congress pulling the plug on the RVN also resulted in the abandonment of Kampuchea and its people to an immense genocide bloodshed by the Khmer Rouge. The Republic of Kampuchea fell quietly on April 17, 1975. This conventional war against NVA infiltration was sadly coined in history with the simple name "Cambodian civil war" as the world did not know about the tragedy in Kampuchea only until it was taken over by the China-backed Khmer Rouge (they were also people of Kampuchea). The near three million lives' massacre did not seem to mean much in the U.S. as the media might already have phased out of this unpopular SE Asian war. Most of the victims were innocent and defenseless civilians or surrendering Khmer Republic military personnel of all ages/ both genders.

Nixon and Kissinger easily brushed off their hands with this one! "C'est un bon débarras!" as the French would say, "what a good rid!" The Soviet-backed Socialist Republic of Việt Nam SRV[314] had supported China-backed Pol Pot's regime until 1977[315], when it turned around when the Khmer Rouge also beheaded Vietnamese immigrants Cáp Duồn in Kampuchea. The SRV overthrew this Pol Pot former ally in 1979 by defeating the Khmer Rouge. This DRV action resulted in China's 1979-1981 retaliation attack in Northern Vietnam[316].

The Kampuchea War (1970-1975[317]) was a short but absolutely bloody one, with Khmer Rouge's genocidal massacres (1975-1979), so it might have been easier to be

---

313   https://www.globalsecurity.org/military/ops/vietnam2-cambodia.htm
314   The Vietnamese might be symbolic as they stress on appellation that will make meaning. As a result, North Vietnam came up with many references for North Vietnam: socialist, republic, democratic, people's, and ended up with many appellations.
315   https://vi.wikipedia.org/wiki/Khmer_%C4%90%E1%BB%8F In Vietnamese: Khmer Đỏ (Khmer Rouge, Communist Kampuchea)
316   War of the Dragons http://www.historynet.com/war-of-the-dragons-the-sino-vietnamese-war-1979.htm 1979 China attack of Vietnam
317   https://www.onwar.com/aced/chrono/c1900s/yr70/fcambodia1970.htm and https://en.wikipedia.org/wiki/Cambodian_Civil_War Cambodia "Civil War"

kept secret and "hidden" from the scrutinizing eyes of the free world. Compared to Kampuchea, the conflict in Laos was not as short as it lasted from 1953 until the end of 1975, but it was still hidden under the intriguing but striking names of "civil war" (like the one in Kampuchea) or "secret war," causing, in my psycho-sociologist analyst opinion, a major confusion of historical facts, especially when it was credited the most only to a small ethnic group, the Miao - Mèo (nowadays Hmong) tribes, leaving out the Kingdom of Laos[318] which was afflicted the most, by the American news media.

Kingdom of Laos

Unlike Kampuchea or Việt-nam, which both had a coastline, land-locked Laos was surrounded by Kampuchea (South), Thailand (West), Burma or Myanmar (North-West), China (North) and Việt-nam (East). Laos acquired its pseudo-independence from France, when France transferred civilian powers (France still maintained military controls) to the neutralist Royal Lao Government (Prince Souvana Phouma) via The Franco–Lao Treaty of Amity and Association in 1953. This transfer caused a serious internal division into three sides, the governing Royalist (King Vatthana) and two rebelling factions: the leftist communist Lao Patriotic Front (Prince Souphanouvong – Pathet Lao Army PLA) and the rightist rival (Prince Boun Oum of Champassack).

First DRV Invasion of Laos

In 1953, PAVN (People's Army of Vietnam was part of the Việt Minh at that time) Gen. Võ Nguyên Giáp invaded Laos with 40,000 PAVN troops, including 2,000 Pathet Lao Army led by Prince Souphanouvong, allowing the prince to occupy Xam Neua where many Miao (or Mèo, until 1970s Hmong) tribes were located. The PAVN continued its victorious advance and stretched out to Luang Prabang (center of map above) and Thakek (South on the map). These victories were actually **NVA diversion tactics "dương Đông, kích Tây» (tackle the West but actually attack the East)** to prepare for the real major

---

318 At least 4 Régions Militaires 1, 2, 3 and 4 in the Kingdom were at war against Pathet Lao and the NVA, instead of one #2 Miao

assault on Điện Biên Phủ that cost France its whole Indochina. This was the time when the U.S. started using Civil Air Transport, later called Air America, to move supplies to support the French in Điện Biên Phủ.

The People's Army of Vietnam (PAVN or North Vietnam Army NVA)'s victory of Điện Biên Phủ (Việt Nam) led to the Việt-nam – French Indochina treaty at the Geneva Conference (Accords) on July 20, 1954. The Agreement on the Cessation of Hostilities in Laos[319] was also signed ending the French rule in Laos.

The Laos War was dwindling down to a ceasefire in 1956 between the French-trained and the U.S. financially-assisted Royal Army (Souvana Phouma) and the Pathet Lao Army (the brother Souphanouvong). Nevertheless, the Socialist Republic of Việt-nam SRV (also called DRV Democratic Republic of Vietnam, or North Vietnam) expanded its support for the Left, the Pathet Lao Army PLA, by sending them, including some Miao-Mèo indigenous people, to North Vietnam for academic and leadership schooling. As the leftists had always maintained their rule over their occupied territory, skirmishes continued to occur in the mountains (this area later became Région Militaire RM#2), while, interestingly, in the administrative city, the PLA was gaining seats in the Royalist Lao joint National Assembly (16 out of 59). This might be the time when the conflict between the Pathet and the Royalist groups was not intense yet.

My father, who traveled intensively to Laos in the late 1940s and with my French godfather Jean Schwoerer ("Directeur Général de la Société des Théâtres d'Indochine") early1950s for business, oftentimes said that the Laotian people were the most peaceful people in the world. "When they (Royal Lao Army) caught PLA prisoners, they would just cane then release them," he smiled... I totally agreed with Father when I started working with the Laotian and Hmong immigrants in Minneapolis Public Schools in 1975. The Vietnamese people might be trickier and more aggressive, while the Lao/Hmong very peaceful and accommodating!

---

319   https://www.mtholyoke.edu/acad/intrel/genevacc.htm the Agreement on the Cessation of Hostilities in Laos

Xam Neua – Laos: Original Headquarters of Pathet Lao Army PLA
Source www.asienreisnder.de

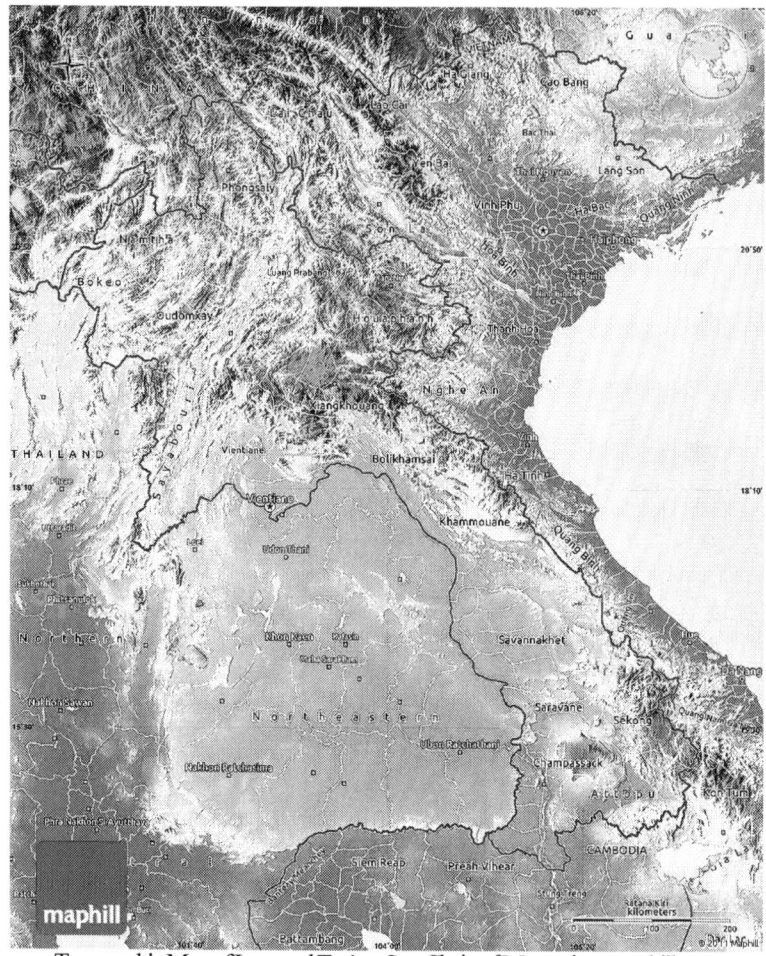

Topographic Map of Laos and Trường Son Chain of Mountains - maphill.com
Source: http://maps.maphill.com/laos/maps/physical-map/physical-map-of-laos.jpg

The Original Lao "Secret Army"

Because North Vietnam increased its support for the growing PLA forces, Royal Lao Government kept supplying the Miao tribes with military aids to combat the PLA who operated in the same areas where the tribal Miaos lived. **Prince Phouma's request for aids from the U.S. had opened up Laos to super power intervention.** American military trainers began to replace the French and the U.S. 77th Special Forces Group started training the **Royal Lao Army to become the first Lao "Secret Army,"** this being a violation of the Geneva

Conference. The U.S. Program Evaluation Office that ran this clandestine operation expanded "twentyfold" **under the pseudo name of Eastern Construction Company**. The Royal Lao Government was trying to resolve conflicts with the opposition (by integrating the PLA in the national army and government); nevertheless, the U.S. intervention caused Leftist Prince Souphanouvong to be imprisoned. The mechanics of war kept escalating. Nevertheless, the Socialist Republic of Việt-nam decided to launch its second major invasion in 1959[320] to test the Lao Army as well as reinforce the PLA.

Gen. Vang Pao and his Miao Secret Guerilla Units in "RM#2"

During the political internal turmoil in the Kingdom of Laos, the CIA discovered young Miao (Meo, later on Hmong) Lt. Gen. Vang Pao of the Royal Lao Army RLA (1963). He was an officer who had started his military career as a lieutenant in the French "anti-Việt Minh commando groupement" and who had gravitated to the rank of general in the Royal Lao Army. Vang was loyal to the King of Laos while being a champion of the Miao people[321]. The Lao Secret Army was originally born from the montagnard Special Guerrilla Units consisting of Yao, and Lao Theung, besides the Miao ethnic militia men[322]. They were trained in Thailand and chaperoned by U.S. CIA and Special Forces in Laos. In the meantime, the RLA also had its own secret army SGU, also trained and chaperoned by the U.S. Special Forces and CIA[323]. RLA as the national forces, the "regular"(professional military e.g. Infantry Army, Air force, Artillery, Airborne...), of course outnumbered the Miao smaller units, considered by the CIA as the "irregular[324]" (militia).

---

320 https://en.wikipedia.org/wiki/Laotian_Civil_War#1953.E2.80.9354
321 https://en.wikipedia.org/wiki/Vang_Pao#Military_career Lt. Gen. Vang Pao of the Royal Lao Army
322 Participating in the Laos War were approximately 50,000 in Royal Lao Army, 21,000 Thai and ethnic montagnard mercenaries and 23,000 Miao militiamen fighting against 48,000 Pathet Lao. Source: Wikipedia
Laotian Civil War https://en.wikipedia.org/wiki/Laotian_Civil_War. A very small secret clandestine force was staffed by American CIA and Thai special forces personnel.
323 CIA Briggs, Tovar, Bruton. Col. Khao Insixiengmay (2018-19)
324 Cash on Delivery CIA case officer Thomas Briggs's book accounting CIA activities with regular Royal Lao and irregular ethnic combatants in Laos. This terminology is also used by Col. D. Courtney, former CIA case officer in Laos.

The CIA was able to form battalions of up to 9,000 men in total, coordinated by 10 U.S. CIA and Special Forces operatives and 100 Thai Special Forces. This operation was maintained strictly secret so that Laos could have its political and military neutrality as prescribed by the 1954 Geneva Convention. When released back in Laos, most of the ethnic and mainly Miao Thai-trained fighters served under the leadership of Gen. Vang Pao (1963-1975) in the mountains of RM#2 in covert operation that was sometimes, if not most of the time, independent from the Royal Lao agenda.

Their major mission was usually gathering intelligence for CIA on Pathet Lao activities and sometimes on the latter NVA advisors as they passed for peasants working on their fields[325]. This army, resupplied by Air America, had spearheaded into and sometimes expanded from Xam Neua (Pathet Lao Army PLA headquarters, North of Plain of Jars) and fought an extension war against the PAVN[326]-assisted PLA in the mountains of Northern Laos (RM#2). It also provided protection to CIA installations there (Radar sites, Na Khang forward resupply base, Long Cheng air runway[327]). Those who were trained in Thailand (note that not all were trained in Thailand) were still not combat ready as the Royal Lao troops who had actually received more intensive strategic and tactical instruction being the core army. Reports from various RM#3 RLA and CIA sources said many Miao SGUs had suffered major casualties when engaging NVA in RM#2 and had to be rescued by RLA from RM#3. Nevertheless, the U.S. particularly relied on the Miao SGU to recover and guide U.S. Navy and Air Force pilots who were shot down in Laos (Barrel Roll) to Thailand: they could pass for indigenous peasants.

The Miao SGU[328] assisted the CIA to protect the radar

---

325   As reported by the CIA, Miao's cash crops were poppy fields and opium trade, which was legal in Laos until 1971. It was alleged that Gen. Vang Pao had Air America transport his tribe crops for trading so he could raise funds to maintain his troops. My brother in law said that RVN Dept of Transportation/ Post Office (Bộ Giao-thông và Bưu Điện) of President Thiệu's Cabinet had intelligence reporting that Opium had been snuck into the South by the NVA via HCM Trail and that POWs confessed they had bought it from the Miao in Vinh for the purpose of corrupting and disabling the "American invading forces and ARVN puppets." It happened during the period towards the end of the American involvement in the war (early 1970s).

326   People's Army of Vietnam, Socialist Republic of Vietnam.

327   Tragic Mountains: the Hmong, the Americans, and the Secret Wars for Laos, 1942-1992 by Jane Hamilton-Merritt (1993,1999)

328   As reported by the Wisconsin survivors of SGU in their Militia fliers (2017).

in the "Lima Site electronic mountains" (as the Miao SGU called it) that was installed and operated by the U.S.AF (e.g. LS-85 TACAN[329] and others, as verified by the CIA and the Royal Lao Spec. Forces). The mission of these stations as run by the U.S. Navy was to pinpoint targets for carpet bombing (B52) and bomb strikes mainly over the Barrel Roll area and the Plaine des Jarres in RM#2 against the PLA and NVA, and, from time to time, 60km around Hà-nội and 19km around Hải Phòng[330] ... to apply political pressure (Rolling Thunder bombing campaign from 1965 to 1968) and in support of the Paris Treaty negotiations (1968-1973)[331].

Contrarily to many claims, Gen. Vang Pao and his SGU never fought in the areas of the Hồ Chí Minh Trail which was located in RM#3 (head of the Trail, per. Col. Khao Insixiengmay and maps) and 4 (the end of the Trail in Laos), as verified by CIA that operated in Laos during the war, as well as by the Royal Lao Army[332] and Special Forces[333] that actually fought there. It should also be noted that the CIA did not work just with the Miao SGUs in RM#2. The Royal Lao Government and Army had intensively teamed up with the CIA (Roadwatch Teams) even more intensively in Regions 3 and RM#4 which received the most carpet bombings in Laos: this was where the HCM Trail was[334], not in Gen. Vang Pao's RM#2. On the eastern side, MACV-SOG and the ARVN Spec. Forces teams (SEAL Frogmen and Airborne) guided strikes using electronic listening devices. Nevertheless, the trail did not stop here: it expanded beyond the Cambodian border and turned into the Sihanouk Trail as discussed earlier.

---

329   https://www.cia.gov/library/center-for-the-study-of-intelligence/csi-publications/csi-studies/studies/95unclass/Linder.html CIA report on Lima Site 85. Phou Phathi area, nicknamed the Rock by the Americans, was the location of the Lima Site Gen. Vang Pao's Miao SGUs was helping with security. It was also where the renowned controversial opium poppy field was grown by the Maio tribes' traditional cash crop. Gen. Vang Pao was alleged for using this crop to fun his war, using Air America (undercover CIA) planes for transportation, this involving the CIA in the controversy. Also, more complete, by U.S. Air force Magazine: http://www.airforcemag.com/MagazineArchive/Pages/2006/April%202006/0406lima.aspx

330   Hà-nội and Hải Phòng were mostly covered by the ARVN Strategic Technical Directorate "jump north" commandos.

331   Buffer zone was 60km from the Chinese border. https://en.wikipedia.org/wiki/Operation_Rolling_Thunder

332   CIA T. Briggs, Tovar, CIA E. Chavez, CIA D. Courtney, Bruton. Royal Lao Airborne Lt. Col. Insixienmay, Mpls MN (circa 2017) Insixiengmay is a highly respected commander of a "4-bataillon Groupe Mobile" in RM#3.

333   As reported by Thomas Briggs, Laos War researcher (Coalition of Allied VN War Veterans)..

334   As verified by CIA sources Laos Briggs and

Most importantly, we need to know that **the Miao was not the only group that had SGU and whom the CIA supported,** as many Hmong Studies and documentary sources (including PBS and PBS TPT television media) had claimed. RM#3 CIA case officer, Thomas Briggs[335] who had been researching Laos CIA activities in Laos, had confirmed that Régions Militaires 3 and 4 also had a major SGU contingency consisting of mainly RLA troops assisted by CIA with similar missions as the Miao SGU in the mountainous RM#2 plus much more... These Lao SGUs, not Miao[336], were especially needed there because this was where the Hồ Chí Minh Trail (name used interchangeably with the "supply line," both referring to the HCM Trail in Lao RM#3 and 4, not Miao RM#2) was located. RM#3 and #4 had suffered the most B52 bombing in Laos (Steel Tiger and Tiger Hound) for this main reason. See incoming B52 Bombing Map of Laos and its 5 Régions Militaires.

Clandestine Operations by the Republic of Vietnam's "Secret Army" in Laos

In 1963-1964, the ARVN Frogmen SEAL[337] commandos who were long-term inserted in North Việt Nam for intelligence gathering (e.g. bombing by B52 Rolling Thunder or by U.S. Navy jet fighters in Hải Phòng, Nghệ Tĩnh, Hà-nội...) were instructed they could "walk into Laos then report to any Miao villages[338]" in order to be connected to

---

335   Lecture in Richfield, MN, in May 2019, and Book Cash on Delivery, 2009, CIA Special Operations During the Secret War in Laos.

336   https://medium.com/@SoldiersWhisper/the-hmong-did-a-lot-of-damage-on-the-vc-supply-lines-4289de7d5b86 or http://www.akleg.gov/basis/get documents.asp?session=28&docid=3115 or https://video.search.yahoo.com/yhs/search?fr=yhs-iry-fullyhosted_003&hsimp=yhs-fullyhosted_003&hspart=iry&p=youtube+hmong+tpt+vietnam+war#id=6&vid=a5ed-6c78c0bcb4cace9cceac6bde1ecf&action=view or https://video.search.yahoo.com/yhs/search?fr=yhs-iry-fullyhosted_003&hsimp=yhs-fullyhosted_003&hspart=iry&p=youtube+hmong+tpt+vietnam+war#id=7&vid=b48a31b-ce31feb66c10c23ea7e4c9826&action=view or the long film The Secret War https://video.search.yahoo.com/yhs/search?fr=yhs-iry-fullyhosted_003&hsimp=yhs- ("we fight the HCM Trail before the HCM Trail became HCM Trail... ") as well as many others

337   Frogmen Units originally trained by MACV-SOG and later reassigned to RVN General Staff - Strategic Technical Directorate

338   STD was trained by U.S. MACV-SOG and run by the RVN Joint General Staff (equivalent of the U.S. Pentagon) that had coordinated all ARVN SEAL commando activities in Laos, Cambodia and North Vietnam since the early 1960s. *The Republic of Việt-nam Commandos by Hieu Vu, 2011*

MACV for extraction transport back

MAP OF UXO IMPACT AND BOMBING DATA 1965-1975

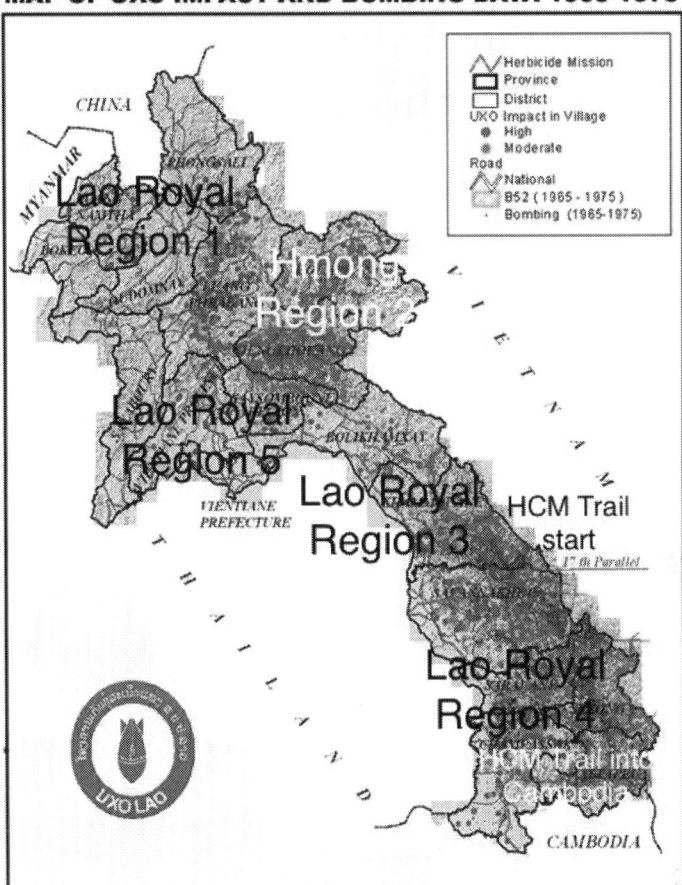

Source: Laos and its 5 Régions Militaires.
Note that the Hồ Chí Minh Trail starts from RM#3,
passes through RM#4 and continues into Kampuchea
https://deanoworldtravels.files.wordpress.com/2012/02/
bombmapopt2.jpg

home, via Thailand. My friend's brother, Bảng Nguyễn, an ARVN Frogmen team leader, had confirmed this in a conversation in 1968. He was killed in the early 1970s while in operation in Lowland Laos while installing listening devices along the HCM Trail. The back of his skull was blown off, as his father told me. As of 1965, the ARVN SEAL Airborne units launched near the border in North Vietnam (e.g. Nghệ An,

Sơn La, Lào Cay…) no longer relied on Miao transitioning for post-op extraction, but, instead, were extracted by helicopter (the same way as U.S. Special Forces) directly from Laos, and transported to the RVN or Thailand Nakhon Phanum Airbase.

A great many ARVN commandos inserted long-term in North Vietnam were captured and incarcerated at Hoả Lò ("Hà-nội Hilton"), then relocated to multiple other prisons spread out in the North to make room for the captured U.S. pilots. One of these was commando Trần Quý who war later resettled in St Paul, Minnesota after 15 of incarceration. His friend Hoàng Thiêm, also a St Paul resident, was incarcerated for 20 years. Both were verified commandos and received a Presidential Award.

In the meantime, the ARVN SEAL frogmen teams (sea commandos) operating near the Gulf of Tonkin continued to be extracted by PTF boats from Hải Phòng, Hạ Long Bay, Móng Cái, Hạ Long Vỷ, Đồng Hới, Nghệ An, Thanh Hóa, Quảng Bình, et. al. Airborne teams deployed in Kampuchea were usually extracted by helicopter or just crossed the border to return to the RVN.

The efforts made by these "secret" teams should have been recognized and their heroes honored, but, unfortunately, their activities and sacrifices were kept secret. They were the bigger part of "America's Secret Army;" however, the news media had either ignored or buried them in the unknown world. Of course, being voiceless freedom fighters, their stories had been silenced forever.

Like Việt-nam, Laos was Politically Divided

As it would happen to Kampuchea a decade later, the domino effects as predicted by Presidents Eisenhower and Kennedy took its toll in Laos. The battles and bloodshed remained absolutely "secret" to the rest of the free world. It was a "secret" involving, not just the 10,000 Miao fighters, but also some 50,000 Royal Lao Army RLA troops and 5,000 Thai special forces mercenaries, fighting a common local enemy, the Pathet Lao Army PLA, and foreign invaders and instigators, the North Vietnam Army.

On the one hand, Hồ Chí Minh really wanted to use

the South Eastern corridor of Laos where the spine of Trường Sơn Mountain opened up (low altitude) from Mụ Giạ Pass in Quảng Bình Province to Ban Karai Pass in Hà Tĩnh Province (Khammouane in Laos, Gen. Insixiengmay's RM#3) for passage via Savannakhet Plain and Sekong Valley, then into Kampuchea to access South Vietnam. On the other hand, turmoil arose, as it was the case of the RVN being ravaged by opposing Vietnamese fractions after having recovered its Independence from France in 1954, when the Lao Neutralist group led by Capt. Kong Le took over the Royalist Capital City of Vientiane by means of a coup d'état[339]. The faction ran into opposition by Gen. Phoumi Nosavan who had family connection with Thailand PM Sarit Thanarat, thus leading to Thai intervention... No matter what happened, the U.S. was supporting the "legitimate" Royal Lao Government, as represented by Gen. Phoumi's group and Thailand-connected government of Royalist PM Samsanith. As in Việt-nam, the situation in Laos was so complicated the Pentagon might have seen it difficult to support a highly divisive system. This explained why Gen. Vang Pao and his tribalist faithful guerila units had become so handy. **They were more united and, therefore, reliable!**

  The Soviet Union intervened by airlifting PAVN (North Vietnam Army) heavy weaponry and gunners to assist the new alliance of the Pathet Lao Army PLA and the Neutralists against the Royalists. The "war" moved nearer to the big city, when the U.S. B26's were brought in to bomb Vientiane. This secret war continued to brew and swell up; nevertheless, it was never recognized by the U.S., and the American people would never know about it.

  The anti-Royalist groups were pushed by the People's Army of Vietnam PAVN (NVA) artillery shelling into the high land (near Plain of Jars) in RM#2[340] bordering the Northwestern part of the DRV where the Miao tribes were located. **This region, originally under Col. Khamkong Bouddavong's command in the 1960s, was strategically transferred to Gen. Vang Pao,** as the Miao lived in RM#2 and had control over the mountains that were infested by the Pathet Lao.

---

339 https://www.airspacemag.com/military-aviation/ravens-of-long-tieng-284722/
340 https://en.wikipedia.org/wiki/Military_Regions_of_Laos#Military_Region_2

Hence, territories were set up for bloody battles, with the expanded PLA resupplied by the Soviet and reinforced by NVA troops. The RLA started receiving fighter planes from the U.S. and Thailand, starting with the "ancient" T-6 Texans reconverted trainers in 1961 and moving upscale to the more agile T-28 Trojans (also used in the RVN Air Force during the same period) in 1963. In the meantime, the Geneva Conference imposing Peace to be brought back to Laos forced both the Soviet and American advisors to leave Laos[341]: the "war" shifted to Việt-nam and Laos War became an American underground operation.

The internal skirmishes between different Lao generals mounting their coups to overthrow the government had kept the "Forces de l'Armée Royale" busy fighting rebels while at the same time dealing with PLA insurgence in the mountains. It had become necessary for the U.S. to bring in B52 bombers to carpet the PLA positions and encampment: the **Barrel Roll** areas (1964, Plain of Jars). This was where **Gen. Vang Pao's army** served the best to rescue downed American pilots and help pinpoint targets for bombing in RM#2 (Barrel Roll, see next map), besides fighting the PLA and protecting the CIA-run Long Cheng Base with its Radar. RM#2 (Hmong) more or less buffered lower half of RM#1 (controlled by the **Royal family, King Sisavang Vatthana**, and the Royal Lao Army) from the heat of PAVN attacks in the Barrel Roll free-bombing zone. RM#5, with the Capitol Vientiane, controlled by **Gen. Abhay,** was sheltered and protected as it was surrounded by the Thailand border on the west, the Royal family's RM#1 on the North, and Gen. Vang Pao and CIA's RM#2 on the North East side (see earlier Map of Laos and its 5 Régions Militaires).

The "Trail" in Regions 3 and 4

After multiple coups and counter-coups mounted by many competing generals, Royalist Gen. Phoumi Nosavan was overthrown. This was followed by the birth of the Hồ Chí Minh Trail in the Lower Land RM#3 (Southern by Trường Sơn Mountains) which was under the control of the Insixiengmay

---

341   https://www.airspacemag.com/military-aviation/ravens-of-long-tieng-284722/ CIA-operated Long Tieng (or Long Cheng) Air Base

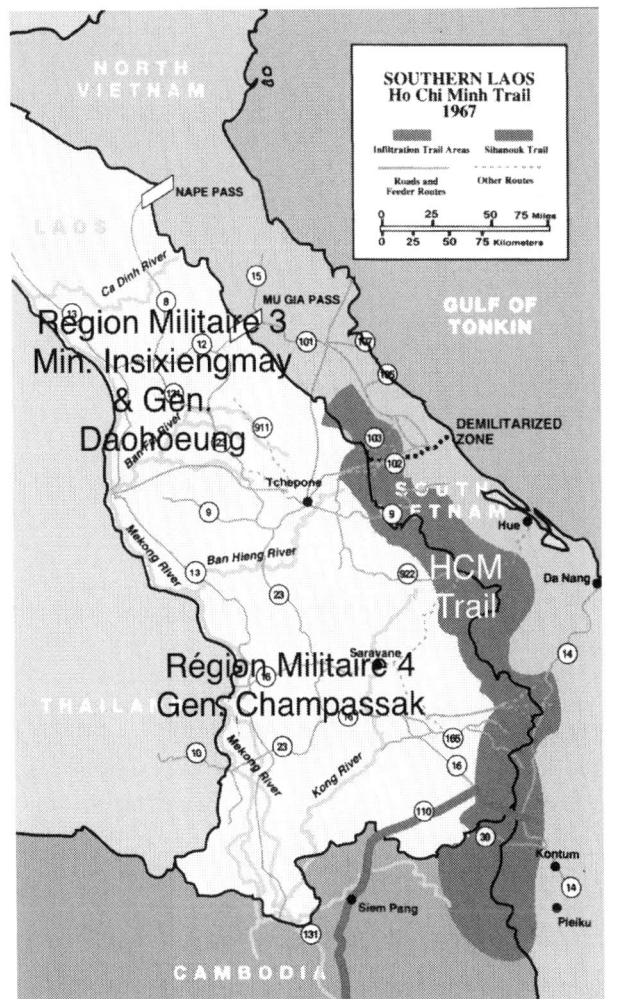

Laos: Hồ Chí Minh Trail in Région Militaires #3 and #4
Source: Hồ Chí Minh Trail en.wikipedia.org

family[342] (Royal Lao Minister Leuam Insixiengmay) and assigned to RLA Brig. Gen Nouphet Daoheuang. This renowned Hồ Chí Minh Trail would eventually lead the PAVN (NVA) and their supplies through RM#4 (controlled by Prince Boun Oum Na Champassack's family) into neutral Prince

---

[342] https://en.wikipedia.org/wiki/Military_Regions_of_Laos#Military_Region_3 and https://en.wikipedia.org/wiki/French_Protectorate_of_Laos#Conflict

Sihanouk's Cambodia [343] (Sihanouk Trail in Cambodia) where the NVA troops could freely access South Vietnam.

The ARVN commandos formed in 1960-1961 by MACV-SOG were launched across the border onto the trail to plant electronic listening devices which would trigger strategic bombing to destroy the incoming PAVN convoys. Bombardments escalated to major sorties operation in RM#3 and RM#4 (**Tiger Hound** and **Steel Tiger**, similar to the ones in the **Barrel Roll** area in RM#2, but much harsher) beginning December 1965 by the U.S. (B52 and jet fighters were employed), the Royal Lao and the RVN (jet fighters) air forces. They eventually expanded into B52 carpet bombing (1966-1972) on the border of Régions Militaires 3, 4 and the Southern part of DRV. Millions of tons of bombs ravaged both Barrel Roll (against PLA, in CIA and Gen. Vang Pao's RM#2) and the Tiger Hound (against NVA on the head of the HCM trail, South of Vinh in Việt-nam, in Minister Insixiengmay and Gen. Daohoeung's RM#3 and Gen. Champassack's RM#4, see map on previous page), while the civil war spread out in Northern Laos in short but sometimes intense outbursts back and forth between the Plain of Jars and Luang Prabang. The Royal Lao Army/Air (mainly) and ARVN clandestine forces (not Gen. Vang Pao's Miao SGU's, as some sources had claimed) had spent a great deal of efforts on the trail in RM#3 and RM#4 against the PAVN movement along this trail between the early 1960s and 1975.

What Happened in Việt-nam was "trailed" from Laos then through Kampuchea

The prelude to the 1968 Tết Offensive in Việt Nam happened in Laos first, when the NVA flooded Nam Bac with many divisions of their main forces. Prior bloody battles had reduced the Royal Lao Army down to one third of full strength. Nevertheless, the latter recovered and continued battling the Pathet Lao Army PLA and PAVN with the support of the U.S. and RVN forces from across the border. Later in 1971, when the U.S. and all allies had started withdrawing their forces from

---

343   Cambodia, as called by the French, was changed to Kampuchea under Lon Nol. Khmer is what the people prefer to be called.

Việt Nam, the ARVN launched divisional Lam Sơn 719[344] from the RVN I Army Corps to assist the RLA combat the NVA, and, especially, to cut off the HCM Trail as well as destroy NVA supply route and troop convoys. Unfortunately, this campaign turned disastrous because of ARVN commander Gen. Hoàng Xuân Lãm's operational incompetence and intelligence leaks to the press media (New York Times) via many stages of ARVN I Corps security breaches[345]. The polarization between Pres. Nguyễn văn Thiệu and VP Nguyễn Cao Kỳ had caused major shaky coordination between I Corps Commander Lt. Gen. Hoàng Xuân Lãm versus Airborne Lt. Gen. Dư Quốc Đống and Marines Lt. Gen. Lê Nguyên Khang, thus allowing the NVA and VC to take the upper hand. Quite possibly, ill feelings about the allies' "betrayal" had caused even more break down between the ARVN high commands and their execution of MACV plan as developed by Lt. James Sutherland[346]. American reports listed between one tenth (source 1) and nearly half (source 2) of the 20,000 participating ARVN troops being killed!

As a test to verify whether or not the U.S. and allies would return to help the RVN (per Gen. Bùi Tín, when discussing Easter Tide Offensive of 1972), the DRV continued to pump more NVA divisions into South Việt Nam for the bloodier Easter Offensive of 1972 (DRV called it the Nguyễn Huệ Offensive). As soon as the Paris Accords were signed in January 1973, the U.S. brushed its hands off of, not just South Vietnam, but also Laos and Kampuchea as well, letting the "big red Hulk" pump up his muscles (with China and the Soviet continued substantial aids) to smash the remnants of the unsupported Republic of Vietnam, Kingdom of Laos and the Republic of Khmer. The war was all ours now, **fighting poorly armed against well-equipped and supplied enemies!**

The consequences for both nations, Kampuchea

---

344 By Military: https://en.wikipedia.org/wiki/Operation_Lam_Son_719#Counteroffensive in English.

345 As researched by **Nguyễn Kỳ Phong of ARVN Strategic Technical Directorate commando groups:** ***Hành quân Lam Sơn 719: Nguồn gốc và khuyết điểm*** http://qlvnch.blogspot.com/2008/06/hanh-quan-lam-son-719-nguyen-ky-phong.html sources and weaknesses, in Vietnamese. Secrets of operation plans were discussed by Lt. Gen. Hoàng Xuân Lãm in public. Because of his incompetence in commanding, Lãm was to be replaced by Lt. Gen. Đỗ Cao Trí, commander of ARVN divisions in Kampuchea incursion (1970), unfortunately, the latter was killed while being transported out of Kampuchea to his new assignment.

346 https://military.wikia.org/wiki/Operation_Lam_Son_719

and Laos, might not have meant much to either Nixon and Kissinger or the Pentagon, but were unequivocally tragic for the innocent people who this American Presidency and its generals had dragged into a violent and deadly tempest. Rapidly, Kampuchea fell from the peaceful Sihanouk reign into a war torn nation which resulted in the devastated Killing Fields atrocities committed by the atrocious murderous Pol Pot regime (1970-1978). Meanwhile, Laos, also a more or less peaceful Kingdom, ended up lingering in international bloody messes they did not even start! The peaceful Miao tribes, rendered as a nuisance to the Pathet Lao by the CIA, were left behind, disarmed, like the ARVN, to be retaliated upon and punished by the conquerors seeking revenge. The most unfair situation was that the voiceless people of these two nations had absolutely no say in the Paris Accords Negotiations, where their fates had been decided by whom they did not even know: strangers by the name of Lê Đức Thọ and Henry Kissinger.

Kampuchea Pres. Lon Nol resigned on April 1, 1975, and was evacuated by the U.S. while most of his cabinet had refused American evacuation and was executed by the Khmer Rouge. Not long later, after the RVN was flooded red in April 1975, Laos King Savang Vatthana abdicated on December 2, 1975 and the PLA took over, and Prince Souphanouvong became president. Gen. Vang Pao had already fled with his 3,600 SGU fighters starting May 13, 1975[347]. By the end of 1975, nearly 40,000 Miaos made it across the Mekong River to Thailand to restart their life like the phoenix being reborn, the free Hmong. Likewise, the Lao civilians crossed the river into Thailand, heading for refugee camps.

Is Vietnam War a Civil War or Second Indochinese War? Neither!

It's in the eyes of the beholder! On the one hand, when thinking sequentially in chronological order and in geographic

---

[347]   https://en.wikipedia.org/wiki/Laos_Civil_War and Book *Sky is Falling by Gayle L. Morrisson (verified CIA oral history)*

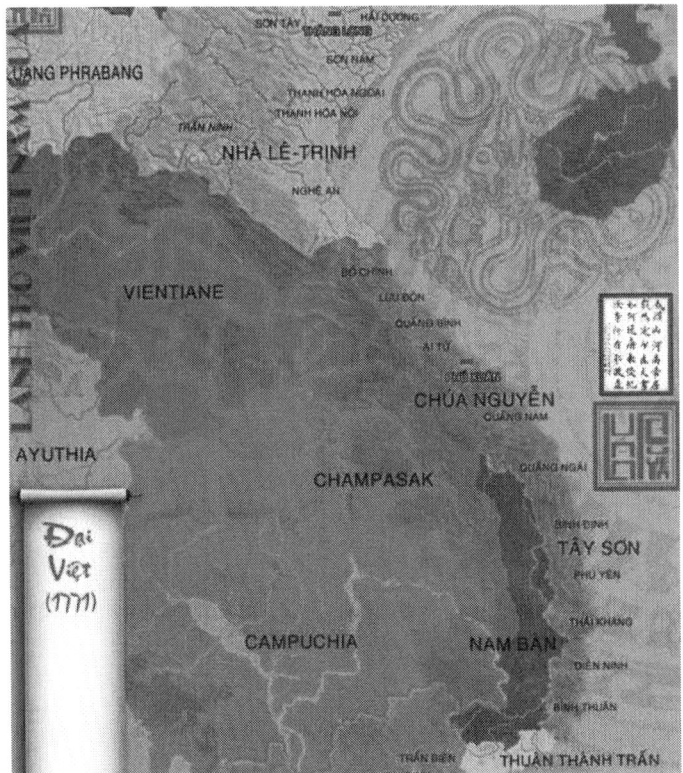

Đại Việt in 1771 being divided into North (Lord Trịnh and King Lê) and South (Lord Nguyễn and King Tây Sơn of another new Nguyễn Dynasty)
Source: https://nghiencuulichsudotcom.files.wordpress. com/2016/07/tayson1.jpg

terms, if we called the Vietnam War "the Second Indochinese War," then the U.S. had violated the Geneva Conference that strictly forbid any political power to intervene in the politics of the old French Indochina. On the other hand, when thinking in anthropological terms, if one called it a Civil War, many Vietnamese Southerners would object to it as North and South had been separate identities, separate sovereignties since the XVIIth Century (as discussed earlier in the section about the Lords and North/ South separation).

Many had argued over the term Indochinese War when they wanted the war to be more inclusive, involving not just Việt-nam, but also Laos and Cambodia. For some Vietnamese, who might have lived a better life during the

French occupation, the terms Indochinese War might be used, but not for others, who thought more in terms of the North and the South being two separate sovereignties since the long war between Lord Trịnh (North) and

Lord Nguyễn (South) in 1627 and 1672, then 1774 and 1775AD[348]. It would be important to note that both Trịnh and Nguyễn Kingdoms were subsequently destroyed by the brothers Nguyễn from Tây Sơn village (led by Nguyễn Huệ, who later on became the great King Quang Trung). This newer identity of the South had given birth to a new feeling of patriotism. Since Lord Nguyễn advanced South in the 1600s and annexed Champa to form the current South Vietnam, a new nation had been born combining both Vietnamese and Khmer (the Khmer-Krom) cultures. The people of the South always considered themselves unique and wealthy having fertile land, as much as those in the North where people considered themselves superior being of the original culture of An Nam. As a result, psychologically, it would be politically inappropriate to refer to the war as a Civil War: they were not the same Vietnamese people in the same country.

The reason why HCM had always maintained it was a civil war was that he wanted to show to the world the U.S. was in the wrong as it was meddling with the internal struggle of the Vietnamese people. It was Nixon who had inadvertently or purposely used the Civil War rhetoric later in his first term in the early 1970s as a pretext to withdraw from the war, indirectly supporting HCM's claim legitimacy when he waged war to liberate the South.

This divisive attitude among the two Việt-nams had actually helped the French implement their "divide and conquer" strategies when they arrived in the 17th Century. Champa-conquered Eastern Cambodia Cochinchina had eventually been acquired, this time by the French. The latter had created the three parts of French Việt-nam (see map), thus maintaining the divided status quo between the "supposedly Vietnamese[349]compatriots. Most importantly, those who believed that American intervention was legitimate and that

---

[348]   https://en.wikipedia.org/wiki/Trinh_Lords in English. https://en.wikipedia.org/wiki/Nguyễn _lords in English, and https://vi.wikipedia.org/wiki/Trịnh-Nguyễn_phân_tranh: Civil War between the two lords Trịnh and Nguyễn during the Mạc Dynasty - in Vietnamese

[349]   At the end of the war, many Southerners had chosen to revert back to Khmer ancestry, while many others maintain Vietnamese citizenship but honor Khmer traditions. Many Champas also maintain their traditions.

the U.S. was truly helping the South achieve its sovereignty as a country politically independent and free from Communism, should have second thought about the term "Second Indochinese War" being attributed to the War in Việt-nam, Laos and Kampuchea. If it were an Indochinese War, the Geneva Conference ruling would have a reason to sanction the U.S. for meddling in these newly liberated sovereignties as well as for violating the clauses of the agreement.

Hồ's team had always propagated, even until the 2010s, the rhetoric that the U.S. was invading and that its sole purpose was to take over the Indochinese colonies from the French. This rational had enabled him to incite people's patriotism and instigate insurgence among the peasants, most of whom (50-80% of the population) were docile and more or less illiterate, and whose heart was filled with hatred against invaders.

Finally, I would say both "Civil War" and "Second Indochinese War" should not be used to qualify the War against Communist Aggression in South East Asia. One has to be extremely keen in psychology and be investigative about this "jeux de mots" (in French: word game) in order to see the game "they" have played with wording to control people's sentiments. **Linguistically-speaking, the Northerners might excel in this field!**

After the French were defeated in 1954, the migration of one million Northerners to the South (Passage to Freedom) had caused a clash between the children of the Trinh and the Nguyen. The Northerners boasted they were the "original Vietnamese" with superior quality and culture[350], while the Southerners that they were the hosts with wealth and opulence which allowed them to always be easy-going and comfortable. Nevertheless, both groups in the Republic of Vietnam RVN (including those in Central Vietnam who were probably the most persecuted by the communists for having a long legacy with the Empire[351]) fled together in search for Freedom and Democracy when the war ended in 1975.

---

350 Linguistically, traditionally, artistically, educationally…
351 Communism hates the Empires

CHAPTER 10

In this Ball Game, Whose Side was the Press Media on?

Unlike WW2, when the news that was considered to be strategically important and that was more or less restricted to need-to-know-only, American people in the 1960s and 1970s were kept more informed of what was going on at the front lines in Việt Nam, whether it was strategic or not, thanks to the daily news updates viewed on television in their living room at home (U.S.). For that reason, news reports that could be detrimental to the success of the war were released, **in the name of Freedom of Speech and the Press,** as they could highly affect military success and warfare security[352].

TV viewers could see war scenes as if they were in a theater watching an action movie in the comfort of their living room. From that point, it was up to their imaginations which, I presumed, had fed anxiety and fear onto each other. Psychologically speaking, the Snapshot and Goleman's Emotional Intelligence theories (as previously discussed), if combined, would explain how upset viewers could be watching those news. The more their emotions brewed up, the more the rivaling news media companies would exploit more sensational ones to beat the opponents, oftentimes with incomplete and instigative rendered information. The same events that had always existed in any past war, e.g. carnage, innocent victims being massacred, or soldiers suffering on the home front… could be sensationalized to the extent they could possibly turn war supporters into a major antiwar opponents. I believed that they actually did! Information had been presented to viewers **selectively** with vivid images and sound which could easily mislead viewers by emotional manipulation. When emotions reached the unbearable limits, the viewers'

---

[352] I have been so shocked nowadays to find out a lot of strategic information being exposed and discussed on the news when or even before a campaign was launched in Middle East during the war against ISIS. How much democracy can we afford without putting our men and women in harm's way?

"Uncle Hồ" Getting Ready for the Kick
By H. Tuong

reptilian brain could not take it anymore, especially when there might be a loved one involved: they rebelled and reacted strongly by fighting back (Goleman's Emotional Intelligence Reptilian Brain's Fight or Flight). It was very simple to understand!

Given two massacres that occurred during the 1st phase of the Tet Offensive of 1968: one happened at the end of January when the communists violated a truce they had agreed on for New Year celebration, and the other one a couple months later. The first one was committed by the VC who had tied the victims up and struck the back of their skull with hoes, or buried, amputated, and beheaded 7,800 innocent civilians of all age when they were still alive. Children were smashed against concrete floor or walls or just stabbed, most of the time in front of parents... The Huế Massacre was more or less ignored by the multiple media networks present. The second massacre happened in Mỹ Lai where 504 villagers were massacred by the U.S. Infantry troops in patrol. The troops were attacked by snipers hidden in tunnels; they lost it and started shooting down anybody in sight, regardless of ages and gender. They happened to have moved out of the strategic hamlets the RVN had built to protect and moved into enemy-controlled zone. The Mỹ Lai incident, on the contrary, was fully covered by media of all types. It had become the only one currently available in English on the internet and in audiovisual documentaries. It was an iconic event that precipitated even more adverse reaction from the television viewers. What would their reaction have been if the Huế massacre received equal treatment by the media, I had been wondering? It was not as momentous as the Holocaust,... but, come on, imagine 8,000 innocent lives being wasted in less than a couple of weeks They were tied up like fish on a chopping board awaiting the club to chop off their head!.

  Meanwhile, the enemy had analyzed these psychological effects carefully, as they had followed up, researched, judged and acted accordingly via their informants in the U.S.… It was the "knowing our foe, knowing ourselves" game they always played. They had "eyes and ears" infiltrated among antiwar protesters, and (who knows!) in the news media or even U.S. government (sympathizers). They knew that divide-and-conquer always worked as they had experienced it with the French during the Indochina years.

  In the meantime, because of emotions that could not be processed rationally, especially when snapshots kept flooding the television screen every day, Americans had fallen victims of their own divisiveness. Pres. Lincoln had warned before that "the house that is divided against itself cannot stand." This situation was in favor of the Democratic Republic of Vietnam DRV. The more the communities were panicking and rebelling in the U.S., the more the DRV Lê Đức Thọ negotiators in Paris and the enemy-fueled antiwar movement in South Vietnam made advances, and the more the White House alienated the RVN, and rushed it into the dead end.

  In this war, support from China, Soviet and the Communist Block (Cuba, Romania, Tchekoslovakia, Bulgaria…) for the Vietnamese Communist Party at the front line was crucial, of course, but, truly, the one that Hà-nội had counted on so much was the "rear" ("hậu cần"): the protesters in Việt Nam and the antiwar movement in the U.S. and the rest of the world. **"Hậu cần" was the weapon of mass destruction that would cause the U.S. to collapse on its own,** as "Uncle Hồ" and his generals had later on proclaimed.

Who could have believed that our own "supposedly rear support" in the U.S. as well as the one in the RVN were destroying us? Could it be possible that the U.S. might have lost the war in Europe and Asia during WW2 if this had happened then? This might be the main reason why the U.S. had always pushed for a RAPID WAR in Việt-nam similar to the one in WW2, so that its "sensitive citizens" would not have enough time to "wake up, feel and counteract."

What follows has to be repeated over and over, so that it sinks into our brain and impregnates our DNA:

NVA Gen. Bùi Tín, the DRV news media official and, by extension, the strategy analyst who was working for the official People's Press Báo Nhân Dân in 1975, strongly believed that **the antiwar movement in the U.S., that was supposed to be the American "rear," was actually supporting its national enemy, the DRV Democratic Republic of Việt-nam (also called SRV Socialist Republic of Vietnam or North Vietnam), by literally destroying its own government (the U.S.) and a few million of the people of the Republic of Vietnam in the name of Peace and Humanities for Vietnam.** *The general further explained that this movement had played an essential role in the Vietnamese communists' victory. With a satisfied smile on his face, during many press releases, he had sarcastically commended the uplifting visits to Hà-nội by public figures such as actress Jane Fonda (as nicknamed by Vietnam Veterans "Hà-nội Jane[353]" after her two-week "tour" in Hà-nội), or former Attorney General Ramsey Clark[354] (visiting North Việt Nam on 8/14/72), or the political "stirrer" Tom Hayden and some other Ministers who, to the people of the DRV, had represented the anti-war "Conscience of America." He especially commented that North Việt Nam was* **"elated when Jane Fonda, wearing a red Vietnamese dress, stated at a press conference during her Hà-nội visit that she was ashamed of American actions in the war and that she would struggle along with us.[355]"** *On the People DRV radio, Fonda denounced*

---
353   https://patriotpost.us/pages/80 Traitor. "Hà-nội Jane" Fonda
354   http://www.history.com/this-day-in-history/former-attorney-general-ramsey-clark-reports-on-his-tour-of-north-vietnam
355   http://www.viet-myths.net/buitin.htm Gen. Bùi Tín was interviewed by Dr. Stephen Young, former MACV advisor in Saigon, attorney and Dean of Hamline

against her very own country being "US imperialism," using the same expression the Vietnamese Communist Party had been propagated throughout the war, while she had the decency and audacity to visit American POW's who were, at that time, incarcerated and tortured at Hà-nội Hilton's.

Gen. Bùi Tín and Army General Võ Nguyên Giáp, the renowned People's Army of Vietnam (PAVN or NVA) supreme military commander, had thanked especially the Anti-war movement that had facilitated North Việt Nam's victory. **"Those people represented the Conscience of America. The 'Conscience of America' was part of its war-making capability, and we were turning that power in our favor. America lost because of its democracy; through dissent and protest, it lost the ability to mobilize a will to win,"** *Tín concluded.*

The Beer Venders set Fans up against their Own Team.

Families that had loved ones fighting in Việt Nam were always anxious to find out what was happening there and whether their relatives were put in harm's way. As coined by CNN in their television documentary series on *"The Sixties"* and *"The Seventies,"* television was gaining its popularity as a new technology[356] that could bring viewers right to the battle field as well as the turmoil that was ravaging the RVN. Unlike WW1 or WW2 (including Korean War), Vietnam War was the first war in which news media networks were allowed to embed reporters with deploying troops. As reported by Minnesota news network anchorman Don Shelby in 2019 in a conference, hundreds of networks were vigorously competing against each other in efforts to report the top priority and the most sensational news. It was like a "survival of the most intriguing" phenomena.

In order to achieve the level of ambiguity needed to attract viewers, the networks did not really think about long term effects of their news on human emotions or on the success of the war campaign. Did they really care? On the other

---

University, St. Paul, Minnesota.

356   *Television: The First Television War by* www.americanforeignrelations/O-W/Television

hand, were any reporters allowed to embed in ARVN troops? Probably not! Due to military and intelligence restrictions, the RVN had its own government reporters that were allowed to reveal only "strategically safe" information; consequently, the world was well sheltered from RVN actions, and therefore, people might think the ARVN did not do enough, or anything at all.

>...NBC frontline news reporter Bill Ryan had reported on U.S. National television, in his own words, when he was covering Charlie Company[357]'s action in the Mekong Delta in mid-1967 that "...Americans do most or all of the fighting in I, II and III Corps areas. The IV Corps South of Sài-gòn area in the rich Mekong Delta is the responsibility of the South Vietnamese Army... There has been talk recently that American units will be deployed in the delta[358]. The terrain is difficult and the South Vietnamese army has had only limited success... Until now only, Vietnamese forces have tried and largely failed around the Việt Cộng in their strong hold..."

If American press was not allowed to embed in the RVN units, how did Ryan come up with such information insinuating to hundreds of millions Americans at home that the ARVN were not fighting or, if they did, that they were failing badly? What would the American people at home think **about their children and spouses risking their life to protect these "incapable Vietnamese?"** Of course, they would be all up roaring and rebelling! Besides, what would they think about the RVN abandoning positions in I and II Corps and running in chaos as soon as the U.S. troops withdrew? Obviously, the National Geographic "Brothers in War" did not show any shadow of ARVN soldiers fighting bravely! If they had not fought bravely and successfully, Quảng Trị would not be retaken and all roads in II Corps[359] would not be cleared and cleaned out so that I, a new 1973 grunt cadet, could travel safely home by unescorted public bus...

---

357   National Geographic documentary Brothers in War on the Mobile Riverine Force MRF Charlie Company of the 9th U.S. Infantry Division operating in Mekong Delta. This film had quoted some sections by NBC reporters. History Net will give more information about Charlie Company http.//www.historynet.com/review-the-boys-of-67-charlie-companys-war-in-vietnam.htm

358   This would be leak of military secret out to the public.

359   At that point in time, I did not know about I Corps security status, but I would definitely be warned by the military authorities if it were not safe when receiving my pass. I travelled unarmed and unescorted.

In my opinion, it was America's over-expanding democracy and freedom of speech that had opened up some cans of worms! During the war, the enemy had certainly loved those worms: they had served their purpose really well! As a matter of fact, this had constituted one of the incompatibilities in warfare between the U.S. and the RVN: the democratic "reveal-everything" policy on the U.S. side versus the totalitarian un-democratic "censorship and limited freedom of press" rules on the RVN side[360].

The truth of the matter was that the ARVN (with no American or allies press embedded) and the U.S. forces (with international press embedded) did not, or could not launch joint campaign, and, probably, had different operational policies and philosophical opinions. Did the RVN bar the American and allies' television press media from embedding reporters in ARVN in the years 1964-1975? If they were not barred, all RVN news and military secrets would be exposed to the enemy's eyes and ears, as the Americans' ones were. Obviously, the RVN would not be able to censor American news as they could with their own Vietnamese reporters. Another nuisance caused by democracy!

Nevertheless, I no longer saw footage of news reels showing side by side operations, as we did of military campaigns before 1964 or 1965. In most of the current documentaries, even though Americans and RVN troops might be participating in the same major campaigns, such as the 1968 Tết Offensive, not much information of RVN post-1965 tactical moves was mentioned, as if they did not fight at all, until the new documentaries that were aired in the early 2000s. **Was there a television press media conspiracy against the RVN?** *It seemed so obvious that the media was leading the public to an understanding that the RVN was either cowardly in their actions (as some network had accused) or that it did not even fight its own war. On the contrary, even recently in 2017, they were making the public believe the enemy was the invincible patriotic heroes.*

Even though RVN commanders and American advisors had tried to maintain good communication, still, they might not be able to cooperate as the allies did when they fought side by side, or at least together in the same campaign, as in WW1 or 2 (as depicted in movies, books and websites). No wonder

---

360   The U.S. imposed the condition that the RVN fight the war exactly in the same way the U.S. did in order to receive military aid.

why this had made the public in the U.S. confused when they were watching their "living room news." The U.S. had won WW2, so the people obviously would still be living in that WW2 paradigm that all allies had coordinated well their action and could work well together. It was not the case of Vietnam War. New snapshots appeared and manipulated the television viewers and paper readers' mind: Where were the ARVN? Why were our loved ones the only ones fighting their war? They did not fight until Vietnamization in the early 1970s...

After all, the taxes Americans had paid were used to build up and support an ARVN of 820,000 soldiers (1960s); it would definitely anger people seeing how their taxes were used, and how their children were suffering uselessly, while the ARVN was not doing anything. Where and what were these great ARVN Infantry, Airborne, Marines, Special Forces, Regionals, Artillery, Cavalry, Air Force, Navy,... doing while the U.S. was sacrificing some 58,000 men and women?... Wouldn't that be so upsetting, and unnerving thinking the ARVN was cowardly and incompetent? I would certainly be enraged myself if I were an American parent or brother then thinking my country was sacrificing for an unfit nation! Of course, Americans would trust the press totally: journalism was supposed to tell nothing but the truth and the total truth, wasn't it? A year later, **Ryan's colleague Ron Nesson added: "Many Americans are confused by conflicting reports about the situation in Việt Nam."** *Of course, the RVN had no say about it... because South Vietnam was voiceless in this war!*

A Los Angeles Time newspaper, as shown in the same National Geographic documentary, had filed an exaggerated report that the VC had disastrously destroyed Charlie and Alpha Mobile Riverine Companies and had lost 134 men in their July 19, 1967 joint campaign. The important missing piece was that they had also taken out two VC companies (250 VC). The interviewees from the original Charlie Company sadly commented that ...

..."such news is made sensational... so the news media can make a name for themselves..." Television, which was a new industry that was supposed to bring out the truth in the name of journalism, had to compete against each other by falsifying or selectively choosing usable news at the expense of soldiers' morale and war outcomes. The more I watched television documentaries from the 1960-70s, the more I realized

that daily news reports on television and newspapers at that time had played an extremely important role in the defeat of the nationalists, the U.S. and allies. Many analysts had also confirmed this. Whether they did it on purpose or not, the press had actually back stabbed their own soldiers, and had caused America to abandon its own faithful allies. Charlie Company 2nd platoon team leader Bill Reynolds[361] commented: **"It was an eye opener that America is not with us! The media had played the information in a wrong way, to make it look like Việt Nam was in the wrong and the only one at fault. They only focused on the negative and reinforced the American people with the worst of Việt Nam[362]..."**

How could the television Press Media determine Win or Lose?

The television press media had influenced public opinion in a negative way: information was left to be interpreted by deduction by spectators. It had consequently enhanced the fear and anger viewers and readers were experiencing. This psychological phenomenon of group and mass horrification had caused the government to become more and more concerned. As "TV-news in the American living rooms" media gained more and more popularity (CNN "the Sixties"), the number of television news press corps grew by tenfold, from 40 in 1964 to 419 in 1965[363]. Sensationalizing the news had become worse and worse as the war intensified. The Tết Offensive of 1968, a clear cut victory, as evaluated by the U.S. and ARVN (and witnessed by myself), was reported back to the public as otherwise. By this time, there were over 600 press corps from all over the world, all being assisted by MACV (US Military Assistance Command in Việt Nam) which had provided them with transportation to the front line...The Tết Offensive was actually the turning point when many reporters no longer supported the U.S. involvement in Việt Nam. Walter Cronkite's, CBS most venerated anchorman and the "most trusted man in America", statement that the conflict has "mired in stalemate" was seen as the signal of major changes in the war. Pres. L B Johnson himself had

---

361   National Geographic Brothers in War.
362   https://en.wikipedia.org/wiki/United_States_news_media_and_the_Vietnam_War
363   http://thevietnamwar.info/media-role-vietnam-war/

stated "If I've lost Cronkite, I've lost Middle America.[364]" Was that the reason why he decided to not run for another term? Nevertheless, LJB must be under major stress and probably invaded by fear, the fear of being himself assassinated. He was perplexed about augmenting troop levels in Việt-nam only after a few years in power.

Pres. Johnson's rejecting Gen. Westmoreland's request for additional 200,000 troops in July 1967 was for the North the signal that the U.S. had already reached its maximal military commitment of the year. "We had stretched American power to a breaking point," as Gen. Tín asserted: "they had no more troops to send over." The three factors "Thiên thời, Địa lợi, Nhân hoà" (timeliness, geographic favor, people's harmony) had been achieved and were definitely in the NVA and VC's favor. Violating the New Year truce did not mean much compared to a possibility of deterring the Americans from entering Laos and destroying the trail, the vein that had been providing the necessary blood to the brain of the NVA and VC's war machine in the South.

Both Westmoreland and Tín admitted that the U.S. war policy in Laos against destroying the vital supply trail was another main reason leading to the nationalists' failure in the war. The U.S. was trying to avoid officially expanding the war outside of Việt Nam, as this move would cause further complication with resentful antiwar groups in the World. Instead, it had chosen a secretive and sneaky way by instigating the Secret War in Laos involving the Royal Lao government and the more or less "privatized" Miao army (not fighting for the Lao Royal government, but as a pseudo-mercenary army[365] serving unofficially a CIA entity via an individual member of the Royal Lao Army, Gen. Vang Pao), as well as another Secret War in Cambodia involving the new Republic of Kampuchea which was given birth by the U.S. Government.

Walter Cronkite, who was considered in 1968 to be one of the most popular reporters, as he had received many nicknames such as the "oracle of America[366]," "the most trusted

---

364 https://www.britannica.com/topic/The-Vietnam-War-and-the-media-2051426 and Walter Cronkite's 1985 documentary series on The Vietnam War.

365 The Hmong army did receive weapons and ammunition from the Lao Royal government in the 1950's at the beginning of the war when Laos was attacked by the People's Army of Vietnam.

366 https://www.nationalreview.com/2017/06/hue-1968-review-key-vietnam-

man in America[367]," or "Uncle Cronkite..." had gained a reputation for being a full-hearted anchorman of the Việt Nam front line. His top sensational breaking news and his devotion to delivering hot news to the viewers had helped him gain confidence and admiration among the people in the U.S. as well as in the world. His analysis of the war responded well to the public that needed compassion which they felt they could not get from the establishment, the compassion of family members that was devastated by the fear of losing their son or husband in Nam. Nevertheless, his quick and forfeiting judgment of the events of the Tết Offensive was doing nothing but giving the enemy major disadvantage: it actually helped the VC and NVA outwit his country!

> *...Who won and who lost in the great Tết Offensive against the cities? I'm not sure. The Vietcong did not win by a knockout, but neither did we. The referees of history may make it a draw. Another stand-off may be coming in the big battles expected south of the Demilitarized Zone. Khe Sanh could well fall, with a terrible loss in American lives, prestige, and morale, and this is a tragedy of our stubbornness there...[368]*

To me, it would be like a television press anchorman telling Gen. Eisenhower and the Roosevelt White House that "the American and allies would have no chance to break through the German strong hold in Normandy to liberate France, given the huge amount of casualties our units had suffered upon landing on the beaches there on June 6th, 1944." If American people were aware that over a million of their soldiers would be dying in Europe, they would have turned off the switch of the war machine even before touching the French beaches.

---

battle/
367   http://www.renewamerica.com/columns/washington/150603 , https://en.wikipedia.org/wiki/Walter_Cronkite , http://www.elliswashingtonreport.com/2015/06/04/uncle-walter-commie-cronkite/
368   Walter Cronkite's commentaries on the Tet Offensive of 1968 http://www.sonic.net/~scds/resources/US-History/1968_Walter Cronkite, Commentary on Tet Offensive.pdf (Feb 27, 1968)

Besides predicting American losses, Cronkite also threw a major blow at the capabilities of the RVN, his country's major ally, thus undermining it by poisoning the mind of the tax payers who had been funding this war against Communism.

> "...On the political front, past performance gives no confidence that the Vietnamese government can cope with its problems... For it seems now more certain than ever that the bloody experience of Việt-nam is to end in a stalemate..."

His conclusion had probably become prophetic as it had strengthened Nixon's and Kissinger's belief that the U.S. would end the war with a Peace with Honor negotiation:

> "...To say that we are closer to victory today is to believe, in the face of the evidence, the optimists who have been wrong in the past. To suggest we are on the edge of defeat is to yield to unreasonable pessimism. To say that we are mired in stalemate seems the only realistic, yet unsatisfactory, conclusion. On the off chance that military and political analysts are right, in the next few months we must test the enemy's intentions, in case this is indeed his last gasp before negotiations. But it is increasingly clear to this reporter that the only rational way out then will be to negotiate, not as victors, but as an honorable people who lived up to their pledge to defend democracy, and did the best they could.

To me, the most serious injustice against the nearly million South Vietnam who died in the war was Cronkite's team ignoring the horrendous massacre of the 7,800 innocent civilians committed by the VC in Huế. News about this major massacre could have made the mind of the frightened and confused citizens wake up and make them understand the U.S. and RVN were fighting to rescue and save innocent lives. Cronkite was there to cover the VC NVA violation of an agreed upon truce that would allow the Vietnamese people to celebrate a peaceful New Year of the Monkey. How could he ignore this massacre? It was so public as it had lasted a few months during the time he was there. He had purposely failed to report communist atrocities, and, on the contrary, had focused on the much smaller one committed to demonize the U.S. Infantry, all resulting in making the public enraged against the nationalist cause, in favor of the Communist Party? Why such a gigantic bias?

Many of Cronkite team's photos and films, such as the iconic one showing the U.S. Marines setting fire to peasant's huts, or other photographers' pictures of the "Mỹ Lai Village[369]" incident, had imprinted long lasting images in viewers' mind that GIs were baby killers, massacre monsters… and caused the American people to mentally massacre their soldiers returning from the war.

How would the GI's family feel when, one day, they saw their son or husband laying or removing land mines (e.g. the film about the Charlie Company maneuvering through the jungle, the explosive squad in action, or the death of a mine sapper…), and, the next day, they heard that he was no longer: he was blown up by enemy ordinance? Who would they hate… the press for delivering the news, the VC who had laid the mine, or the government for having sent their loved one to his death in a "non-sense" war? Not their trusted press, not the VC who were too far to reach! Obviously, they would blame it on their government and the incompetent country it was supporting.

I strongly believe such direct news coverage, maybe even uncensored in the name of Freedom of Press, should not be allowed during the much bloodier great wars. The series Presidents at Wars recently shown on television[370] explained that, during the U.S. Campaign in Europe during WW2, the only film clips, besides Army propaganda reports shown to

---

[369]   Photographer Ron Haeberle's iconic photos of Mỹ Lai Village massacre victims were widely used by the press media. Company C of the 23rd Americal Infantry Division was involved in this massacre.

[370]   Presidents at War 2-episode series on TV

the public by the Pentagon, were the 400 some movies made by Hollywood. Viewers particularly loved the series with Pres., then movie actor, Ronald Reagan. Watching war movies had become a favorite pastime, besides going to the bars or dancing places. Reagan was known for being excellent at telling stories of the "heroichome" boy soldiers who were fighting for a good national cause, rescuing and saving Europe and Asia. There was no embedded press media then, and, if any, reporters had to submit themselves to a strict censoring policy, for the sake of preserving military secrets. The Operation Overlord high command of U.S. Europe Invasion, Gen. Dwight Eisenhower[371], had specifically ordered strict secrecy. "The enemy is watching" and "they are listening" signs were posted everywhere.

On the contrary, there were between 500 and 900 different networks spread out everywhere during the late 1960's in Việt-nam reporting whatever truth they chose on their war coverage, such as Cronkite's of Tet Offensive as described above, to compete against each other for program sensationalism and individual reputation. While the American viewers were devouring Vietnam War news in the U.S., the enemy's eager and restless eyes and ears were scrutinizing and analyzing the American people's emotions and reactions, so their high command back home[372] in Hà-nội could assess their troops' success or failure in the South, and make adjustments to their tactical plans, as confirmed later by supreme NVA commander Gen. Võ Nguyên Giáp and NVA political cadre Gen. Bùi Tín. Hồ Chí Minh had pointed out, as his forces could not rely upon its human and fire power, "the People" could win only by outwitting the enemy: "Know our Foes, know Ourselves, one Hundred Battles, one Hundred Victories." "Knowing our Foes" did not apply solely to intelligence collected from the front line (using sympathizers and collaborators), but it did back at their home (U.S. or RVN), particularly when using enthusiastic but ignorant participants (inadvertent "blabbers") or infiltration. In this case, thanks to their news media. Making a move in this chess game is so predictable on the nationalist part!

As a result, Americans, who had been getting more and more anxious at home, would totally freak out seeing,

---

371 https://en.wikipedia.org/wiki/Operation_Overlord
372 Hồ Chí Minh and his political and strategic cadres Trường Chinh, Lê Duẫn,...

not Hollywood movies as they did during WW2, but real live captioned and interpreted "Nam" scenes every day in the comfort of their living room. They were bombarded by Mỹ Lai types of film and photos[373] and, as a reaction, condemned their sons and husbands for atrocities they had committed against "innocent civilians" who they wouldn't imagine could be a disguised enemy.

Did the American public and the world know that Mỹ Lai victims were living in unauthorized zone controlled by the VC and that those snipers killing GIs actually came out of tunnels in their village? Just a few weeks before Mỹ Lai, approximately 7,800 defenseless Huế civilians (real civilians, mostly students, teachers, priests, doctors, civil servants, housewives, infants…) of different generations were gathered by VC and NVA, muffled up and used shovels to kill their victims by striking blows to the back of their skull, then buried them in mass graves during the Tết Offensive of 1968[374]. Did viewers see these atrocities? CBS (Cronkite), ABC, LIFE magazine… were there covering news… and they should have informed their viewers for fairness sake! And the cans of worms kept getting opened up and the list kept growing longer…

Did news viewers in the U.S. know about the 2,791 innocent civilian victims of the Cao Đài religious group were ordered by Hồ Chí Minh (Việt Minh) in 1945 to be executed in Quảng Ngãi by "slow beheading" (slashed around the neck until beheaded), alive burying, shooting by firing squads or tying up and throwing into the ocean[375]? What about 30,000 to 80,000, many were even members of the Communist Party, were persecuted and executed in Quỳnh Lưu, Nghệ An[376] in 1956 for being land owners, as ordered by Hồ Chí Minh's People's Court (Land Reform Law); this causing hundreds of other thousands to rebel[377] against Hồ … and nearly 300 anti-communist montagnard civilians were also incinerated to

---

373   Corpses of elder, women, children, GIs burning huts…
374   https://en.wikipedia.org/wiki/Massacre_at_ Huế and interviews of Huế survivors in Minnesota by Partners of Defense Scott Walker and myself (June 6, 2017)
375   Quang Ngai 1945 massacre in Vietnamese language: https://sites.google.com/site/hanhtrinhvechanlydao/61-vu-tan-sat-tin-dho-cao-dhai-giao-tai-quang-ngai-thang-8-va-9-nam-1945
376   Quỳnh Lưu massacres in Vietnamese https://ongvove.wordpress.com/2011/11/11/ cuộc-nổi-dậy-quỳnh-lưu-nghệ-an-1956-2/
377   Quynh Luu, Nghe An 1956 massacre in Vietnamese https://ongvove.wordpress.com/2011/11/11/ cuộc-nổi-dậy-quỳnh-lưu-nghệ-an-1956-2/

death along with their whole hamlet, by VC in a flame thrower/grenade in a revenge attack in Đăk Sơn, Sông Bé in 1967[378]...

...Vietnam Veterans Bill Laurie reported a well-documented list of massacres committed by the VC NVA that he knew about in his June 25, 1965 St. Louis Post Dispatch article. He also succinctly pointed out how biased the news media, e.g. Los Angeles Times, were when purposely misleading their readers[379]...

I vividly remembered reading in the Sài-gòn newspaper that 48 civilians, including 18 Americans, were killed by VC bombs while eating in Mỹ Cảnh Restaurant in 1965[380]... and that from 50 to 100 civilians of various ages were bombed when crossing a Farm Fair bridge near Sài-gòn in the mid-1960s... or the dozens of passengers (multiplied by hundreds or thousands of repeated times that I remembered as they were reported in the Vietnamese news in Sài-gòn in the 1960s and 1970s) riding city buses... 100 villagers of Dục Đức, Đà Nẵng were massacred by NVA in 1970[381] with hundreds more wounded and 800 houses burned... Thousands of civilians, mostly women and children, were bombarded by NVA artillery while they were evacuating southbound from Quang Tri during the 1972 Easter Offensive... My lycée buddy Chí's house received a SAM missile on the roof, almost killing his whole family while they were sleeping. Shrapnel hit him on the face causing him to almost lose his eyesight.

Did the American news fans know terrorism during Vietnam War happened every day, as often as having our meals three times a day "như ăn cơm bữa?" If they did not know about it, it was because these pieces of news did not serve news media's purposes. To make their news sensational, reporters had done their best to bend to the mass interest: shut down the war at any price.

---

378   Vietnam Veteran Bill Laurie's report http://vnafmamn.com/VNWar_atrocities.html and Wikipedia https://en.wikipedia.org/wiki/Đăk_Sơn_massacre. Bill Laurie also reported major biases in the news media reporting atrocities (e.g. Los Angeles Times), praising the enemy and putting down the American and RVN. http://namrom64.blogspot.com/2013/12/hinh-xua-bac-cong-tham-sat-dak-son-song.html   RVN Archives and blog in Vietnamese.

379   Bill Laurie *Bias in Vietnam War Atrocities' Reports.*

380   Vietnam War atrocities in English http://vnafmamn.com/VNWar_atrocities.html

381   http://anhoaproject.co.uk/pdf/%20The%20Duc%20Duc%20Massacre.pdf and http://www.home.earthlink.net/~ducducvietnamfriends/an_unknown_massacre Dục Đức, Đà Nẵng massacre of 1970

Of course, these massacres were reported in Vietnamese by Vietnamese news media; as a result, the public in the U.S. and the world had even less chance to know even if it scrounged around for news. Nevertheless, the free world media could have had their multilingual agents do more research in Vietnamese... Perhaps it was not to their benefit to make such efforts because they knew the viewers or readers were not interested in Vietnam War any longer and would block out of their mind horrendous snapshots

. This also constituted a weakness of the RVN in making use of their propaganda machine, especially abroad. Perhaps the RVN did not know how to take advantage of these atrocities to show, **in English (or network with English-speaking reporters)**, to the supporting allies' "rear line" (such as the people of the U.S.) that they were fighting against evils and that **the people of RVN deserved the support of the free world, in the name of Peace and Humanities.** This was the reason why I had commented earlier that RVN psychological warfare was inadequate[382] as it had failed to expose to the world so many impunitive genocidal crimes against humanities committed by the communists! The weak were voiceless, and the voiceless became even weaker!

Interestingly, Mỹ Lai Massacre (committed by the U.S. troops during the heat of the Tết Offensive) seemed to be the only Vietnam War atrocity that was fully and frequently covered by the American, as well as world press and television media during the time of the war, or even on the internet in the present time. Today's website was populated with information to demonize the American G.I.s, even nowadays. Nevertheless, not many, if not none, of the public blog, not even the daring and graphic Pinterest[383], had any documentaries about the massacres committed by the VC and NVA that I had cited above...Whether directly or indirectly, purposely or unintentionally, the American press media had shared their seemingly-selected findings with the world to "villainize" their own soldiers and, at the same time, to give the antiwar movement a reason to condemn the war, and to "heroize and

---

382   Even nowadays, Vietnamese complaints are still written and posted in Vietnamese on the internet or social media. Why in Vietnamese and not in English if they want the world to know? Vietnamese people know about atrocities already, by word of mouth!

383   Pinterest display materials that include the most diverse and complete documentary photo collections of important events in the world, such as horrendous pictures of the WW2 Holocaust caused by the Nazis or the Nanking China executions of civilians by the Japanese

angelize" the enemy. Of course, the images of massacres shared (e.g. Mỹ Lai) were real, and the stories presented were true, so the networks had become increasingly popular thanks to a newly born info-war system. Nevertheless, the truth of the matter was that only a tiny side of the whole truth, the side that would benefit the most to the networks, was presenting to the angry public the demonized side!

In the meantime, the Vietnamese nationalists who could have used communist massacres in their psychological warfare propaganda to boost its cause did not really know how to exploit information: complaints presented in Vietnamese language in the Vietnamese press media were heard and seen only by the Vietnamese, the voiceless people. Even nowadays, reports of communist atrocities had been presented on websites or social media only in the Vietnamese language, so how could we expect the world to know and have compassion? Consequently, the freedom fighters had lost and continued to lose the information war. They had totally lost at the main battle field, as referred to by the communist as "the rears[384]" in both the U.S. and South Vietnam turf. **The "rears," not the frontline or battle fields, had actually determined winning and losing.**

Even the NVA had acknowledged their horrendous inhumane killings: As told by NVA 8th battalion recon squad leader Trần Đức Thạch[385], approximately 600 villagers of all ages of Tân Lập, Long Khánh were massacred just a few days before the Fall of Sài-gòn, when the NVA avenged their losses caused by the ARVN 18th Infantry Division as the latter was withdrawing from positions during its ordered strategic "repositioning[386]." Thạch was later jailed for revealing the truth. The surviving ARVN officers of Tân Lập were executed by the VC with a shovel to the skull, so that no one would stay alive to tell the truth. Of course, this horrible news was never broadcast on the living rooms television in the U.S., otherwise the Americans would understand why we fought against the communists! Until today, no information in English could be found on the internet and social media either about the 65,000 formally[387] reported deaths of former ARVN and RVN civil

---

[384] The opposite of front line, referring to the people behind the front line who are supposedly supporting their government and army.

[385] YouTube in Vietnamese: <u>Thảm Sát Tân Lập - Tan Lap Massacre 1975</u>

[386] In Vietnamese https://baovecovang2012.wordpress.com/2013/04/21/tham-sat-o-lang-tan-lap-xuan-loc-hai-trieu/

[387] The figure would be higher if all deaths were reported.

servants executed by the communists in the so-called "re-education" camps between April 1975 and 1983. No one would know about the secret disappearance of inmates, of course, as well as those who died of illness or prolonged injury after they were released.

Eventually, everything in the American living room television news seemed to show there was no good cause in fighting the war, as W. Cronkite had so boldly determined it to be "stalemate" after his 1968 tour of Tết Offensive. It could have been different if people in the free world had known how barbaric the enemy had been and why innocent should have been saved; on the contrary, they were graphically informed that their own troops were the inhumane "guys." No wonder why those at home watching the news had so strongly objected to the draft, refused to be sent to their guaranteed death, and insisted that U.S. troops be brought back home. The Americans had just discovered that Life was indeed beautiful, especially that people became more cognizant of Freedom[388], of their human rights and equality, and that their future should not be as bleak as it looked through the Vietnam War glass… It was exactly as if the U.S. was going through its own Cultural Revolution.

Much later after the war had ended, one way or another, Cronkite and Fonda had offered "some kind of" explanation about the well-known statement about the war in 1968 (Cronkite, not an apology[389]) or apology for having sided with the enemy and trashing ones' government and army. Along the same line, as her husband had attested, Jane Fonda[390] had also apologized, but she still maintained that she was right or did not regret what she had done during her visit in North Vietnam. No matter how much one explained or apologized, the harm had already been done and being superficially apologetic would not heal anything, as their mistakes would continue to affect the victims the rest of their life. If one were civil and humane,

---

388   A song that was performed at the Woodstock concert of Mid Aug 1969 said it all. The lyrics had one word repeated over and over: Freedom, Freedom, Freedom, Freedom,…

389   https://www.wnd.com/2018/01/cronkite-admitted-blowing-it-on-his-famous-commentary/#! Cronkite admitted he had over-stated when claiming the war being a stalemate in 1968, but thought he had nothing to apologize.

390   Hà-nội Jane, while apologizing and admitting the Hà-nội photo-op was a "huge, huge mistake," but still tried to find excuses, blaming on innocence and being taken advantages of by DRV press. http://www.vanityfair.com/hollywood/2015/01/jane-fonda-regrets-hanoi-jane Vanity Fair, http://abcnews.go.com/Entertainment/video/jane-fonda-apologizes-1972-vietnam-trip-28380629 ABC News

one would know that apologies meant nothing without any concrete redemption and restitution. Like termites, the pain would continue to gnaw into the victims' sub-conscience, and to subtly harm their rehabilitation back into the society from re-belonging through self-realization and actualization stages (Goleman's Hierarchy of Needs); besides hurting American veterans, those harms had also indirectly affected millions more lives in Southeast Asia; so, how could one expect a mere apology could help heal a whole generation?

Among the back-stabbers, there are also hundreds of thousands more who are ganging up with each other in the U.S. to bully returning veterans. Meanwhile, relentless misleading information continues into current days[391] to drive the Vietnam Veterans and, maybe, their next generation descendants into internalized shame. It is the new era of identity search, researching and fact-checking what their fathers and mothers have truly done in Nam that had made them evil.

The communities in the U.S. definitely need to initiate and propagate concrete social and rehabilitation programs such as public recognition, contribution to veterans' mental and physical health care, concrete support for survivors and gold star families, etc... Even the government itself has not concretely honored veterans and recognized their sacrifice during the war until the first "Welcome Home Vietnam Veterans[392]" in 2009[393]. Consistent nation-wide and substantial events ought to be celebrated and commemorated on a yearly basis; otherwise, it will just be plain hypocrisy! **For that betrayal, "we might forgive you, but we would never forget. If you are really repenting, do something to prove it!"**

---

[391] Some network is still propagating a mysterious agenda to glorify surviving VC and NVA while showing pitiful U.S. troops and coward and incompetent ARVN fighting in vain. How will the latter's descendants feel about such misleading images? Atrocities committed by the Communist are still hidden, of course!

[392] Vietnam Veterans refer to American troops who have served in Vietnam War. This might extend to other ally countries such as Australia, Rep of Korea, New Zealand…, but does not include South Vietnam, Kingdom of Laos or Republic of Khmer. Politically-speaking, VN Vets are American G.I.s.

[393] In 2009, Minnesota and California were among the first states that enacted March 30 as the Welcome Home Vietnam Veterans day. Nevertheless, it seemed to be celebrated a few times, then fell back into forgotten memory. Observation has been inconsistent throughout the country http://www.ncsl.org/research/military-and-veterans-affairs/vietnam-veterans-day-legislation-and-statutes.aspx

Who else is the Enemy?

So far, we have been describing the enemy as the Việt Cộng VC (Mặt Trận Giải Phóng Miền Nam or the National Liberation Front NLF), the North Vietnamese Army NVA "Bộ-đội Bắc Việt" of the Democratic Republic of Vietnam) and their allies (China, the Soviet Union, Cuba as support, Pathet Lao Army PLA involved in a parallel war and Khmer Rouge towards the end of the war). Furthermore, if we look through Sun Tzu's lens "knowing our foes, and knowing ourselves,…" we will be surprised.

The nationalists have usually looked closely into "Knowing Foes and Knowing Friends" while collecting intelligence; however, they might have failed to look at themselves, and to see if they are really fit to lead. If one is not fit, one should elect and support those who are, even those who belong to a rival party. This wisdom is vital to any success in war, as preached by Sun Tzu. He has even said "We should keep our friends close to us, but we should keep our enemy even closer…[394]" The enemies of the U.S. in this war are not just the Northerner (DRV) and Southerner (VC) communists in Việt Nam, but they are also Americans in its own backyard. These people have been misled by the media and by their own government. For the RVN, the enemy can be the South Vietnamese citizens' closest friends and neighbors, the people of their own kind (clan, party, creed…), their relatives, the RVN own media and members of the government. **The true enemy is within! That is why it is called a Guerrilla War. That is why it might be more complicated than the other great wars in the past.**

As people become aware of the rights their own individual or clique/ group have, they will not allow those rights to be inhibited by the establishment. Their reptilian brains (Goleman's Emotional Intelligence) will kick in and guide them into a fighting mode, even before they realize what they do might harm their own **greater common cause**. In some cases, in the name of Democracy, some will test the system and find out they can oppose anything and whatever is imposed on them and that they believe is oppressive. They realize they can live freely and object to the national military draft that is a threat to their well-being. They believe that others, whether friends or foes, also have those rights and cannot be subject

---

394 *The Art of War by Sun Tzu*, 5th Century BC China.

to injustice or cruelty. They ignore that those people might be their deadly enemy or barbaric people who may be massacring other innocent people.

As we have already discussed "eye of the beholder" and "snapshot" in an earlier chapter ("Psychology: Eye of the Beholder"), what some see as right might not be right to others: the truth in the eye of the beholder. Some might believe in what they are capable of emotionally perceive; however, it happens that the perceived image is just one of the snapshots of reality, not the complete and real truth. Differences in perception cause polarization. It becomes worse when the reptilian brain ("Psychology: Emotional Intelligence") interferes in rational thinking and causes chaotic panicking possibly leading to inappropriate or harmful decision making.

With that in mind, we can now look at events in RVN in the big picture point of view. The behind-the-scene enemy, referred to earlier as the sympathizers or the undercover collaborators, look and behave exactly like friends or victims. Taking advantage of the new democracy, they keep feeding the naïve environment with supplementary stimuli in the name of political and religious repression. The propaganda vitamins that give the rising left a reason to victimize the enemy and demonize the defenders! Divide-and-conquer has always worked for the winners. Now, it is being well used by the underground. "Keep your Friends Close, but Keep your Enemy Closer" is what it entails to if we want to distinguish one from the other; this being a major lesson one ought to learn. The political and propaganda machine, as unveiled by Gen. Võ Nguyên Giáp and Bùi Tín who has eluded to having happened in the rear in the U.S., is actually also occurring by ten-fold in South Vietnam. Political turmoil have stirred up chaos in the second republic, when the Buddhists and the intellectuals rise to protest against Pres. Thiệu's administration. It eventually makes the people in Việt-nam as well as in the U.S. and Europe believe that Thiệu's government is repressing its people; this being probably another reason why the U.S. has decided to abandon Việt-nam.

Differences in viewpoints and policies have probably started to propagate since the antiwar movement has sprouted during the Vietnam War and grown into political chaos in the 2017 America[395]. Even nowadays, they have caused a more

---

[395] The left front was coming to life since the mid-1960s, as analyst Dr. Stephen Young, former Dean of Hamline University in St Paul, Minnesota, has

and more aggravated war between the conservatives and the liberals, between the republicans and the democratic clans... It is so awkward that we, the free world (and the RVN), all want Democracy, and yet, it is that same Democracy that has turned around and bitten back at us. It is like a double-edged blade that has, with history, when, people become more and more diversified, then polarized, grown into a multi-edged knife that is undoubtedly slashing up every fish in the pool, thus internally and externally hurting each other. Nevertheless, one cannot win, whether in the RVN or in the U.S., when "the house divided against itself cannot stand" (Abraham Lincoln[396]). A lesson should be learned that, one will always lose any battle when one cannot stand united under the same flag, the flag of Unity and Belief. In order to achieve such unity, leaders at all levels must eliminate all urges for personal profit and all citizens must know being divided will do nothing but help the enemy win the war.

Democracy is great as it honors every individual's rights; nevertheless, the truth of the matter is that individuals in the system are diverse, unstable, unpredictable, and somewhat unconventional; therefore, the established policies will change with their will, guts, and emotions, and, as a result, might become disoriented or maladapted. With Democracy, people's self-realization is expanded in a way that contradicts and rebels against the conservative common good-centered traditions. It will make the people polarize into small groups and divert them from pursuing the same basic goals, as they might see things differently.

JFK's "Ask not what your country can do for you, ask what you can do for your country…" 1961 inaugural quote now does not seem to apply any longer, as people start refusing to do things their country, the "establishment, expects them to do, believing that the expectation is wrong and that patriotism is to be questioned. For the establishment, the new idea gets in the way of national security and defense, and, therefore, it becomes a weapon against our own self and against the country common cause. That is what has happened in the U.S. towards the end of 1960s into the beginning of 1970s. This is the era

---

recently coined in his June 18,2017 article *Bitter fruit is the result of our half-century derangement in Minneapolis Star and Tribune Sunday paper, Opinion Exchange Section. It has kept the Vietnam Derangement Syndrome bleeding until today and has further delegitimized traditional American values and identity.*

396   President Lincoln's House Divided Speech: http://www.abrahamlincolnonline.org/lincoln/speeches/house.htm

when national policy is doubted and starts collapsing to give in to a populist expansion.

Celebrities were stabbing with imputatively metaphorical knives into the backs of their own soldiers, while the politicians stood up against the government and antiwar brothers and sisters cursed those who were serving. Many draftees tore up draft cards, and confused returning veterans trashing their earned medals ... Meanwhile, many of the young ones had resorted to psychedelic escape to avoid confronting their fear of reality and disappointment. Citizens fighting against their own had become the normal community activities.

Upon returning home, the war veterans, many mentally and physically crushed by Agent Orange and other airborne chemicals as well as the trauma of their own experiences, were considered demons and in many instances had to unfairly suffer the wrath of anger and ingratitude from their own compatriots... Pres. Lincoln might be right when he says "America will never be destroyed from the outside. If we falter and lose our freedom, it will be because we destroyed ourselves." America was way too distant from Việt-nam; nevertheless, Americans' humaneness had allowed the enemy to influence their heart against their own government. They would not lose their freedom, but they had nearly destroyed themselves. A few million of their allies and their soldiers became the victims who ended up being more or less physically and mentally obliterated, **this being a total Social Injustice.**

Democracy is so precious, and yet, so dangerous! **If the "other" enemies were not the press media, the U.S. government, or the antiwar movement, then it would be Democracy itself!** The main enemies, the VC and NVA, say it so well about their supporters, who actually are the "other" enemies" of Freedom and Humanity, this being summed up by two NVA generals as follows:

NVA Gen. Võ Nguyên Giáp, supreme commander of the DRV, in an interview by CBS in 1989, had commented on the role of American television media in the U.S. defeat: *"do not forget the war was brought into the living rooms of the American people... the most important result of the Tết Offensive was it made you de-escalate the bombing, and it brought you to the negotiating table. It was therefore a victory...* **the war was fought on many fronts. At that time, the**

most important one was the American public."

Furthermore, NVA Gen. Bùi Tín, a colonel and officer of North Vietnam political news media in 1975, had also added a comment that "both Jane Fonda and John Kerry[397] (who became Secretary of State under Pres. Obama), or activists such as Tom Hayden (the master-mind of the Students Democratic Society), are essential to the People DRV strategy in support for the our rear line: *'everyday, our leadership would listen to the world news over the radio to follow the growth of the American antiwar movement...'*" It did sound as if the Vietnamese Communist Party was "growing a cattle or crops" and was waiting to harvest benefits. "Know their Foe...."

Britanica.com described Tom Hayden's *vigorous activities*[398] *that might have helped him earn Fonda's heart: As the American presence in Việt-nam escalated, Hayden began organizing* civil disobedience *protests against the Vietnam War. In 1965 he made the first of many trips to Việt-nam, including a controversial 1972 trip to Hà-nội with his future wife,* Jane Fonda. *In 1968 Hayden played an integral part in the protests at the* Democratic National Convention in Chicago, *eventually being beaten, gassed, and arrested along with* Abbie Hoffman, Jerry Rubin, *and others as part of the* Chicago Seven, *who were charged with conspiracy, impeding police officers, teaching other protesters how to make incendiary devices, and crossing state lines to incite a riot. After five years in the judicial system, Hayden was acquitted on all counts.*

The result was as expected as the cattle grew well: the pressure from the antiwar movement expanded towards the end of the war. The antiwar protesters were even supported by the participation of the distressed medal-trashing Vietnam Veterans in the antiwar movement. The support turned more intense in 1972 with the Easter Tide Offensive that had proven the U.S. would never return after they had withdrawn. It was actually a test showing to the DRV the crops are ready to harvest: all signs showed they were winning, as U.S. troops continued to withdraw in spite of this offensive being much bloodier than the one during Tet 1968. The American negotiating team even grabbed any solution demanded by Hà-nội negotiator Lê Đức Thọ betraying its main ally and sacrificing more lives. At any cost, Nixon and Kissinger had to flee the war "with Honor,"

---

397   Patriot Post: Traitor: Hà-nội Jane Jane Fonda http://patriotpost.us/pages/80
398   https://www.britannica.com/biography/Tom-Hayden

and Hà-nội to win the war.

In the last moments before the Paris Treaty was signed, Nixon/ Kissinger tried their best to convince Thiệu to have his RVN team resume their participation at the Paris Accords negotiation table by promising to restart carpet bombing Hà-nội if the communists violated the treaty. Thirty some secret re-assuring letters Nixon were sent to Thiệu[399]. They were kept secret even Secretary of Defense James Schlesinger did not know about their existence. The latter was the person who would technically have to implement such an "assurance" in case Nixon could not fulfill his promises. Nevertheless, Thiệu had no choice as the RVN was so dependent on American aids and so desperate for any shred of hope it could still be saved. He allegedly told his National Security Council that, "if Kissinger had the power to bomb the Independence Palace to force me to sign the agreement, he would not hesitate to do so." What followed happened really fast: the agreement was signed by all four parties, then the DRV violated it, and Nixon was "watergated," while American people were still enraged and divided, the U.S. economy was in disarray, leading to Democratic legislators (e.g. Idaho U.S. Sen. Frank Church[400]) voting against all aids and preventing the U.S. from returning to its agonizing ally (War Act). Schlesinger also forfeited on presidential promises (to restart bombing if the enemy violated the treaty)... most of the football team had already left, and the only one player remaining on the field was gasping for a last breath of air, the Republic of Vietnam! (see the graphic "Paris Accords Main Clause- Comparison by Dr. H. Tuong" in the next section <u>Paris Accords, the Twisty Treaty</u>).

---

399   Excerpts from http://www.newrepublic.com/article/79923/promis-es-promises: The Palace File
by Nguyen Tien Hung and Jerrold L. Schecter (Harper & Row, 542 pp.)
400   Frank Church: Viet Nam War and Church Committee http://en.m.wikipe-dia.org/wiki/Frank_Church

CHAPTER 11

The Fans join in to boo at their home team

Back in early 1960s Việt Nam, after many coups d'état, most of the people were uneducated peasants, and consequently, they were confused about democracy and did not know which ones were the true leaders to follow. Meanwhile, the people of the U.S. were also confused, not just about the war, but also about the new discovery of their rights and the meaning of life. The nationalist Vietnamese were struggling with their newly acquired Independence from China, France and Japan, their newly acquainted Western Democracy and their long time thirst for Peace. Just like wild animals, after being raised in controlled captivity, they definitely would not know which directions into the jungle to take when they were released. Freedom and Democracy sound palatable, but they would require new knowledge, clairvoyance, tolerance, mutual respect, acceptance, self-realization, and, perhaps a willingness to abandon some old outdated knowledge and practices from the past in order to embrace new values. It would be a difficult journey, especially for people such as the Vietnamese, who felt more comfortable living in the past (because some of their past was nurturing and reassuring, or that they had survived it) more than for the future (because the future had always been unpredictable, deceptive and futile, if not bloody). Failure to cope with the future could lead them to self or mutual destruction, which had indeed happened in the South (political turmoil, coups d' état) after the French accepted Việt Nam's total Independence in 1954. Perhaps that was the reason they avoided risk-taking adventures and would rather hide in secured environments, such as the mono-cultural mono-lingual communities Little Sài-gòn in California… (to be discussed in chapters to come).

Democracy Divides - Part 2

The Vietnamese society, whether it was in the North, Central or South, had always been pluralistic, if not polarized, not just because of the diversity with over 50 ethnic groups, but because of individual cultural pride and segregation inflicted by the French's Divide and Conquer strategies used during a century of colonization. Consequently, the newly Democratic-converted sides of the RVN were more or less..., I'd say, very divided. Individuals or groups believed that, with democracy, they could have their way to do or say whatever they wanted, and believed their way was more better than others'; this had led the country into internal self-mutilation between different factions of religious groups, political clans, military **cliques, French versus American-trained personnel,** creed or regional origins (North, Central or South-born, or subdivision into even smaller demographic units, e.g. which part of the Mekong Delta, of Central...). The situation had one more time confirmed my theory about how the colonial past divide-and-conquer experiences had affected the unity in both 1st and 2nd republic.

As much as I could remember, even among us French-educated, I could feel a certain subtle, but high level of pride, among those cultivated and educated in an elite French Lycée, compared to the "lower level" American, and even "lower" Vietnamese high schools. As reflected in more than one conversation with my French-educated brother Anh Bảy, the American university-educated leaders seemed to look down at the French-educated counterparts for not being innovative and fast thinkers. The Vietnamese-educated leaders were considered by both the French and American ones as slow and routine loners, who could not work in teams and were usually unable to cope with sudden moves.

Furthermore, the moderate socio-political research groups in France must have discussed Indochina in their political arena and had come to a conclusion, as pointed out by my young History Geography professor[401] in our classe de 11ème at Lycée Jean Jacques Rousseau in 1967-68, that the colonial mentality had set up the society so that the colonized countrymen would do their best to put each other down in order to elevate, be known and appreciated by the colonials so they could gain power for themselves in the colonizers' system. In his 1970 <u>Pedagogy of the Oppressed,</u> *Paulo Freire examined the relationship between the colonizers and the colonized, "oppressors" and the oppressed", and concluded that, once the colonized were able to recover their Freedom and Independence, they were actually perplexed in not knowing*

---

401   I forgot his name, but vividly remember his received a prestigious History degree of "agrégé," something like a Masters Degree but even more prestigious than Masters, from Sorbonne University, equivalent to U.S. Harvard

what to do with it (do not know what to do with each other), and that they even fear it, especially when everybody was still fighting each other, thinking one is better than others... "Freedom is acquired by conquest, not by gift. It must be pursued constantly and responsibly. Freedom is not an ideal located outside of man; nor is it an idea which becomes myth. It is rather the indispensable condition for the quest for human completion[402]" Perhaps this was the reason why Pres. Kennedy, in the late 1950s when he was still a senator, had coined that, after having recovered their Independence, the colonized "must support their struggle (be on their own, so that they can process their emotions and grow out of any complex of inferiority) and any intervention by the U.S. will be bound to be futile" (see Chapter 9). Once Independence has been established, the colonized oftentimes experience a period of self-rediscovery while exploring the newly found Freedom: as a matter of fact, it could be very difficult to achieve self-realization after exiting from a long period of well-structured (well controlled) regime. It is like a frog being freed from a small vase and discovering that the sky is vast and unlimited, not like what it has been looking at through the opening from the bottom of the vase.

What a feeling being lost in the newly acquired Independence! It was clearly a feeling the people could not internalize foreign ideologies; however, they kept letting themselves affected by it. It started making more sense to me now rethinking what the French educators had coined in psychological notation about the Vietnamese "mentalité des colonisés." Encouraged and empowered by their French "chaperones," they became arrogant thinking they knew everything and were better than anyone else. This could have been the reason why there had been a great many coups d'état, political rebellions, religious protests and uprisings, every several years since 1954 under Pres. Diệm (betraying his Emperor Bảo Đại) through the end of the 2nd Republic in 1975 (Buddhist uprising against the RVN government). Perhaps so many events had happened that the Vietnamese never had time to recuperate from.

The transition from being dominated by foreigners into a pseudo-Democracy tidal wave was so abrupt the people were confused and drowned in their Freedom. The RVN was

---

402   https://en.wikipedia.org/wiki/Pedagogy_of_the_Oppressed Paulo Freire's Book Pedagogy of the Oppressed

seriously divided within. So many military politicians wanted to be leaders, thinking they were "more right" and could lead better than the sitting ones. So many conflicted against each other as they believed they were better and knew more than the others.

On the contrary, the enemy, who was totally submitted to the totalitarian Communist Party, and who strongly **believed** (as if it was a religious faith) that they were going to rescue and liberate the South, and had abided to communist rules. In reality, the communists knew they had no choice: they were strictly monitored by their leaders and by each other. Failure to obey and exert their loyalty to the Party might result in being publicly executed[403] or their family being harmed, one way or another, or both. When they were taken prisoners by the RVN, they could see the big difference: that the RVN would treat them humanely and would not do any harm to their family as a punishment for "insubordination or Treason to the People" as the Party would. Many NVA and VC POWs had declined being returned to North Vietnam during the post-Paris Accords Exchange[404] in 1973: They would rather commit mass suicide than to go back to the DRV.

The Political Divide

If we look back into history from the days of the State of Vietnam (beginning of Ngô Đình Diệm in power) until the end of the 2nd Republic (Pres. Nguyễn văn Thiệu), we would see so clearly how unstable the South was. Those involved in the turmoil might not see it because they were deeply committed to the cause they believed was right: fight the opposition, overthrow the tyrant. When the French was about to be defeated, every, or every other year, a new prime minister was appointed by Emperor Bảo Đại. The latter himself was betrayed by the last one, PM Ngô Đình Diệm.

Independence came in 1954, and Diệm took power after the 1955 referendum-election (viewed as fraudulent as per U.S. sanction) and had to face rebellion from the Emperor's loyal

---

403 As discussed earlier, executions were used to warn others more than to punish the individuals.
404 To be discussed later in 1973 – Returning NVA and VC POW to the Socialist Republic of Vietnam

groups, particularly Gen. Nguyễn văn Hinh[405]. The uprising of the armed religious groups, Gen. Trịnh Minh Thế's Cao Đài and Gen. Lê Quang Vinh (Ba Cụt)'s Hoà Hảo, and the organized-crime boss Gen. Lê văn Viễn (Bảy Viễn)'s Bình Xuyên[406] contributed to the decade-long heavy government destabilizing efforts of both political and military systems of a newly born nation[407]. Everyone was trying to secure their positions and power in the new nation. The failed 1960 coup attempt by both Lt. Col. Vương văn Đông and Col. Nguyễn Chánh Thi had inaugurated a series of other "coups d' état."

1963 came with the uprising in Huế of the Buddhist group; this movement was repressed by the sitting president's siblings Ngô Đình Cẩn and Ngô Đình Nhu, and viewed by the U.S. as a religious repression. The civil turmoil grew out of hand and ended with the U.S.-backed 1963 coup which opposing military leaders took advantage to execute both Pres. Diệm and his brother Advisor Nhu. No one knew who executed them! Gen. Dương Văn Minh was one of the main generals leading the 1963 coup. A year later, Big Minh's junta government was overthrown by his "companion d'armes" (from the days of the French Union), Gen. Nguyễn Khánh, and other generals such as Trần Thiện Khiêm, Nguyễn văn Thiệu, and others, all believing they deserved better posts in the junta.

In 1964, Gen. Khánh exiled Gen. Khiêm, his ally in another coup against Big Minh, to Washington D.C. as Ambassador. Other generals, Trần văn Đôn and Lê văn Kim, were added to the mess with their proposals to defeat the NLF non-militarily[408]. At the beginning of 1965, Gen. Khánh himself was deposed and forced into exile as Ambassador-at-large by the wolf pack. It was actually Gen. Khiêm's clan who led the coup, with Gen. Lâm văn Phát and Col. Phạm Ngọc Thảo who did not take power. Then Gen. Nguyễn Chánh Thi and Air Marshall Nguyễn Cao Kỳ came into the picture. Phát and Thảo were sentenced to death in absentia: the former stayed in hiding and was pardoned, but the latter hunted down and executed.

---

405   http://www.zoominfo.com/p/Nguyen-Van Hinh down-played by the U.S.
406   https://en.wikipedia.org/wiki/Trình_Minh_Thế, https://en.wikipedia.org/wiki/Lê_Văn_Viễn, and https://en.wikipedia.org/wiki/Ba_Cụt Le Quang Vinh
407   http://www.cornellpress.cornell.edu/book Cauldron of Resistance and http://muse.jhu.edu/chapter/895393 The "sect" crisis of 1955 and America's miracle man in Vietnam
408   https://en.wikipedia.org/wiki/1960_South_Vietnamese_coup_attempt, https://en.wikipedia.org/wiki/1964_South_Vietnamese_coup and http://military.wikia.com/wiki/1965_South_Vietnamese_coup

1966 followed with more Buddhist uprising in Đà Nẵng and Huế against Pres. Thiệu's Sài-gòn regime. The protesters took over the public radio station and threatened to expand into an anti-American movement by the people. The Marines had to be brought in to reclaim control. The ARVN was totally out of control and divided: Sài-gòn-loyal Marines versus rebels-supporting ARVN, all due to allegiance to their leaders who were themselves divided[409]. The U.S. army intervened and chaos outgrew into U.S. versus ARVN rebels led by ARVN Col. Đàm Quang Yêu.

*It is incredible what the Self (psychology term) could transform people into! In the meantime, the NVA and VC were probably laughing and enjoying the RVN chaos. They might have been the ones instigating the Buddhists using **Belief** to control **Politics**! When people had minimal education level, it was easy for them to mix up religions in politics and the enemy obviously could take advantage of their "confusion" to instigate them into dissidence and insurgence.*

It was believed, even though he disavowed involvement, that Gen. Thi had something to do with the rebellion. Also, we could not forget to mention the serious disagreement between Thiệu and Kỳ which I had discussed in a previous section: Buddhist rival Kỳ could have instigated the Buddhists against Catholic Thiệu. Perhaps the Tết Offensive of 1968, Tết Mậu Thân Year of the Monkey, had helped the RVN internal turmoil become more controllable as Vietnamese people were able to open their eyes and realize who the real enemy was.

The Religious Divide

The internal bleeding was not caused just by political factions. It had also manifested religiously as I had pointed out earlier, mainly between various Buddhist groups against both Republic establishments that were identified Catholic (both Diệm and Thiệu were Catholic) and that were supported by the Christian Unites States of America. Would it be possible that the Buddhist movements were communist or aspiring communist, or were the Buddhist groups taken advantage of by the Party? Many present-day bloggers, socio-political analysts, and documentary film producers had suggested that the

---

[409] http://www.historynet.com/the-1966-buddhist-crisis-in-south-vietnam.htm

Vietnamese Communist Party had succeeded in manipulating and instigating Buddhist movement to start Peace and pseudo-human rights uprisings against the RVN government. Others had suggested that the Peace-loving Buddhist movement, during both Republics, had actually been led by the communist infiltrators, if, in fact, they were not communist groups themselves.

Even though my family was "passively" Buddhist (we did not go to temples regularly as the active members would) and that most of my "précepteurs" (school pre-tutors), uncles, and myself included[410] had participated in Buddhist monks' and students' uprising and demonstrations against the government. I had been perplexed, since the day I joined the ARVN. about whether or not Buddhism was used by the Việt-cộng VC. Was I brainwashed in either case? I was not really sure. Like any other "fawn," I was young, light-hearted, naïve and confused.

Nevertheless, the Communist Party members who were preaching human rights and the need to fight for independence from foreign invaders sounded were doing their best to show they would be the true leaders. They had already proven they could successfully extinguish the century-long French colonization. They appeared to be more patriotic than the rest (sympathizers and collaborators?), especially when they preached what the Buddhist disciples had been longing for, Peace and Humanity. These ultimate goals were also supported by the Buddhist reactive organizations, such as Ấn Quang Buddhism[411] (also Unified Buddhist Sangha of Việt-nam[412]), which was the more politically-active group led by Superior Monk "Thượng Tọa" Thích Trí Quang and Venerable Monk "Đại Đức" Thích Nhất Hạnh (Huế and Đà Nẵng), totally independent from the conservative mainstream group[413] Việt Nam National Temple Block "Việt Nam Quốc Tự" (also "Viện

---

410   Vietnamese Buddhists protested against Cambodia President Lon Nol-incited "Cap Duon" campaign in the late 1960s beheading Vietnamese and throwing heads down the Mekong River

411   Ấn Quang Temple in *Buddhist Group Joins Sài-gòn Opposition – The New York Times September 15, 1974* www.nytimes.com/1974/09/15/archives/buddhist-group-joins-saigon

412   *Buddhism in Vietnam Today – History and Current Events* www.thoughtco.com/buddhism-in-vietnam-450145

413   Conservative Buddhists condemned Communism as atheistic, denounced military governments in all oppressive forms, and rejected any political influence. Ấn Quang Temple was even more militant and reactive against the RVN establishment. https://en.wikipedia.org/wiki/Unified_Buddhist_Sangha_of_Vietnam and http://www.vietnamwar.net/Buddhism.htm

Hoá Đạo") led by Superior Monk "Thượng Tọa" Thích Tâm Châu (Sài-gòn).

Both republics of Việt Nam having been ruled by Catholic presidents caused for many years discontentment among many Buddhist believers. The Buddhist anti-government movement in Việt Nam, reinforced by antiwar intellectual groups, actually fit in well with the anti-war groups in the U.S. and European countries. The latter also preached Peace and Human Rights.

On the one hand, Venerable Monk Thích Nhất Hạnh's spiritual teaching and Peace activities during his exile in the U.S. and in France beginning 1966[414] had definitely conveyed the Buddhist, and particularly the Vietnamese Buddhists' wishes to end wars. They might have as well ignited among the American people some major doubts and suspicion about the American policy in Việt Nam. Furthermore, Sup. Monk Thích Trí Quang[415]'s political activities, also **in the name of Humanity and Peace (Buddhist) and of Religious Freedom and Human Rights** (American Democracy – leftist antiwar) in Việt Nam, were so militant that they had caused VP Nguyễn Cao Kỳ's secret police to arrest him as a communist suspect. This had further lifted him up to the level of a Peace hero, and landed a color photo of him on the front page of the Time Magazine[416].

Vietnamese Buddhism had eventually split into two distinct groups. Some RVN rightists believed Vietnamese Buddhism was actually communist or, at least, pro-communist, while the more compassionate leftists, including U.S. government and many Vietnamese columnists[417], had asserted that Buddhist activists, such as Sup. Monk Thích Trí Quang, were not communist, even though they had actively protested against the Vietnamese Republic until its last agonizing breath in 1974-75. Had this American experience with Vietnamese Buddhism spiritually opened up the horizon for the antiwar

---

414　https://en.wikipedia.org/wiki/Thích_Nhất_Hạnh #During_the_Vietnam_ War in U.S.A and France 1966-1975

415　Monk Quang was identified by French "Sureté" report to have joined the Indochinese Communist Party in 1949 against the French colonials; this information was later confirmed by DRV Deputy PM Tố Hữu in 2000. https://en.wikipedia.org/wiki/Th%C3%ADch_Tr%C3%AD_Quang

416　April 22, 1966

417　In Vietnamese: https://thuvienhoasen.org/a7145/con-nguoi-that-cua-thuong-toa-thich-tri-quang-dao-van-binh blog on Buddhist Lotus Library written by blogger Đài văn Bình (11/14/2010) *The real person of Thượng Tọa Thích Trí Quang*

movement to, not just to fearing the war, but rather to shutting the war down? Besides, all lives mattered, whether they were nationalists or communists, didn't they?

On the other hand, the rise of the Social Justice movements in the U.S. supporting anti-discrimination, Human Rights, women's liberation, gay and lesbian GLBT liberation, along with the vision of the long lost Peace at the end of the tunnel, the promise of individual's Freedom, and Freedom, and Freedom[418], and spiritual expressions (East Indian influence) had openly embraced the participation of peace protesters along with many Buddhist monks in demonstrations.

Nevertheless, as a result my family being Buddhist at that time, I believed that, if the Buddhist movement in Việt Nam were not communist, it would have definitely been seriously taken advantage of by the communist sympathizers and the Communist Party. It had significantly increased communist popularity among the non-politicized population, especially when Buddhism was the dominant religion in the country. Some documentary photos on the internet had shown Asian[419] Buddhist monks, in the name of Peace and Humanity, demonstrating side by side with NLF flag-carrying[420] (Red-Blue flag with a centered yellow star) protesters in antiwar demonstrations in the U.S. Blogs suggested that these could be communist infiltrators who had been welcomed as Peace combatants[421] along the same side as the young idealistic Americans against the booming Social Injustice movement in Việt-nam... Many of these (monks?) in the U.S., without any doubt, had links with the Vietnamese Communist Party[422].

Things continued to gradually change for worse towards the end of the war, when Buddhist groups supported

---

418   The word Freedom was chanted as the main lyric by artists performing at the 1969 Woodstock Concert. The Long March for Freedom also symbolizes Americans' aspiration for Freedom (CNN <u>The Sixties</u>).
419   Wearing grey-brown Vietnamese "Big Vehicle" Buddhist robes?
420   Perhaps protesters did not realize the NLF flag symbolizes Vietnam War being a civil war and that the U.S. was actually invading Vietnam, exactly the way Hồ Chí Minh wanted the world to perceive.
421   <u>Kissinger Sold Out Blog by Xóm Cồn Ha Uy Di, Nha Kỹ Thuật, ARVN Tactical, Nov. 2009.</u>
422   By the same token, some VC who were found hiding inside Sài-gòn Ấn Quang Buddhist Unification temple.

by the educated university student mass intensified their anti-government activities. The rebellion seeds were deeply sewn into the soul and heart of the people who were all aspiring for Peace, a word that was loved by Buddhists and taken advantage very well by the enemy. Nevertheless, the dormant Buddhist movement surfaced again later during and after the Easter Tide Offensive of 1972, demanding Sài-gòn regime of Thiệu to resign. It was a crucial time as the country was highly at risk as all of the allies were swiftly withdrawing from the war and American aids were being drastically cut.

The fact that many VC were found harbored or hiding in the Buddhist Temple Ấn Quang (leftist group headed by both monks Thượng tọa Thích Trí Quang and Đại đức Thích Nhất Hạnh[423] since the 1960s uprisings) in Sài-gòn proved that many monks and followers, knowingly or not, were actually supporting, if not being themselves members of the Communist Party. It was so obvious the Buddhists were working hand in hand with the advancing enemy, possibly not politically the Communist Party way, but in the name of Empathy and Humanity[424] the way Buddha had taught. Here again, the communists had cleverly and skillfully disguised themselves as the Red Riding Hood grandmother, the humane killer.

One could argue that both Presidents Diệm and Thiệu being Catholic were used as a reason by the Communist Party to seriously set the RVN government and the predominantly Buddhist population against each other. They were using the same tactic: "the Americans are here to take over Việt-nam" compared to "the American-supported Catholic establishment is oppressing the peaceful Buddhists."

After the fall of Sài-gòn, Ven. Monk Hạnh was barred from returning to the DRV, while Sup. Monk Quang was placed under house arrest there. In 1981, the DRV consolidated all Buddhist organizations into the Buddhist Sangha of Việt Nam and placed it under state control. As of now, after forty some years recovering from the Liberation of the South, the Buddhist people have just realized they had been duped again by a fake Democracy. Once more, they are now rising again, against the current DRV government, this time fighting the communist government, not for Peace, but definitely for the Human Rights and Freedom for which they had fought before against the

---

423   https://vi.wikipedia.org/wiki/Giáo_hội_Phật_giáo_Việt_Nam_Thống_nhất Unified Buddhist Association

424   …also, possibly with the same goal as the antiwar movement in the West.

RVN. Peace or Freedom sounds simple, but is the toughest choice to make, ever!

Was it just a coincidence that the Buddhists had protested **in the name of Peace and Human Rights** against both the 1st and the 2nd Republics, both regimes being headed by Catholic presidents? Or, in 1963, when Rev. Monk Thượng-toạ Thích Quảng Đức self-cremated in protest against Ngô Đình Diệm's Presidency oppression[425]? This Huế Central Việt Nam political conflict turned into a religious one between the sitting Catholic Regime (Ngô Đình Nhu and Ngô Đình Cẩn, Pres. Diệm's brothers) against the dominant Buddhist population. Subsequent self-immolation protests of more Buddhists took place in the late 1960s against Prime Minister Nguyễn Cao Kỳ and Head of State Nguyễn văn Thiệu's military rules. Why were many VC found and captured inside Buddhist Ấn Quang Temple? Even in 1974, just a few months before the Fall of Sài-gòn, when the RVN could barely fend itself and had to abandon major military strong holds in I and II Corps, major protests were staged by Buddhist monks and led by Monk Thich Tri Quang demanding Pres. Nguyễn văn Thiệu to step down.

A strong belief in a better level of Dharma[426] and self-sacrifice for Social Justice had indeed helped Buddhist activists boost their courage and make them put self sacrificially in harm's way, thinking they were fighting to help the People rid of tyranny. Perhaps, this was the main reason why the Communist Party had embraced the Buddhist teaching to divide the Buddhists away from the "Catholic establishment." Again, "divide to conquer" really worked as it could liberate the Buddhist People (and subsequent groups) from Christian tyranny in the name of Freedom of religions.

---

425   http://news.bbc.co.uk/onthisday/hi/dates/stories/may/31/ BBC on Buddhist burning to death and http://www.history.com/this-day-in-history/buddhist-immolates-himself-in-protest History Channel on Buddhist immolates himself

426   Wikipedia: In certain contexts, *dharma designates human behaviors considered necessary for order of things in the universe, principles that prevent chaos, behaviors and action necessary to all life in nature, society, family as well as at the individual level. Dharma encompasses ideas such as duty, rights, character, vocation, religion, customs and all behaviors considered appropriate, correct or morally upright. See* The War, Chapter 2. https://en.wikipedia.org/wiki/Dharma#Definition

CHAPTER 12

Paris Accords, the Twisty Treaty

Returning to the Paris Accords events, Thiệu's refusing to recognize the VC Provisional Government had totally changed the format of the Paris negotiations, from four to two-way, just between the U.S. and the DRV. This had allowed the Nixon-Kissinger team to monopolize nationalists' decision and deal directly with the DRV politburo Lê Đức Thọ (as I had discussed in details in previous chapters). All along, Thiệu had definitely tried to make a point that the war was not a Civil War, but one between two separate countries, the South versus the North (thus justifying U.S. intervention[427]), contradicting the long-time DRV claim that Vietnam War was a Civil War between Vietnamese and Vietnamese. The latter argument would demonstrate that the U.S. was at fault when intervening by invasion in an internal matter and sovereignty of a country; it would also prevent the U.S. from giving foreign aids to the RVN.

On January 27, 1973, the accords were signed by all four sides. It was agreed upon that a cease-fire would end the war throughout Việt Nam, that the United States would withdraw all troops, and close all its bases within 60 days. In return, the DRV agreed to release all U.S. and other prisoners of war. Both sides would withdraw all troops from Laos and Kampuchea (or Khmer, former Cambodia under the French and Prince Sihanouk). All current local troops, including the RVN, NVA and VC, would remain on stand-by, and the 17th Parallel per 1954 Geneva Conference (Accords) would still be the demilitarized zone DMZ. The government of the RVN would not be dismantled; nevertheless, it had to recognize the insurgent VC National Liberation Front provisionary government as well as accept the fact the VC and the NVA could keep all the territories they had taken; besides, the communist side could resupply from the North and its "big brothers" China and the soviet, while the South would lose all aids from the U.S. Of

---

[427] President Diệm wanted the opposite, that it was a Civil War between North and South to prevent American intervention that would give the North a reason to liberate the South from American colonization.

course, all forces supporting the South during the war had to withdraw as soon as the Accords went into effects.

This clause oppressively placed the RVN in a totally disadvantaged situation. As naïve as the world could be: Would the Paris Accords be solemn enough to keep the North and the VC from cheating? What Thiệu had

"PEACE WITH HONOR"
By Pres. Nixon

Paris Accords Jan 27, 1973
By SRV Lê Đức Thọ and U.S.Sec. Henry Kissinger

United States and all Allies	Socialist Rep. of Vietnam SRV
Will *totally withdraw* from Vietnam	All troops *can stay and occupy territories*
Will *cut all aids (Case-Church 1973)*	Soviet-China *full aids continue for NVA*
No replacement of equipment	*Can replace used weapons/troops*
RVN (South) will stop all wars	NVA and VC will stop all wars
RVN Government will continue until Unification	National Liberation Front Provisionary Government is recognized
POW's will be released	POW's will be returned to SRV

Pres. Nixon secretly promised Pres. Thiệu he would resume aids and Linebacker III B52 Bombardment if the enemy violated the Treaty. Of course, his promise would never be carried out (Pres. Ford) because of Watergate and 1973 War Power Act.

March 1974:
NVA launched Ho Chi Minh Campaign...
RVN ran out of ammo and had no aid...

Paris Accords Main Clause- Comparison
By Dr. H. Tuong

feared the most had happened. Immediately after the U.S. had totally withdrawn by January 1973, the Hồ Chí Minh trail became the official extremely busy Hồ Chí Minh highway, as fresh conventional troops started moving in legitimately, with new and even more sophisticated heavy equipment and weaponry. They need more troops to occupy more land as quickly as they could: per Paris Accords, they could keep these territories.

Death by Strangulation… by Our Own Friends!

People DRV's violation of Paris Accords was totally ignored: Sec. of Defense Schlesinger's June 1973 duly recommendation for resuming bombardment was rejected by Congress; he was severely criticized. The most cynical part was that, **on the one hand, the 1973-74 U.S. Congress, significantly more Democrat than Republican (Democrat Majority Senate% 59D/40R and House% 56D/43R[428]), had immediately banned all funds appropriation (the Case: Church's bill) and cut off all military aids to the RVN. Remember that the NVA and VC would still receive their supplies and, supposedly, replacement equipment from China and the Soviet!**

The RVN was like one's body being totally invaded by spreading final phase cancer tumors, but no medicine or cure would be administered. In the meantime, Nixon and Kissinger got what they had wanted, **Peace with Honor (or not!)** to keep their re-election promise, at the expense of a few more million innocent lives. It was a death sentence to the RVN and all South Vietnamese!

While the Paris Treaty was being negotiated, as reported by RVN Gen. Cao văn Viên[429] in his Memoir:

"…Gen. Murray had been telling the Vietnamese, "you are going to get what we promised." They would receive the one-for-one replacement of equipment under the Paris Accords, **as would the NVA and VC.** *On February 13 (1973), Gen. Viên*

---

428   This is the reason why the majority of the ARVN veterans support the Republican leaders no matter how wrong or incompetent they can be. Information on Congress is from Wikipedia U.S. Congress Summary.

429   Lt. Gen. Cao văn Viên, one of the only two four-star commanders of the Vietnamese National Army in the 1950s and recipient of the U.S. Legion of Merit - Commander in 1969, was Chief of RVN Joint General Staff in the 1960s. He had planned the Laos incursion Lam Sơn 719 in 1970. After the ARVN was able to recover most territories taken by the enemy during the Easter Tide Offensive of 1972, he strongly believed the ARVN was capable of expelling the NVA across the DMZ given that it received adequate supplies. He passively supported President Thiệu on the latter decision to abandon I and II Corps https://en.wikipedia.org/wiki/Cao_văn_Viên https://military.wikia.org/wiki/Cao_văn_Viên. https://www.goodreads.com/book/show/6147049-the-final-collapse One of his books was the Final Collapse (2005) as part of the Indochina Monographs Series written for the U.S. Army Center of Military History.

ordered restrictions on all types of weapons. Within a four-month time lag between ordering and shipping, the supply line dried up by April, 1974 and the "system was never to recover". **Why? U.S. Congress had cut off all aids as the treaty was finalized!**

There was a growing list of critical shortages, including ammunition, medical supplies, and funds for the subsistence of South Vietnamese troops. Infantrymen, accustomed to carrying six hand grenades into battle, were issued only two. Mortar and artillery rounds for the defense of outposts were limited to four rounds a day and all harassing fire was halted to save ammunition. More than half of all armored vehicles were out of operation and 20 percent of all aircraft were grounded. The use of firepower to save lives, the American way of fighting, was being eliminated, and by the spring of 1974, the Vietnamese realized that "Vietnamese blood is being used as substitute for American ammunition." The casualty rate increased drastically so that by the end of June, during a period of increasing North Vietnamese violations of the cease-fire, there were already 19,000 dead for the year and more than 70,000 wounded. As a result of supply reductions military hospital did not have enough medicine or bandages. In many cases, bandages were washed and reused. **Meanwhile, the enemy received even more equipment, human power and supplies... Why? The HCM was then wide open because no one controlled it!**

In his memoir, Gen. Cao văn Viên concluded:

"Thus the South Vietnamese soldier of 1974-1975 marched into combat with the deep concern that his ammunition might not be replenished as fast as it was consumed and that, if wounded, he might have to wait much longer for evacuation. The time of abundant supplies and fast helicopter-lifts was over...To reduce that aid so drastically and so abruptly ended any chance of success and generated panic among the people and armed forces of South Viet Nam while encouraging the communists to accelerate their drive to conquer by force." **That was why I and II Corps had to be abandoned**[430]**!**

---

[430]　President Thiệu decided to abandon I and II Corps https://en.wikipedia.org/wiki/Cao văn Viên https://military.wikia.org/wiki/Cao văn Viên. https://www.goodreads.com/book/show/6147049-the-final-collapse One of his books was the Final Collapse (2005) as part of the Indochina Monographs Series written for the U.S. Army Center of Military History.

Janette Maring (researcher)
Facebook Blog - Vietnam War: Fallen Leaves

Just a few weeks earlier, the U.S. had significantly increased the bombing of Hà-nội (Xmas Bombing[431]), to bring the DRV back to the Paris table. Before this, they were "pouting" and had withdrawn from negotiations! On the other hand, in order to have Thiệu's team back to finish the last phase, sign the Paris Treaty, Nixon had to reassure Pres. Thiệu that the U.S. would not really abandon the South. Thiệu's team was pouting, too, and had withdrawn from negotiations when they found out both Kissinger and Nixon were cheating and secretly negotiating in secret with the enemy. On many occasions, he had promised Thiệu the U.S. would resume bombing the North if the latter violated the treaty. Whether his intention was true or just to bypass Thiệu's anger so he would have his team return to the negotiations table and sign the papers, Nixon and Kissinger had succeeded in finalizing the Paris process, an achievement that allowed Tricky Dick to be re-elected for a second term, and Kissinger to achieve glory (Nobel Prize nomination). Nevertheless, Thiệu knew the behind-the-scene reality: the U.S. would never return even if the RVN needed to be rescued if the North and the NLF reveal their true aggressive face: violate again an agreement that they had signed[432] and that was witnessed by the whole world.

Towards the end of Paris negotiations, the relationship between the two "big lovers", the RVN and the U.S., turned more and more from sour to "stinky," until it completely disintegrated when "spouses Nixon and Kissinger" were caught red-handed cheating on Thiệu: the former had flirted and slept with Thọ, Chou and Mao" in 1972. Thiệu had good reasons to refuse participating in the Paris negotiations that were actually a tragic joke! Even though his family was allegedly accused of corruption[433], Pres. Thiệu, the "American

---

431   Christmas Bombing Linebacker II https://www.historynet.com/the-christmas-bombing.htm and https://www.history.com/this-day-in-history/nixon-announces-start-of-christmas-bombing-of-north-vietnam

432   NVA and NLF first violation was the Tết Offensive of 1968 when they attacked 150 RVN cities in spite of New Year Truce. The second time was in 1974 against the Paris Accords they had signed a year earlier, when they launched the Hồ Chí Minh Campaign to attack the South. The latter had been abandoned by its allies.

433   There was an allegation that he and family was involved in an illegal trade of luxury goods and perhaps drug trafficking, as denounced by his right hand

clan man," still received some recognition by nationalists for being daring enough to confront his former supporters, the United States. He really ought to be commended for his "audacity" of standing up against the "Big Brothers" when he realized the RVN and its people had been deeply betrayed.

After all, as Brother Anh Bảy had commented, the U.S. tended to support leaders who were corruptible as it was always easier to work with leaders to whom you gave "special benefits." Pres. Diệm[434] might have become a victim of the U.S. because he was hard-headed and refused to abide by the U.S. war plan, even though he was considered by Pres. Eisenhower to be "the Churchill of Việt-nam", an upfront, more or less "clean[435]", clairvoyant, and a real patriotic leader. In the 2nd Republic, Pres. Thiệu was supported by the U.S. probably because he and/or his family were corruptible, or that he was more pro-war than his predecessor. He had agreed to fight the war as the American military had wanted: launch a war of attrition against an enemy that used the "not-so-heroic guerilla hit-and-run tactics." As a matter of fact, a great many RVN upper echelon officials, most of whom being high-ranking, received "gifts[436]" from subordinates military as they had control over the latter's life and death. As a matter of fact, a few of these leaders even attained power by networking (e.g. Lam Sơn 719 Campaign in Laos, I Corps in 1970, per NewsMax documentary).

Poverty and abuse of power gave birth to corruption, this perhaps being one of the reasons why the U.S. abandoned RVN. This traumatic social disorder and injustice had propagated through many layers of RVN military and civilian

---

Vice President Nguyễn Cao Kỳ

434   The nationalist Vietnamese had seriously conflicting feelings about President Diệm. Many strongly believed he was a real patriotic hero and defender of the Republic against Communism, which is true, while others disliked him because his siblings were oppressors, this being true also.

435   … not like his siblings who were totally corrupt tyrants! President Diệm, as observed my many researchers, including those who disliked him such as Seth Jacobs (*Cold War Mandarin: Ngo Dinh Diem and the Origins of America's War in Vietnam – 1950-1963 2006), had to admit he was not corrupt. Col. Lý Trong Song, as President Diệm's bodyguard, revealed that the president, a Catholic but also true Confucian (with all five virtues Humaneness, Loyalty, Respect, Wisdom and Trust) who had led a very modest life..*

436   Gift giving, according to Asian culture (a Chinese tradition adopted by Vietnamese), was an act of showing gratitude and respect towards benefactors. There was a very fine line between gift giving and corruption, so fine both sides did not see it as bad, as both sides would benefit from it, without realizing that traditional gift giving had caused a major social calamity as those who could not afford this practice were the ones suffering injustice the most.

hierarchy. Asian's tradition of gift giving could outgrow into corruption easily, usually through the wife! Brother Anh Bảy, who was perhaps "clean," sometimes mentioned he had always kept an eye on his wife to prevent corruption. "We'd accept 'Tré,' sausage from Huế, nothing else!" he quoted in a joke.

... By extension from this concept of corruptibleness, is it still for the same reason the U.S. nowadays (in the XX-XXI centuries) gives substantial aids, billions and billions of U.S. dollars per year, to a great many countries, especially in Africa, Central and South America, the Middle East, even when they are at Peace or do not need aids? Are foreign aids used to buy status quo and to make it easier for the U.S. government to work with them? Aids are even given to rich countries, which is unbelievable, when there are plenty of poor and hungry American people in the country. Words also go around that leaders from those countries oftentimes give donations to U.S. officials for their election campaigns... Money is used to buy Peace! Money is used to buy political support or endorsements! Money is used to buy control over sovereignty and favor land grab, as in the case of China... Are American tax payers aware of this trick? It's another conspiracy theory we need to consider... Politics has become an equal- give-and-take game, where deals are made via endorsement. The game in Việt Nam during the war is actually a matter of life and death. The game China is playing in Africa, SE Asia, South America, Middle East...

Now going back to the Paris Accords discussion, one might realize that the American move was actually a self-defeating forfeit, so that Nixon's presidency could gain another term. It allowed the National Liberation Front VC and the North Việt Nam Armed forces to continue to occupy territories where they were already established in the South. Pres. Thiệu as well as the communists knew the Americans would not return to the war if the enemy violated the treaty and restarted their invasion. The bloody 1972 Easter Offensive that occurred just before the Paris Treaty was signed had proven this. The U.S. ignored it and kept withdrawing troops and cutting off military aids in spite of the attack.

More aggressive than the Tết Offensive of 1968, this 1972 Offensive had allowed them to take even more land that they could keep according to the Paris Accords, and, by the same token (Kissinger's Paris Accords), to strengthen the

NVA VC positions in the newly-occupied territories using replacement troops and continued supplies from the North[437]; besides, it was a preempted move to probe a possible final attack that would totally dissolve the Republic of Việt-nam and subjugate its people to a totalitarian strangulating regime.

In the meantime, unfortunately, the RVN, being deprived of all new supplies and replacement equipment, would have to lose more territories. That was why Pres. Thiệu had to resort to the last solution: to pull all ARVN troops from I and II Corps in order consolidate forces on defendable territories (strategical repositioning). How could an ill-equipped army defend a long territory and its millions of population? Had Nixon and Kissinger foreseen that they had negotiated in Paris would allow this calamity to happen? Or were they so desperate they had to play such a low hand to please the opposing side so the latter would grant them a quick and easy way out of the war? Of course, to reach the end, win the second term election, they had not hesitated to use all means, including the most unethical and inhumane ones: betraying friends (the Kingdom of Laos, the Republic of Khmer and the Republic of Vietnam) and sacrificing millions more of "meaningless" lives.

An international control commission representing both communist and non-communist blocks was established, made up of Canadian, Hungarian, Polish and Indonesian, these being countries that did not participate in VN War. Its purpose was to keep peace and oversee the execution of the Paris accords while waiting for South Việt Nam to have a democratic election that would "determine its people's self-determination." This commission was nothing but a weak token. So the war continued, this time completely out of balance.

The only thing I knew was, from a conversation with Brother Anh Bảy (he was in the Second RVN Cabinet), that Thiệu was very angry at "Mỹ phản bội" (US betrayal). The latter saw no way out but to consolidate the ARVN and reposition its forces ("di tản chiến thuật, or strategic evacuation") where they could defend. When bullets for ARVN individual soldiers' M16 rifles were rationed to the maximum, I realized then it was becoming drastically serious. And, by an association of thoughts, I started to understand why, in 1974, we had to pull from our old stockpiles outdated ammo the U.S.

---

[437]   via the Trails in Laos-Kampuchea that then grew into highways

had given to the French Army in 1954 (the Battle of Điện Biên Phủ).

I did feel really disoriented and disappointed when I discovered at the shooting range we were using old WW2 ammo! I thought the U.S. had given them to us as part of the WW2 surplus. As a matter of fact, as naïve a new member of the ARVN as I was, I was still not able to connect the dots... that our allies had long abandoned us before I was drafted. No ally meant the end of supply or equipment replacement! It was the shortage of ammos that made us resort to outdated French Army supplies. No wonder why we were ordered to restrict shooting as a 5.56 bullet (standard caliber for M16 or current M4 rifles) would cost U.S.$16 each against the budget, as we were told! In the meantime, the enemy had received not just more equipment and supplies than before, in spite of Paris Accords limiting it, they even received newer updated tanks, not the old T54 but the more modern T55 and PT76, unlimited RPGs (at least one per squad of 10 or 12), missiles... It was really the end of the RVN and the ARVN.

Fighting a war against a well-equipped enemy with practically no updated equipment and resupplied ammunition was like running a marathon with our blood supply to our brain being squeezed and cut off, paralyzing our whole body. No matter how good RVN commanding officers were, no matter how hard and courageously our soldiers fought, the lack of supplies and tactical support (artillery, air…) would certainly demoralize the armies. Nevertheless, contrary to what major film documentaries had claimed otherwise[438], the world should admit the ARVN had fought their best with the very little tools they still had. The proof of valorous ARVN success was that all major artery highways were cleared in 1973, a year after the Easter Offensive and the year when there were no more ally forces. I knew it because I was on them, between Bình Định in II Corps and Sài-gòn in III Corps, traveling back and forth on civilian buses and military convoy: there was no sign of NVA and VC activity. If the ARVN did not have control of those areas, I would have been caught and executed.

Many Vietnam Veterans asked me what would happen if military aids were not cut. The ARVN Chief of General Staff,

---

[438] TV reporting that ARVN taking the War over from withdrawing U.S. Army was not good…..

Gen. Can văn Viên, had stated in his Memoir that he strongly believed the ARVN would be able to repel the communists out of the South and even push them Northbound all the way to Hà-nội, should military aids not be discontinued. Nevertheless, in my opinion, the ARVN might need to fight more assertively using the same warfare the enemy had been using, in spite of Geneva Convention: torture, exemplary execution[439] to counterbalance enemy's cruelty. Based on the cities being liberated and the national routes successfully cleared and under ARVN control as I could see while riding on unescorted civilian buses between II (Nha Trang boot camp) and III Corps (Sài-gòn) in 1973, I believed our forces were strong enough to repel the enemy without American boot-on-the-ground, if fully equipped and supported. It would also be necessary that B52 carpet bombing on the Hồ Chí Minh Trail and Hà-nội were continued with the assistance of the CIA Roadwatch program and the efforts of Royal Lao in RM #3 and 4 and Miao SGU in RM#2. Given the fact that the communists had been "cheating" during the war, they would continue to cheat and manipulate the naïve illiterate Vietnamese; thus preventing the war from ever ending unless the enemy was annihilated. Once a war was started against a passive aggressive enemy whose guts were full of hatred, one had to expect an eternal war. If one loved Peace, at any cost, it had to be upheld right from the beginning, otherwise there would never be Peace with Honor the way Nixon and Kissinger had wanted!

The Generals

As the Paris Accord negotiations winded down to the final stage, things were not going well in I and II Corps, as I had reported in Chapter 3. Many strategic troop evacuation (nicknamed "reposition") and reassignments of military commanders were ordered by Pres. Thiệu. One such commander was Lt. Gen. Ngô Quang Trưởng, Commander of Corps I towards the end of the War. He was the most respected by both ARVN and U.S. troops and commanders, as well as by Pres. Thiệu's establishment, for being talented, straight forward, fair and uncorrupted. Replacing his incompetent predecessor Lt. Gen. Hoàng Xuân Lãm who had almost lost Central I Corps, Trưởng was considered by the U.S. army to

---

[439] Not massacre as used by the enemy!

be a top general[440] for being able to stabilize Central after the massive Easter Offensive of 1972[441].

Gen. Creighton Abrams, U.S. Forces commander in Việt Nam 1968-72 had told his officers that Gen. Trưởng was competent enough to "command an American division." Gen. Norman Schwarzkopf, U.S. commander of the U.S. Forces in Desert Storm in 1991, had commented about Gen. Trưởng as being the most brilliant tactical commander at that time when he was fighting in the renowned Battle of Ia Drang as a Major, advisor to Lt. Col. Trưởng[442]. Lt. Col. George W. Smith, advisor to Gen. Trưởng when assigned to the ARVN 1st Infantry Division, had made a powerful comment that really revealed Trưởng's personality: tough, disciplined, dedicated to his military profession, and, unlike his contemporaries who had climbed the ranks through political influence, nepotism, or just using cold hard cash (bribery). A general who had earned his stars on the battlefield! "He was viewed as a self-starter, without a hint of corruption or ego. He was regarded by the Americans as unquestionably the finest senior combat commander in the South Vietnamese army[443]." Indeed, wherever he was assigned, either in IV Corps or I Corps, the first thing he would do was to clean out nepotism, "ghost (on payroll but staying home as civilian) and ornamental (on payroll, present on roll call, but not doing military duties)" soldiers, these being the major facets of ARVN military corruption[444]

With great bitterness, Gen. Trưởng had to withdraw troops and abandon the provinces in Central Việt Nam, as ordered by Pres. Thiệu in his plan to consolidate forces around defendable territories; this retreat had caused major troop demoralization. Was the president's order to abandon Central I and II Corps in March 1975 the result of an emotional decision

---

440   http://www.historynet.com/the-most-brilliant-commander-ngo-quang-truong.htm
441   Đại lộ Kinh Hoàng (the Route 1 of Terror). YouTube Video clip.
442   Vietnamese monthly magazine *KBC, issue #5, Westminster, CA: article about Trung Tướng Ngô Quang Trưởng and 2007 Washington Post article by staffer Patricia Sullivan Ngo Quang Truong: South Vietnamese Army General* http://www.washingtonpost.com/wp-dyn/content/article/2007/01/24/AR2007012402276.html
443   Vietnamese monthly magazine *"KBC," Issue #5, Westminster, CA: Article Trung Tướng Ngô Quang Trưởng*
444   President Thiệu's cabinet and many American advisors believed that a great many high ranking officers in the ARVN ranking from Captain and above were corrupted. Military personnel who tried to avoid combat assignment would buy their way out and become "ghost or ornamental" soldiers.

(anger at the U.S. Congress resolution to cut off aids, and at Nixon and Kissinger for playing dirty tricks) or a panic attack? Nobody knew. The only thing we could all see was that columns of ARVN troops and civilian refugees plugging up the South bound main routes in another disastrous "di-tản chiến-thuật[445]" (strategic evacuation). It became one more time the blood Route I, the Road of Terror (Đại lộ Kinh Hoàng), a repeat of Easter Offensive of 1972, as columns of troops and refugees were bombarded by the communists everywhere they went. Gen. Phạm văn Phú, commander of II Corps, and his units suffered the same confusion when their troops were ordered to withdraw from II Corps. Obviously, the enemy had known many details of this evacuation plan, perhaps because of intelligence leakage from families as they were communicating with their military members. It was worse than the horrendous retreat of the French army and allies in Dunkirk at the end of May, 1940, when they barely escaped being overrun by the advancing Nazi forces. This major failure had caused serious demoralization among the ARVN units, as soldiers had to witness their families and defenseless civilians being mercilessly butchered by enemy shelling during the evacuation.

Co-dependence on family can cause operational problems. This makes substantial difference between communist and nationalist forces. The NVA soldiers, coming a long way from home, have no loved one to worry about: their family, except those participating in war campaigns in the South, is in no harm's way. The VC have already been brainwashed that all, themselves as well as their family, have to sacrifice for the liberation of their Ancestral Land. As for the nationalist side, like the NVA's, allied forces such as Americans, Aussies, South Koreans, et.al., are far from home; therefore, they do not have to worry about their loved ones being in harm's way either. The ARVN soldiers in special units, such as infantry, airborne, rangers, marines, commando and Special Forces, et. al., who are classified as the "regular forces," are also away from home, so they worry much less about family safety than the larger chunks of Regional Forces Địa Phương Quân and militia Nghĩa Quân forces (these were considered "irregular forces"). The two latter usually station and operate within a shorter distance (between 5 to 50 km) from their home within the same province. The advantage of using Regional Forces to defend urban areas is that these

---

[445] The first one happened during the Easter Offensive of 1972, evacuation from Quảng Trị Province.

units are local and can identify the "invisible" collaborators and sympathizers more easily than the "regular." During the 1974-1975 strategic reposition (abandon I and II Corps), maneuvering of troops turns out significantly chaotic mostly because of interaction with the panicking civilians (regional forces families) who were trying to evacuate[446] with their military family members: regional-force soldiers attention, priorities and devotion are distracted and divided.

Brother Anh Bảy thought Lt. Gen. Trưởng was quite "obedient" to Pres. Thiệu. He could have resisted and proposed an alternative plan to Thiệu's "di tản chiến-thuật" strategic evacuation. We regretted that our nephew (Biên Hoà side), Lt. Gen. Đỗ Cao Trí[447] [448] had already died in a helicopter accident in Kampuchea. If Trí were the Vietnamese "flamboyant" Patton, Trưởng would have been the mellower version. Trí was more assertive and hard-headed, so he would not have followed Thiệu's order to abandon Central Vietnam without getting into a tantrum and going about his own way. He could have ordered a counter-attack and opened up a blood route to at least save those evacuating from the Imperial City of Huế, the provinces of Thừa Thiên and Quảng Ngãi, as well as the City and naval base of Đà Nẵng: they were strongly believed by commander Trưởng to still be defendable.

Nevertheless, in his autobiography[449], Trưởng had commented that Pres. Thiệu was "in disarray not letting his army commanders know what his intention or war plan was." Whether Thiệu was mad at Nixon/Kissinger's hypocrisy (betrayal while pretending they really cared for the RVN), or discouraged at U.S. Congress decision to abandon its ally, nobody knew for sure besides speculating about it. Abandoning Central would certainly demoralize soldiers, but what else could he have done? We were at a dead end and he was a perfect Confucian Gentleman "kẻ sĩ," an invincible docile servant!

---

446   To leave family behind means to let them be executed by the advancing and invading communists. It is nothing like the Dunkirk in France during WW2 when soldiers are evacuated without civilian families.

447   https://en.wikipedia.org/wiki/%C4%90%E1%BB%97_Cao_Tr%C3%AD

448   Trung Tướng Đỗ Cao Trí Article in Vietnamese published by Vương Hồng Anh in Chiến-Sĩ Cộng Hoà magazine

449   Vietnamese monthly magazine KBC Issue#5: *Vì sao tôi bỏ Quân Đoàn I?" (Why did I abandon I Corps?) by Ngô Quang Trưởng. Westminster, California (2003)*

Army General Đỗ Cao Trí[450] was one of the highest ranking generals of the ARVN. He was respected, not just by the ARVN, but also by the allies. Born in Biên Hoà, he was related to my first marriage in-law's side, so we knew him quite well and received political updates often. Trained by both the French (Auvour Infantry School in France) and the Americans (Command and General Staff, Fort Leavenworth, Kansas, and Air-Ground Operations School, Fort Kisler, Washington), Airborne Gen. Trí became one of Pres. Diệm's right-hand military commanders during the infamous Huế Buddhist Crisis and, later on, participated in the 1963 successful coup against Pres. Diệm's. He was promoted to Lieutenant General, commander of I Corps in 1963, then II Corps in 1964 in Central Việt Nam. Being suspected for being involved in subsequent coups, he was lifted from command by Prime Minister Nguyễn Khánh in Sep. 1964, and was forced into retirement in 1965 by Pres. Nguyễn văn Thiệu, who was at that time Chief of State of the Military Junta. Thiệu sent him to South Korea in semi-exile to serve as ambassador of the RVN, until after 1968 Tết Offensive when Thiệu recalled him and made him commander of III Corps to counter-balance VP Nguyễn Cao Kỳ (pro-American) power. Gen. Trí, nicknamed by the U.S. news media as the "Patton of the Parrot's Beak," launched the incursion into Kampuchea in 1970 [451] side by side with the "secret operation" by the U.S. army. For him, it was a two-prong campaign: to annihilate the NVA and VC's base (where they resupplied and received their pre-invasion training), and to avenge the Vietnamese immigrants in Kampuchea (new appellation for Cambodia) who were beheaded by the Khmer locals in their "Cáp Duồn" campaign[452]. Records as cited in the Vietnamese magazine "Chiến-sĩ Cộng-hoà", Issue #2, Aug. 2009 showed the ARVN from both III and IV Corps (Infantry, Rangers, Artillery, Armored) had joined in this operation, independent from the U.S. campaign; however, up until recently, American documentaries did not cover any RVN activity, including its participation in Kampuchea. Nixon had tried to cover up U.S.

---

450   https://en.wikipedia.org/wiki/Đỗ_Cao_Trí Gen. Đỗ Cao Trí's biography

451   Vietnamese periodic magazine *Chiến-sĩ Cộng-hoà, Issue #2, Aug. 2009* Garden Grove, CA.

452   As mentioned in earlier sections, Cap Duon was instigated by U.S.-backed Cambodian Premiere Lon Nol based on propagandas that Vietnamese had invaded and taken part of Khmer territories in the far past. It was propagated among angry nationalist Khmer so they would behead Vietnamese-Cambodian immigrants and throw the heads into the Mekong River. Lon Nol, who was prime minister in the early 1960s under Prince Norodom Sihanouk.

participation in Kampuchea (but he could not), as it was a violation of the sovereignty of an uninvolved country (neutral). Nevertheless, it was an ARVN victory, as there was a major display in front of the House of Parliament in Sài-gòn (now Grand Opera) of that action.

Unfortunately, Gen. Trí was killed when his helicopter was shot down in Feb. 1971. His death left many controversial theories among those who loved and knew him. Was he assassinated by VP Kỳ for siding with Pres. Thiệu and being a threat to American-backed (CIA?) the VP? Or was he killed by the U.S. for being close to rebelling Pres. Thiệu, the latter having shown signs of defiance against Nixon's policy in the Paris Accords negotiations? Gen. Trí was promoted Army General (four stars) posthumously.

The informed nationalists began realizing the Republic of Việt-nam and the National Liberation Front (VC) were nothing but Nixon and Kissinger's and the People's Democratic Republic of Việt-nam (North)'s pawns in this political chess game, and that the U.S. government and the DRV were the only two parties that had actually had the control over all the pieces. Who had really won the war? Who had lost? Who were the real Heroes and who the Villains? As the Truth would always be in the Eye of the Beholder, as there were always things we as human beings with diverse experiences could not and would not agree on. Only rationally thinking based, not on mouth-to-mouth stories, , commentaries, gossips or emotionally-diverted and unfounded information, but on more concrete factual verifiable data (e.g. by including the newly declassified documentaries, taped interviewed of decision-makers such as R. McNamara, Nixon, Kissinger, Thiệu, et.al.) could shed some brilliant lights over this complicated war that had divided so many hearts and souls.

CHAPTER 13

Win or Lose?

The War ended abruptly, as abruptly as it was started in 1964 with Pres. L. B. Johnson's Bay of Tonkin incident, through B52 bombing events in the North (Operation Linebacker I and II) leading to DRV returning to the bargain table finalizing of the Paris Accord Agreement in 1973[453]. Nevertheless, continued violations of Accord in 1973 through 1975[454] with increasing infiltration of NVA forces via HCM Trail, then the abandonment of the Army Corps I, then II by the RVN (as ordered in 1974-75 by Pres. Thiệu) caused by U.S. Congress to cut military aids and support[455] to the RVN, and the internal sabotage by various pro-communist factions (including the Buddhist protests)… all led to the inevitable fall of Sài-gòn in 1975.

Taking Advantage of the Rear Debauchery and Decadence

Indeed, towards the middle through the end of the war, as the U.S. was going through major internal socio-political turmoil and a recession economy at home, most Americans wanted to ignore or forget Việt Nam and refocus on their needs at home. Their survival was all that mattered, as they started realizing that it was people's life they should be concerned about, lives they ought to live, their self-satisfaction and love they ought to enjoy. Besides, there were many rights they did not know they had, and that thought they ought to explore more, demand and defend, such as their Civil and Human Rights, Women's Liberation, Gay and Lesbian's Rights, Freedom of Expression, Freedom of the Press… In the midst of expanding psychedelic and debauched life styles, some were even doing their best to probe to see how far they could push in

---

453   https://en.wikipedia.org/wiki/Operation_Linebacker_II and https://en.wikipedia.org/wiki/Paris_Peace_Accords
454   January 11, 2013 The 1973 Paris Peace Agreement Reconsidered published in Counter Punch - Tells the Facts, Names the Names blog by Gabriel Kolko – January 11, 2013
455   In the meantime, NVA and VC were allowed by Kissinger-negotiated Paris Accords to continue receiving supplies from China and the Soviet Union.

their protest against the establishment. The antiwar movement stemmed out of compassion and humanitarian feelings, in response to appalling images sent back from the front line in Việt Nam by the press media.

...One of my dear friends, a native of Minnesota, described her appreciation of how much Freedom, enlightenment and ecstasy she and her friends had discovered during the three days of the 1969 Woodstock Music and Art Fair. They could not believe they could be entitled to such an immense Freedom and such a way of life that had been hidden from them. In my opinion, the Woodstock "phenomenon" was the prelude to an American Socio-Cultural Revolution: "We were like in paradise... there was nothing but love, freedom and complete ignorance of the present or the future. We had sex all day, or enjoyed weed and nurturing companionship. Who cared about the war! We had to live the day, as tomorrow might not be the same." Indeed, who cares about the war! Half a million had attended this peaceful and orderly event[456]*, and, even though the old society frowns on them, they had discovered that everything they had thought forbidden was actually legal, and that they deserved it all...*

On this same note, I was wondering whether the Vietnamese communist propaganda machine had anything to do with helping promote a lifestyle of debauchery with their infiltration efforts. After all, they had more or less been "accommodating" the use of drugs among American GIs in Việt-nam, using children and women who believed they would be patriotic by helping the enemy sell weed to corrupt the wellbeing and morale of "invaders," and, at the same time, take advantage of the trend to bring in some income for their poor existence. Marijuana had been a foreign thing to Vietnamese people, who preferred opium, until they found out that American G.I.s loved it (even though it grows in the

---

[456] http://classicrock.about.com/od/photogalleries/ig/Woodstock-1969/ Woodstock 1969 Photo Gallery: "Half a million strong", "Billion year old carbon", "Reading, napping, and rapping", "Heat, rain, mud." http://classicrock.about.com/od/newreleases/tp/The-Woodstock-Diaries Woodstock daily diaries Aug 14-18, 1969 and the 2009 movie *Taking Woodstock (Amazon.com) illustrating this event.*

wild in Việt-nam) [457] and [458]. *Opium could be possibly sold to the incoming NVA in Vinh, DRV, North of the head of the HCM trail, by Miao farmers to the NVA[459] who resold it in the South to the G.I.s and South Vietnamese users via children and peasants; it could also be possibly sold by the CIA as per the RM#2 Miao's accusation. The opium issue had eventually become a major political controversy[460]. Nevertheless, U.S. Drug abuse seemed to culminate in the last years of their services in the War. If the communist had pulled the string to manipulate both civilians (politics) and the military (strategies), they had actually succeeded in corrupting the internal infrastructure of the rear support (American youth in the U.S.) as well as mental wellbeing of the front line. "Nội công, ngoại kích[461]," indeed! This was a communist strategy learned from the two Opium Wars that had incapacitated many Imperial armies and the Qing Dynasty of China during the wars against the Britain (1839) and France-Britain (1856)[462].*

The question concerning drug abuse still boiled down to be whether or not "Uncle Hồ's" team had anything to do with drug and weed use among American GIs in Việt-nam and the younger generation in the U.S. in the 1960-70s? Perhaps watching the 3-day Woodstock events in 1969 had given them the idea of using the 1800s Opium War tactics to corrupt the U.S. forces. It had surely helped them fight the war both in "front" (front line) and on the "back" (the support back

---

457   Opium was more popular with the rich Vietnamese during the French occupation, as it was legalize then, probably to corrupt anti-French movement. Intelligence, as discussed at our Gia-Định Sector Headquarters (counter substance-dependency reports – Bureaux Phòng 2 & 3) reported that Opium was brought in by NVA troops via HCM Trail for two purposes: for injured NVA troops (in place of Morphine) to reduce their pain, as for passing it on to American and RVN troops to cause dependency. Cheap Opium was sold by Miao tribes (their principal crop in High Laos mountains) in Laos and Vinh, North of the HCM Trail. Nevertheless, opium was not as popular as weed.

458   https://isthmus.com/opinion/opinion/vang-pao-drugs-and-the-cia/ Gen. Vang Pao and the Hmong opium crops during the Vietnam War

459   The NVA also assumed that opium would be popular among Americans.

460   The Hmong vs. CIA controversy on opium trade during the war had alleged that the CIA was involved in the opium trade with the Hmong, and had made them addicted to it. However, the bottom line was that Air America had used the planes to transport Gen. Vang Pao's crops out of Long Cheng poppy fields. Also, TPT PBS asserted that Opium was traditionally used by the Hmong as a medicine against most illness. .https://www.nybooks.com/articles/1990/11/22/heroin-laos-the-cia/ and https://en.wikipedia.org/wiki/Allegations_of_CIA_drug_trafficking

461   Sun Tzu's tactic of attacking the American government "rear support" in its own turf, while assaulting its forces at the front line in Vietnam.

462   https://www.britannica.com/topic/Opium-Wars and https://en.wikipedia.org/wiki/Opium_Wars Britain and France used opium to weaken China in the XIX Century.

home). More will be discussed in later chapters.

My brother-in-law who served as a Minister in Pres. Thiệu's Cabinet stated that NVA POWs had confessed they were the ones bringing into the South tons of opium they had bought, as I had mentioned earlier, from the Miao peasants in Vinh before they crossed into the Laos towards the Hồ Chí Minh Trail. This information verified the plan the NVA had developed using Opium to corrupt the American and the South Vietnam armies. I did not find out until later that the Miao's specialty was growing poppy crops in Laos, now verified as Gen. Vang Pao's RM#2[463]*. What an interesting controversy: Gen. Vang Pao selling opium to our common enemy, the NVA, to weaken the U.S. and RVN Army! How twistier could war politics turn out?*

The general question would be whether or not drug abuse issues were[464] *the only factor that had caused the war to become unpopular? New researches showed from findings via converted former VC and NVA members that underground infiltrators were not the only destruction agents involved, but that even "friendly" backdoor behind the scene opportunists were also successfully pulling the instigative strings, not just in Việt-nam but also in the U.S.*

In the meantime, even though they were excited about receiving updated but used equipment handed down by the departing U.S. Army, the **nationalist Vietnamese could not believe they could have been betrayed** so quickly and easily by their long-time trusted ally the United States.[465]. It would take them a while before they could imagine or visualize the impact of all the allies withdrawing; nevertheless, as reported via my brother Anh Bảy, the Sài-gòn government strongly believed its army would be able to handle replacing the allies. The ARVN had been fighting their own war anyway; it would be just a matter of upgrading the ARVN from 800,000 to

---

463   CIA and Vang Pao controversies relating to opium: https://www.revealnews.org/article/pao-drugs-and-the-cia/ , https://www.nybooks.com/articles/1990/11/22/heroin-laos-the-cia/ Marc Eisen confirmed on heroin deals between Vang Pao and South Vietnamese syndicates https://isthmus.com/opinion/opinion/vang-pao-drugs-and-the-cia/

464   CNN series on the 1960s pointed out that this was a new era when Television was becoming the main attraction for American families and that networks were extremely competitive to push themselves forward. The winning ones are the ones reporting the most on the life of GIs: how they struggled, how they survived and how they died.

465   *Khi Đồng Minh Tháo Chạy* in Vietnamese (When Our Allies All Fled) by Dr. Nguyễn, Hưng Tiến - 2005

1,200,000 strong to fill in the gaps. Perhaps this explained why I found everybody remaining calm in Sài-gòn as if nothing had happened.

Who could tell the future? It might even make it easier for some commanders, as there would be no one else to breathe "side-tracking recommendations and rules of engagement" on their neck. Nevertheless, adjusting to the changes would never be easy by all means, especially when the U.S. had also cut off all aids and allowed the enemy, via the Paris Treaty, to remain in the South and continue to occupy taken territories. Life went on, anyway, as the RVN had no choice, really no choice! With the newly acquired equipment, they might be able to handle the commies. Another omen was that the latter were also allowed the resupply of used "equipment" and perhaps "replacement of unusable personnel!" **It was the Agony of the South, indeed!**

A leak that unveiled the Truth of the Matter: The Pentagon Papers[466], leaked by Daniel Ellsberg (and accomplice Anthony Russo) in Oct. 1969 and fully declassified in 2011, actually revealed a sad truth that, to Pres. Kennedy and, especially, Pres. Johnson, South Vietnam was nothing but a pawn manipulated by the CIA since its earlier existence under the name Office of Strategic Services. It was the CIA that helped set up Pres. Ngô Đình Diệm as the first President of the RVN in 1955, and it was the same CIA that helped rebel generals plot their coup that overthrew Pres. Diệm in 1963. The Defense Department under Pres. Johnson determined the reasons why the U.S. was persistent about its involvement in Việt-nam were:

- *"70% – To avoid a humiliating U.S. defeat.*

- *20% – To keep [South Vietnam] (and the adjacent) territory from Chinese hands.*

- *10% – To permit the people [of South Vietnam] to enjoy a better, freer way of life.*

- *ALSO – To emerge from the crisis without unacceptable taint from methods used.*

- *NOT – To help a friend"*

---

[466] The Pentagon Papers: https://en.wikipedia.org/wiki/Pentagon_Papers#Internal_affairs_of_Vietnam , http://www.history.com/topics/vietnam-war/pentagon-papers

The Tonkin Incident was preempted as soon as the South was able to stand up on its feet to stir up an escalation that led to the bloodiest and most costly war in history, involving, not just Việt-nam, but also Laos and Kampuchea.

EXCERPT FROM DECLASSIFIED "TOP SECRET" DRAFT[467]

3/24/65 (first draft)

ANNEX-PLAN OF ACTION FOR SOUTH VIETNAM

1. U.S. aims:

70% --To avoid a humiliating U.S. defeat (to our reputation as a guarantor).
20%--To keep SVN (and then adjacent) territory from Chinese hands.
10%--To permit the people of SVN to enjoy a better, freer way of life.

ALSO--To emerge from crisis without unacceptable taint from methods used.
NOT--To "help a friend," although it would be hard to stay in if asked out.

2. *The situation:* The situation in general is bad and deteriorating. The VC have the initiative. Defeatism is gaining among the rural population, somewhat in the cities, and even among the soldiers--especially those with relatives in rural areas. The Hop Tac area around Sài-gòn is making little progress; the Delta stays bad; the country has been severed in the north. GVN control is shrinking to enclaves, some burdened with refugees. In Sài-gòn we have a remission: Quat is giving hope on the civilian side, the Buddhists have calmed, and the split generals are in uneasy equilibrium.

3. *The preliminary question:* Can the

---
[467] https://www.mtholyoke.edu/acad/intrel/pentagon3/doc253.htm

situation inside SVN be bottomed out (a) without extreme measures against the DRV and/or (b) without deployment of large numbers of U.S. (and other) combat troops inside SVN? The answer is perhaps, but probably no.

4. *Ways GVN might collapse*:

(a) VC successes reduce GVN control to enclaves, causing:

(1) insurrection in the enclaved population,
(2) massive defections of ARVN soldiers and even units,
(3) aggravated dissension and impotence in Sài-gòn,
(4) defeatism and reorientation by key GVN officials,
(5) entrance of left-wing elements into the government,
(6) emergence of a popular-front regime,
(7) request that U.S. leave,
(8) concessions to the VC, and
(9) accommodations to the DRV.

(b) VC with DRV volunteers concentrate on I & II Corps,

(1) conquering principal GVN-held enclaves there,
(2) declaring Liberation Government,
(3) joining the I & II Corps areas to the DRV, and
(4) pressing the course in (a) above for rest of SVN.

Even though approximately 1.5 Million Americans[468] had actively participated at the frontline in this war, in one way or another, a vast majority of American civilians at home, including those with family members and relatives who were participants in the war, did not really know much about what was happening. If they did, it would just be conflicting and

---

[468] Vietnam in HD: On the Frontline (2016) by HistHD History Channel

confusing bits and pieces, particularly "Why" and "How" it was fought. As a matter of fact, as far as the RVN was concerned, this was the time when America might not even care about what would happen to its abandoned allies in the aftermath.

Benefitting from the Allies' Withdrawal

Although it might be more or less true that the high command of the ARVN was plagued with, on the one hand, corruption[469] and, on the other hand, incompetence and mutually-undermining rivalry between multiple old-time leaders from the "ancient" French-colony era, the "little soldiers[470]" and officers were credited by eye-witness U.S. Veterans for their bravery and valor at the battle field.

It was also unfair for the news media to expect the ARVN to continue fighting full-fledged an enemy that was armed to the teeth after "Vietnamization," as if it were still receiving full equipment, artillery and air support and supplies as in the late 1960s. Besides, the enemy could now parade legitimately and openly in their occupied territories instead of being pushed out as they would before Paris! The RVN was now like a patient's body plagued with cancer, as a result of Kissinger's low-hand moves in the Paris card game, secretly colluding with his counterpart and enemy North Vietnam Lê Đức Thọ.

Even before the Paris Accords became effective, aids started being cut and the RVN had to "frugally ration ammunitions." Actually, as I could vouch for this when ammo and air and artillery support started to be restricted as soon as in 1972-73 as a result of political rhetoric from the U.S. Democrat-controlled Congress and the weakened White House. As a consequence, full military aid cuts lead to ARVN troops' uncertainty; nevertheless, ally's withdrawal did not actually

---

469 ... as rumored that a great many corrupt officers between captain (company commander) and colonel (battalion or division commander) or even higher were taking bribes to maintain "ghost soldiers," the "lính ma" (on roll but never present), or belonging to the "network" with upper echelon in order to receive favors (promotion). These were the causes of incompetence.

470 In the context of this book, "little soldier" refers to the ARVN soldiers who are in the lower ranking echelon, including second class "binh nhì" to first and second lieutenant (and maybe some exceptional captain) "thiếu, trung và đại uý" who execute war plans and usually have limited operational authorities.

lowered morale and fighting spirit... Still, while all of the allies were getting ready to disappear, the ARVN was determined to continue to fight, and the enemy to muscle up and to engulf the South. The "Hội Nghị Diên Hồng" call for patriots to defend their Ancestral Land was sounding again: "How can we fight without resupply?" The answer from both the communist and the opposing nationalist fighters sounded out in unison "Sacrifice!"

> "Toàn dân! Nghe chăng? Sơn hà nguy biến!"
> To the people: Can you hear me?
> Mountains and Rivers
> (Ancestral land) are at risk!
> "Hận thù đằng đằng! Biên thùy rung chuyển"
> Hatred is piling up! Borders are quaking.
> "Tuông giày non sông rên vang tiếng vó câu"
> Stamping Mountains and Rivers
> (enemy), cavalry is
> thundering.

During the time I was being transported in long convoy across provinces (end of 1973: Nha Trang through Tuy Hòa Pass to Bình Định II Corps Sector), deploying into battle fields or being dropped from Huey-copters (1974) with my Infantry platoon, or patrolling in a jeep along rural areas with my boss Lt Col. Quý (1975)..., I never heard anyone panic, or even moan and groan out of control, in spite of we were under attack or someone got shot at. Everyone was smiling and conducting normal daily business at the front line. The civilians were doing the same thing.

The ARVN was indeed able to keep their morale up, and continued to fight bravely to successfully liberate cities and townships occupied by the enemy during the 1972 Easter Offensive, not in disarray or chaos as claimed by the American television press media as I found out later. At that time, as we were ordered to restrict the use of ammunitions, preserve weapons and equipment, we were all aware that U.S. military aids were being drastically cut. Huey helicopters, instead of landing on the group to launch our platoons, would not go lower than 8 to 10 feet off the ground to release troops they needed to take off right away and not get shot down. We were also ordered to preserve equipment to the maximum, and, if needed, to sacrifice personnel in order to recover lost weapons[471]*... Nevertheless, it had become so obvious when we had started to ration and count bullets, and shells, and to salvage parts from used hardware (M16 riffles, M79 grenade launchers, M72 antitank, artillery pieces, tanks, Huey's...). It felt as if there were no more formula for mothers to feed their infants, and the latter had to live on meager rice juice.*

We knew the republic was in danger, but we did not freak out. What really mattered for us was, not a lack of human power, but rather whether or not we had enough supplies and functional equipment to defend a long country that used to be protected by many armies of the world. In the meantime, the enemy whom we captured was armed to the teeth, and even T55 and PT76 (new Soviet amphibious) tanks, with abundant artillery, mortar and antitank B40[472] ammo newly supplied by the Soviet and China via the expanded HCM Highway. All of those were legitimized by the Nixon and Kissinger team's "arrangement" at the Paris Accords.

Whether Winners or Not, They are the Heroes.

Now, more than forty years after the Vietnam War has ended, the outcome still remains a mystery to many Americans. Many are still wondering "Who won the War?" or "What is the difference between the Geneva Conference of 1954 and the Paris Accord of 1973?"

As quoted by an actual "two tours" veterans: *"Vietnam Veterans, even though time has somewhat helped them make peace with their existence at home, still feel bitter and estranged as they are not welcome back home as heroes as the WWI and WWII combatants had been*[473]. *Instead, they have for years been treated as "baby-killers, hut-burners and monsters," have trash thrown at, or are disrespectfully cursed at. Moreover, in addition to the very public physical and mental abuse Vietnam veterans have to endure, many must struggle*

---

[471] Americans and allies did the opposite: sacrifice equipment in order to save lives.

[472] The original RPG from Russia and China.

[473] Documentary films show glorious parade welcoming WW1 and WW2 soldiers back with jungles of people cheering for them while confetti rained from the high rise. Vietnam veterans have never had such a welcome back!

and continue to struggle with Post Traumatic Syndromes (PTSD). This has been a damning indictment against the returnees, who, in service of their country, are blamed for all the ills and social woes that the war has brought. Americans serving in Vietnam are not the ones who have made rules of engagement and war policy. As patriots who serve their call for duties, just like the WW1 and WW2 soldiers, they only do what they are asked to do and serve their nation and their people (many Americans do not have to), and yet they are blamed for all the terrible things that have incurred in the war nonetheless! We will never know the numbers of the individual returnees who have suffered personally from the collective guilt Americans have felt about the Vietnam War. They have never expected such a return home and are struck-shocked by it.

In addition to being treated unfairly, many have been suffering mentally and physically because of Agent Orange which they were soaked in during the tours in Nam their country has asked them to serve. Even though some recognition has been made now (e.g. the National Welcome Home Day or Vietnam Veterans Day on March 29), efforts to psychologically support the Vietnam Veterans, in my opinion, is still superficial from the government and those who have maltreated them and many Vietnam Veterans see a "too little too late" move by such efforts, since there is some confusion as to who benefits from such celebrations."

A vouch for the G.I.s accused as "baby-killers:" "Ironically, never before in the world have there been expeditionary forces that are so much in love with children! Just Google search on the web for photos of American G.I.s in Việt-nam and you will be flooded with photos of American troops giving treats to Vietnamese kids, playing with them, helping them and their parents and grandparents do heavy chores, rescuing them from harm's way... As I am writing right now and searching for 'pictures, American GI, children' on the web to develop PowerPoint presentation, I can confirm that 99% of the pictures that come up are of troops loving, entertaining, and caring for South Vietnamese children. It just brings tears to my eyes seeing those photos and thinking: "Who has the decency to accuse these most kid-loving people for 'baby-killing'?

Never before in American history, and nowhere else in the world had war heroes been treated this unfairly, by mostly their own people who themselves had neither served nor ever served. Furthermore, deep inside, these people could not believe or accept the idea they, not the soldiers, had lost the war or helped the enemy win it. The U.S. did not lose the war because of the enemy superiority, but rather because of the confusion the people back home had about the war and the lack of a well-planned psychological warfare strategy. They ought to recognize and honor these veterans as they had done their best with very little support from the "rear." The next sections would explain how and why.

Win Lose, Lose Win

Towards the end of the war, when the Paris Accords negotiations were in a deadlock, the Nixon and Kissinger government could not wait to step away from the war "with honor," so that normal life would resume in America and people would think "we did not lose the war: we left it before it ended." The bottom line was, as revealed by documentaries "Trials of Henry Kissinger" and the declassified Nixon audio tapes, that winning the second term was much more important to Nixon than abandoning and betraying a close ally. He was hoping that an instant Peace, coined with the magic word Honor, would immediately re-establish the balance between the three success elements "Thiên Thời, Địa Lợi, Nhân Hoà" (Celestial Timeliness, Favorable Environment and Mindful Harmony) that would allow Pres. Nixon to win his second term election and Sec. Kissinger to maintain his influential and powerful position… That was all it mattered to Nixon and Kissinger.

The truth of the matter was that winning, losing, or even breaking even would really depend upon the eyes of the beholder. No matter what Nixon and Kissinger tried, for sure the majority of the Americans, as dedicated sports fans, had known deep inside their soul that the U.S. had actually forfeited in the middle of a major game. A proud country that had never lost any war had to face failure this time, overwhelmed, unpopular, exhausted, depleted, both physically as well as mentally.

In sports, Americans' favorite hobbies and entertainment, or in any kind of competitive game, whether it was forfeiting, choosing to surrender, giving up, or refusing to continue before the game ended because of one's inability... all of the above meant losing the battle:

...

Cambridge Dictionary[474] *clearly defines "to forfeit" in American English:" "To give up or lose something because you cannot do something that the rules or the law says you must do... Ex: to give up or lose something because you cannot do something that the rules or the law says you must do: to give up or lose something because you cannot do something that the rules or the law says you must do: She had to forfeit the tennis match after she fell and hurt her wrist....'*

Oxford Dictionary[475] *defines "to forfeit:" "to lose or give up (something) as a necessary consequence of something else."*

Legal Dictionary[476] *defines "forfeiture:" "the involuntary relinquishment of money or property without compensation as a consequence of a breach or nonperformance of some legal obligation...*

Basketball[477] *rules published by FIBA, a forfeit and default are two different things. A forfeit occurs if:*

- *Fifteen minutes after the scheduled starting time, the team is not present or is unable to field five players ready to play.*

- *Its actions prevent the game from being played.*

- *It refuses to play after being instructed to do so by the referee.*

A forfeit results in loss for the offending team, a score of 20−0, and in tournaments that use the FIBA points system, 0 points in classification of teams ...

474	http://dictionary.cambridge.org/us/dictionary/english/forfeit
475	http://www.oxforddictionaries.com/definition/english/forfeit
476	http://legal-dictionary.thefreedictionary.com/Forfeiture
477	https://en.wikipedia.org/wiki/Forfeit_(sport)

If I am not mistaken, the world always thinks of Winning or Losing as the obvious outcomes such as: card or board games, sports activities, duel or challenges, arguments, and war campaigns. It has always been understood that the sole purpose of a match between two individuals or two groups of individuals is to see who will win the contest... until Vietnam War.

Turning the "Peace with Honor" Switch on

The U.S. was not the only country fighting for South Vietnam. South Korea (Republic of Korea ROK), Australia, New Zealand, the Philippines, Thailand, and Taiwan had fought in this war, too. **Everybody withdraw as if a war can be turned on and off with a switch.** I have always wondered what the citizens in those countries think about the outcome of the war. Do they ever question the "discrete" and rapid withdrawal[478] initiated by the U.S.? For many years, I keep wondering how they feel about their country abandoning its ally? Or do they feel the same way as the Americans do, relieved that their loved ones are coming home, "having successfully escaped the cataclysmic hell of Nam?" After all, they have participated in the war to assist the U.S. One can presume that they are not the primary one responsible for the withdrawal. The truth of the matter is that it is all about survival of the fittest: save oneself first and not to care much about the others. It is very sad but funny to realize that, in a sudden, the whole football team is reduced to one player playing against a full strength opposition!

I was drafted into the ARVN when the Paris Accords were finalized. Interestingly, I never heard Sài-gònese discuss or panic over our allies having withdrawn from the war until I was told by my in-laws when they were discussing Pres. Thiệu's order to abandon I and II Corps in our family gathering (end of 1972). How could I have missed the news about the allies withdrawing? They would come back, wouldn't they? Were we thinking it would be just like the cases of Dunkirk, or Gen. MacArthur in the Philippines? The Americans had never abandoned their allies, had they ever?

---

478   *Khi Đồng Minh Tháo Chạy* in Vietnamese (When All of Our Allies Fled). Nguyễn, Hưng Tiến, Ph.D. (2005). San Jose, California: Hứa Chân Minh Publishing.

Then the newspapers began publishing stories and photos of chaotic military convoys mixed with columns of refugees and civilians bombarded by VC and NVA on the highways. It seemed to me changes had happened so fast, overnight. How could 536,000 U.S. soldiers, 50,000 S. Koreans, 8,000 Australians, and thousands of other allies just disappear in a blink of an eye? That was unbelievable! How did the people in Sài-gòn[479] feel about the absence of allies then? I didn't really see people behave differently! Why did the RVN government not make a really big deal about it, at least to alert its citizens of incoming cataclysm? Or was I so busy studying and making a living that I did not even notice political and strategic changes? Or was I, just like many other civilians, unaware or unsensitized to war politics? Why did Vietnamese civilians not panic? It was strange that, when I spoke with a lot of people at the university, no one even brought it up! Not even during our summer military training (for male students) when we spent weeks in the same classes and barrack, or out on the practice field.

The politicization of civilians in Sài-gòn at that time was obviously at a really low level and psychological warfare on the part of RVN particularly weak! Everyone minded their own business, as I did; therefore, no one was up to date on the daily news. Life went on as if everything was normal. I was certain the government had long felt the RVN would sooner or later be abandoned by its allies long before it happened.

Now that I thought about it, probably the reason why the U.S. and the allies could claim they did not lose the war when they had actually "forfeiting" (I believe this word is appropriate for me to use): the war had not ended yet. There was still one player in the game, playing to the death against a whole team of opponents. That could be how the phrases "Withdrawal with Honor," or "Peace with Honor" would make sense and would appease **most Americans** so they would vote for Nixon. Anyway, they were tired of the endless "futile war" and wanted nothing but to clear their mind of it. One thing for sure, no one wanted to lose a war. Yes indeed, **America had never lost a War. Neither did America's Free World Coalition!**

The purpose of the wordings was solely to provide them with a reassuring relief and consolation so they could

---

[479] Perhaps people in other cities where the allies stationed knew the latter were withdrawing from Vietnam

make Peace with their mind: that Peace was coming and their friend the RVN and its people would finally achieve Peace. Yes indeed, **Make Peace with Oneself!** Probably, deep inside, Nixon and Kissinger knew the RVN would not last long without resupply and the communist would win as they would continue to resupply, and replenish their forces. As they were long gone, it would not matter to them whether or not the South would be defeated two years later and over a million more lives would be wasted[480]! The Việt-nam nightmare had already been wiped out of their memory…

Anyway, everyone had to admit it: it had been a very difficult and confusing war… but it was obvious someone had won, and someone else had lost the war. It sounded like I was whining, moaning and groaning about the injustice that the People of the RVN and its army had endured, but I would be hypocrite if I did not. I felt most obligated to complain repeatedly, at least for the innocent voiceless dead and the disabled millions others who could not moan and groan, as well as for the U.S. military who, upon returning home, had been treated with shame, disrespect, and dishonor by their own People for whose Freedom and Democracy they had sacrificed. May they all **Make Peace with their mind**, too. Whenever the war was discussed, the same questions continued to resurface: "who had really won Vietnam War?", and "who had really lost it?" or, to avoid the words "lose or surrender, "who are the heroes, the villains and the victims?"

As the Vietnam War winded down, there were actually two major frontlines, if we did not want to count the "indirect ones" that were going on with the antiwar movements in the U.S. and in Vietnam: the bloody one in Việt Nam where all parties were doing their best to annihilate the opposition, and the policy one behind the scene where the U.S. was making deals with the Soviet Union and China. The former related to the deadly endless bloody confrontation between opposing marionette forces, while the latter was actually a living room or meeting room dialogue where the puppet handlers on both sides were manipulating the strings with glasses of wine in their hand. They both had absolutely nothing to lose. If they had a dispute, they could just negotiate and mediate. On the contrary, the puppets were getting hurt or killed, and their bloodshed and sacrifices could never be negotiated by a Peace with Honor.

---

480   Final phase of the war and aftermath refugees.

If Vietnam War were just a skirmish between groups of fighters, then perhaps mediation and negotiation might be successfully used to resolve differences. This is what we do in secondary school settings whenever there is a disagreement or verbal fight[481]. The administrators always arrange a meditative meeting to trouble-shoot or problem solve. Both sides "swim up the stream of events and go over what has happened, the "who-did-what's," from the very beginning, ... and, usually, all sides will blame it on misunderstanding, misjudgment, someone else's instigating... Finally, both sides will make peace with Honor, promise it will not happen again, work out restitutions and life start all over again. That is Peace with Honor...

Perhaps I am trying to justify what both Nixon and Kissinger have done trying to save the U.S. from a shameful defeat, but this mediation-type of argument still does not make any sense and cannot clear them from their enormous unpatriotic and inhumane mistakes and their purposeful wrong-doing. One cannot self-mediate a war one has started and escalated oneself, resulting in millions of lives lost and subsequent human suffering. Land destruction, Agent Orange, permanent physical as well as mental disabilities (post traumatic syndrome), family separation, vindication...! It is unforgivable!

No matter what rhetoric both Nixon and Kissinger had used, everything still boiled down to one reality: a presidency had started a war which another could not handle, so one negotiated a way out and tried to cover it up using dirty tricks, through secret negotiating with the enemies and playing a low hand cards, **in the name of flowery wordings "Peace with Honor" or "Withdrawal with Honor,"** and at the expense of its ally's existence. It was truly "surrender in disguise." It was truly Abandonment and Betrayal. Nixon, Kissinger, the U.S. Congress, specifically the United States of America, had definitely lost the war because of the two men in power's selfishness and thirst for personal power. Clearly, Kissinger had already made up his mind when he wrote in his memoir in 1969 when Pres. Nixon visited Sài-gòn: "We were clearly on the way out of Việt Nam, by negotiations if possible, by unilateral withdrawal if necessary.[482]" That was just a year after

---

481   Physical fight always results in a three-day suspension.
482   Excerpts from http://www.newrepublic.com/article/79923/promises-promises: The Palace File

the Tết Offensive of 1968!

Nevertheless, things are more complicated than just withdrawing from the war. Those who had really committed to it (American G.I.'s are included), and who have seen companions-in-arms sacrifice their life will not allow their leaders to turn their back and "brush dust off from their hands[483]. That is exactly what Nixon and Kissinger have done to the Vietnam War men and women, whether they were Americans, Allies or South Vietnamese.

When Kissinger was negotiating with his counterpart Thọ, he foolishly, if not cowardly, agreed on the Paris Accords that **North and South Vietnamese forces were to hold their occupied territories in the South.** He had submissively played with a-lower-card-hand and actually agreed to allow the NVA to maintain their current positions with full renewable strength and replenishable weaponry and supplies, while the ARVN was using its reserve and recycled equipment from pre-1972 aids. **Resupply military materials**[484] for the enemy meant continued abundant aids from the Soviet Union and China **to the extent necessary to replace items consumed in the course of the truce**[485].

It was not just a simple unconditional withdrawing of the U.S., but it had at the same time given the NVA 100,000 main force combatants[486] the control of their occupied land in the South and deliberately handed them the machete to cut the throat of the South. How could Kissinger done that? It was as if Pres. Lincoln had accepted the South's terms of surrender while allowing it to continue lynching all slaves and giving Gen. Lee full control of the South! Why was he in such a rush in concluding the war? He knew that Ambassador Bunker had affirmed with Pres. Thiệu (June 3, 1971) in his presence that the NVA would not be allowed to stay in the South, and yet, he had accepted Hà-nội's Aug. 1, 1972 demand that, not just their forces be kept in the South, but also that Pres. Thiệu resign[487].

by Nguyen Tien Hung and Jerrold L. Schecter

483   In Vietnamese "phủi tay" meaning "I am out of it."
484   ...while the RVN had to survive with no supply as all aid had been cut
485   https://en.wikipedia.org/wiki/Paris_Peace_Accords Paris Accords Provisions
486   Excerpts from http://www.newrepublic.com/article/79923/promises-promises: The Palace File
by Nguyen Tien Hung and Jerrold L. Schecter
487   As reported in Mường Giang's Blog Xóm Cồn Hạ Uy Di posted in Sep,

The NVA and VC, I was sure, had to laugh really hard, not at Kissinger's inept diplomacy[488], but at the U.S. succumbing to Hà-nội outwitting it by using its own People's heart! Could the most powerful country in the world be defeated by the poorer and weaker one? **It was all about Uncle Hồ's "Eternal Resistance."**

Eternal Resistance "Trường Kỳ Kháng Chiến" and the Anti-war movement in the U.S.

...In an interview by Minnesota attorney and human right activist Stephen Young, former NVA Gen. Bùi Tín[489], now in exile in France, had quoted[490] Hồ Chí Minh's statement[491] that the communists "don't need to win military victories, but we only need to hit them until they give up and get out." This reflected their philosophy of "Trường Kỳ Kháng Chiến" (expression that could be translated into "Eternal Resistance") which might be influenced by China Chairman Mao Tse-tung's 1966 Cultural Revolution[492] or Lenin's Permanent Revolution. Perhaps the Vietnamese Communist Party was well aware that, because of the periodic change in the U.S. government and of, especially, the American people's change in opinion, the U.S. would not do well in an "eternal war." It also realized that this powerful country was weakened economically and socio-politically since the late 1960s and that the government was experiencing a major socio-economic crisis. Hồ's strategy of

---

2009 by the RVN Strategic Technical Directorate

488   Or was Kissinger too distracted by Israel tension with Egypt in the pre-empted Yom Kippur?

489   He was the NVA Thượng Tá Colonel who, as the highest-ranking NVA officer present in Sài-gòn on April 30, 1975, had accepted the unconditional surrender from the RVN interim government of Dương văn Minh in the Independence Presidential Palace. From _How North Vietnam Won the War by Stephen Young._ http://www.viet-myths.net/buitin.htm.

490   _Gen. Bui Tin Describes North Vietnam's Victory_ Article Excerpts of The Wall Street Journal (Aug 3, 1995) interview of NVA Gen. Bui Tin by Minnesota attorney and civil rights activist Stephen Young, posted late by Sgt Grit Staff on Sgt Grit Blog pages. **http://www.grunt.com/corps/scuttlebutt/marine-corps-stories/gen-bui-tin-describes-north-vietnams-victory/**

491   In Vietnamese http://www.baomoi.com/bac-day-khang-chien-phai-truong-ky/c Hồ Chí Minh strongly believed that Resistance had to be long-lasting when making plans fighting the French: "Trường là dài, tức là đánh bao giờ địch bại, địch 'cút', thể mới là trường". The enemy wanted to fight a quick war, quick victory, fast solution, so the Party called for a long resistance, "because our homeland is narrow, our people are few, our country is poor... If the French were thick orange peel, we will need to take time to file our nail sharp, so we can defeat them wide open."

492   http://www.nytimes.com/learning/general/onthisday/bday/1226.html

the mind turned out to be simple: be persistent and just wait for the enemy to collapse on its own blood.

To win a war, one must strike from all directions: front, back and flanks, from outside in (attack from perimeters) and from inside out (rear support, rear war "hậu cần"), physically (using forces) as well as mentally (psychological warfare), regardless of time and space availability. Use any means, anyone, any age, any gender, friends or foes… to fight them. **This was transcended from Hồ Chí Minh's "eternal resistance" when promoting a restless war of attrition.** *"Because our Ancestral Land is long, our resources are limited, but our generations are eternal."* This correlates with *"Thiên thời, Địa Lợi, Nhân Hoà"(See Sun Tzu).*

Interestingly, Gen. Moshe Dayan, during his 1966 "study-visit" in Việt Nam, had sensed the problems of mismatching the <u>Art of War</u> teaching between the U.S. Army goals and action and the communist ones, which he had reflected in his articles about the politics of war[493]. The U.S. policy in Việt Nam seemed to be ineffective as a "war of containment[494]," to prevent the spread of Communism. This containment policy had probably deviated war efforts from the original war of attrition one when it was started in the early 1960s. This subtle change of course, in my opinion, might be due to Nixon and Kissinger's strong belief they could solve the war by negotiating.

Liberate or Invade?

It is clear that, on the one hand, the People's Democratic Republic of Việt-nam DRV has won the military campaign thanks to their well-funded unlimited supplies from their allies that have weaponized their "eternal resistance"" endurance and strong held-belief that the Americans are there to invade Việt Nam. It is plain to see that, on the other hand, the RVN has no choice but to lay down their weapons and surrender when equipment wears down and supplies run out. While the allies of the DRV (mainly China and the Soviet Union, and Cuba, North Korea) have not giving up on them but, on the contrary, have even bumped up reinforcement, all

---

493   http://www.historynet.com/moshe-dayan-sounds-the-alarm-in-vietnam
494   https://history.state.gov/departmenthistory/short-history/containmentand-coldwar

the countries that have supported the RVN have, not just fled, but also turned off the blood supply to its brain in the middle of the game. Nevertheless, winning or losing might not matter as much as the millions of innocent lives that are sacrificed before, during the war and thereafter, and millions more that continue to succumb to a total disaster. The main question is: **why does the North, which has been trying to trick the world into believing that NVA are actually the local Southerner NLF (fighting under the NLF flag[495]), still want to liberate the South when the so-called invaders have been long gone?**

The twisty part remains entangled in Nixon and Kissinger's expression "Peace with Honor", as they hope it will work tricking people into believing that the U.S. has not lost the war, so they will vote for them. It insinuates that none of the U.S. and allies has been defeated yet, and that they have left Việt Nam in order to "allow Peace with Honor to be fulfilled through negotiations." Even the world is tricked by the word Peace everyone loves, the peace they have been pushing for when a Peace Nobel Prize is given to both Kissinger and Thọ…

Is it possible that a major near-world-war-caliber Vietnam War can be resolved in the same way the skirmishes between two groups of high school students can?... by problem-solving, giving-and-taking (actually more give than take for the voiceless and powerless nationalists) and mediation?... with the expectation that all sides will come out of mediation and "chill" as if nothing has happened… Quite impossible! Vietnam War is not a little skirmish, or not even a "conflict" as many analysts have coined: it actually costs over a million lives. Besides, the Paris Accords Agreement will not mean anything when the bullies have not got what they want. They will try again until they achieve their ultimate goal: win at any cost. Eventually, they will cheat and attack again until final victory! It is not fighting for Independence any longer because the alleged "American imperialist invaders" are long gone. Inevitably, the bullied one, the RVN, will fold into fetal position and give in when all aids dry up. In the name of what is the DRV attacking the RVN this time? **Liberate the South from what? There has been no more "invasion for two years since Jan 1973,"** if there was one… Did they turn it into a civil war (still led by the NLF) between the protesters,

---

495   Blue and Red flag with a yellow star in the center, as seen on most NVA vehicles, tanks that entered Sài-gòn on April 30, 1975.

intellect and Buddhist? Was that why the Buddhist had stirred up the civilian population during the last few months, fighting the 2nd republic "injustice and oppression[496]?"

How could the South have prevented the spread of cancer when, by nature, that cancer would eventually spread again? Perhaps, the only way to cure Cancer to the root might require an atomic bomb!? For innocent civilians, it would be just like to push the reset button and restart their life with reincarnation. For the bad guys, it would be their total annihilation…

Both Moshe Dayan and Bùi Tín concurred in one point: "the key to victory relies on the breaking of Hà-nội's fighting spirit…by keeping up the heavy bombing of North Vietnam… Hồ Chí Minh will not be able to withstand for long." When training his anti-French troops, Hồ had asserted that, "as Việt Nam is poor, militarily weak and geographically small but elongated, in order to defeat the French, the people have to prepare carefully for war, a total preparedness, to fight an extended long war of resistance 'Trường-kỳ Kháng chiến[497]', the Eternal Resistance." In order to defeat Hồ, Dayan and Bùi Tín had commented that the nationalists could have won the war should they have fought harder, at maximum power, at all battles, in a strictly war of attrition, the way Pres. Eisenhower and his clan had wanted both Presidents John F. Kennedy and Ngô Đình Diệm had done in the early 1960s. Tín asserted that, should Pres. Johnson had granted Gen. Westmoreland's request to enter Laos and block the Hồ Chí Minh's Trail, Hà-nội could not have won the war.

Stephen Young: *How did Hà-nội intend to defeat the Americans?*

Bùi Tín: *By fighting a long war which would break their will to help South Việt Nam. Hồ Chí Minh said, "We don't need to win military victories, we only need to hit them until they give up and get out."*

Stephen Young: *Was the American antiwar movement important to Hà-nội's victory?*
Bùi Tín: *It was essential to our strategy. Support of the war from our rear was completely secure while the American*

496
497   http://www.baomoi.com/bac-day-khang-chien-phai-truong-ky Trường-kỳ Kháng chiến eternal resistance in Vietnamese

rear was vulnerable. Every day our leadership would listen to world news over the radio at 9 a.m. to follow the growth of the American antiwar movement. Visits to Hà-nội by people like Jane Fonda, and former Attorney Gen. Ramsey Clark and ministers gave us confidence that we should hold on in the face of battlefield reverses. We were elated when Jane Fonda, wearing a red Vietnamese dress, said at a press conference that she was ashamed of American actions in the war and that she would struggle along with us.

Stephen Young:: Did the Politburo pay attention to these visits?
Bùi Tín: *Keenly.*

Stephen Young:: Why?
Bùi Tín: *Those people represented the "Conscience of America." The "Conscience of America" was part of its war-making capability, and we were turning that power in our favor. America lost because of its democracy; through dissent and protest it lost the ability to mobilize a will to win.*

The general also admitted that the National Liberation Front NLF (South Việt Nam VC) was actually a People's Democratic Republic DRV set up: "There is only one Party and one Army in the war to liberate South Việt Nam, the Vietnamese Communist Party, led by Chairman Hồ Chí Minh.

The truth of the matter was that the People's Democratic Republic of Việt-nam DRV had continued to infiltrate, with not just "replacement supplies, equipment and weapons," but also with new fresh troops coming from the North as well as new and more modern equipment from China border. The U.S. Congress's shutting down all aids meant strangling the RVN to death. I am simply proving that the U.S. government had actually taken advantage of the Americans' confusion about the war to by-pass them and negotiate an unconditional withdrawal more or less in favor of the NVA and VC, and, at the same time, kill the RVN, its closest ally in this war.

While the U.S. and Allies had been withdrawing from the war for nearly a year, the Paris Accords were quickly turning against South Vietnam. Military cuts by U.S. Congress caused the ARVN to become weak and the RVN undefendable. The accords clause that allowed the VC and NVA to keep their occupied territories was causing major nuisance as the enemy occupied more and more and converted more peasants into "patriotic" pests.

December 1974 marked a historical turning point when both the Soviet Union and China made new vows with Hà-nội. Was this a follow-up from their secret back channel negotiations with Nixon and Kissinger during the Paris Negotiations?

The "special favors" Nixon and Kissinger had granted the enemy in Paris were now back-firing:

...SOVIET ARMED FORCES Gen. Viktor Kulikov visited Hà-nội to endorse the DRV 1975 offensive plan Chiến-dịch Hồ Chí Minh Campaign to finish off South Vietnam. Additional military aids were agreed on: 6,000,000 Tons of aids, and new heavy weaponry, tanks and fuel.

... PEOPLE'S REPUBLIC OF CHINA deployed 50,000 engineering troops to the DRV to keep its transport system operational to assist Hà-nội rearm. Chinese troops were also deployed to the North to replace the NVA divisions that had been launching into the South.

(Janette Maring - Việt Nam War – A Memoir of the Unsung Heroes – July 2017)

This historically unprecedented betrayal resulted in a successful 2nd term election win for Nixon, and continued power for Kissinger as the U.S. Sec. of State: they appeared to be bringing Peace to Việt Nam, and Honor to the U.S. For these two men, the end justified the means. In order to get what they had wanted, to be re-elected or elevated to higher level of political influence, they had bent to what the misled antiwar movement had requested.

Will there be a consequence for these two men? Probably not! Nixon has already passed away, but Kissinger is still alive and trying to continue to politically influence politicians (e.g. 2016 Presidential Campaign)…

CHAPTER 14

The Confusion of the American Public about the War

The ARVN don't fight?

I grew up through the war and had seen many friends die since the early 1960s. I had witnessed many victory celebrations and confiscated weapons exhibits by the great ARVN. I had seen in public RVN government newsreel film shown at the round-about park "Bùng bình Nguyễn Huệ" (originally "Cây Liễu" in front of Sài-gòn House of Congress near where I grew up) reporting the victorious activities of the ARVN divisions such as infantry, airborne,... fighting in battles. Among those were Gen. Đỗ Cao Trí's 1964 victorious "heliborne" Đỗ Xá Campaign in II Corps that took place just before the Tonkin Gulf incident[498]*. Later on, subsequent propaganda documentaries shown to the public actually portrayed ARVN military campaigns, including the 1970 invasion of Kampuchea. They all showed only ARVN soldiers in action. I never saw American troops fighting in any RVN Breaking News reel, until I watched the American television after I came to the U.S. as a refugee!*

How could this happen? Where were our allies fighting? I had seen ally soldiers, predominantly American or Australian, wearing different uniforms in Sài-gòn. My martial art instructors (Tae Kwon Do and Hap Ki Do) were Korean body guards for their generals... I have heard about the Bi Dul Ki (Dove), the Hac Long Black Dragon, or Pat Long White Dragon, and other Long's (Dragons), or White Horse divisions, and how fierce they could be... but I never saw these soldiers in action in any RVN or American filmed news. Sometimes, daily Vietnamese news report might mention their campaigns, and I knew about them mostly because of my connection with the Korean Tae Kwon Do and Hap Ki Do world. I did not see U.S. troops (and Aussie or Republic of Korea allied) in action until I resettled in the U.S. and watched documentaries on American television channels: I had never,

---

[498] This incident *gave the U.S. the official "go" to get directly involved in the War (President L B Johnson*

ever seen what the American people had on a daily basis in the U.S., as shown in documentaries today. It seemed as if there were two different worlds, two separate truth in this war that contradicted each other, thus making people even more confused... Definitely, these American documentaries showed no ARVN soldiers fighting at all, not until the early 2000s when I vividly remembered seeing the first film about the 1968 Tết Offensive that included a section on side-by-side operation of the ARVN Marines in harmony coordination with their U.S. counterpart around Huế Citadel. Would the American viewers have misunderstood that the ARVN had not actually fought in the war because of lack of media representation? In my opinion, Nixon and Kissinger's term "Vietnamization" could have caused even more bad reputation for the ARVN!

After having watched a great many series on Vietnam War on History Channel, National Geographic to name a few of everyone's favorite..., I realized some American G.I.s may have thought the ARVN did not really fight in the war, or, if they did, they might be connoted as lazy, incompetent and not willing to fight. The very simple reason was **that they fought their own war, and never saw ARVN soldiers fight with them until "Vietnamization"** (of Vietnam War), a term created by Nixon and Kissinger that had mislead people back home into thinking the ARVN had not fought before.

Many times (perhaps, as my wife could verify it, 95% of the time) had I been asked by Vietnam Veterans whenever I told them I had also fought in the war: "So you must be Hmong?' or "You speak Hmong then, don't you?" Did they think the Vietnam War was fought only by the Americans and the Hmong? Some of my colleagues, highly educated principals in Minneapolis, had even asked me to help facilitate their Hmong parents meeting because "you fought in Vietnam War, so could you help me?... because you can speak Hmong."

These incidents really caught my full attention and raised flags when highly educated professionals misunderstood that Asians fighting in Vietnam War were Hmong and not Vietnamese. Who could Minnesotans think had defended the Vietnamese Ancestral Land through thousands of years dominated by the Chinese, French and Japanese? The Annamites did it! Who had been fighting in the French and Japanese wars? The nationalists had fought on both camps and had formed elite forces and they were all Vietnamese and the Vietnamese montagnards. The nationalists had been part of the Việt Minh, too, but a great many Vietnamese did not know that. Who had been fighting the Vietnamese communists before the U.S. arrived? The army of the State of Vietnam did it! Who had been fighting Vietnamese communists during and after the time the U.S. Army was involved there? It had always been the ARVN fighting since the 1950s until the end and beyond (ARVN resistance).

A puzzling question had crossed my mind when I heard Nixon's term Vietnamization. Not only that it struck at my discontentment seeing how severely the ARVN was looked down at, but it also made me think about the questions that many veterans might have when they heard that term. **Where had the ARVN been "until the U.S. passed on the war to them" in 1971** (Laos incursion "Lam Sơn 719")? None of the American television news media had shown them in action. Their news reports were just about the U.S. battlefields fighting by themselves. As a matter of fact, documentaries had even alluded to the fact "**the ARVN had participated for the first time in the battle field**" in Lam Sơn 719 after having been trained in the Vietnamization program… "**as graduation from training.**" This was definitely an insult at the sacrifices the Vietnamese soldiers had made since they fought against the Chinese a few thousand years ago!

For years, the U.S. news media had even commented that they had been busy launching "coups" against their own government instead of defending their land, thus leaving the war to the U.S. troops! Had such a negative image of the RVN government and its army been presented to television viewers throughout the war? All along, the people of the RVN never thought they would be portrayed as such! It was a seriously unfair accusation to prove the incompetence and laziness of the ARVN! The worst part was that no ARVN knew about this (they did not watch documentaries in English) to defend themselves; nevertheless, they had been tagged as lazy, coward and incompetent, while being funded through the war by American tax payers. I am not whining here! I'm just trying to claim social justice for the dead and for the voiceless people!

It really hurts when I viewed the newly revamped <u>Vietnam in HD</u> film series by HistHD in 2016 showing an American G.I. making the comments "there are some (ARVN)… don't get me wrong… really crack outfits, but, the rest are...I don't want to have any part with..." or Joe Galloway's statements "it's (ARVN Laos incursion Lam Sơn 719) a disaster...Vietnam is crumbling... it's supposed to prove that South Vietnam is competent and capable enough to take over American role... a Vietnamization graduation exercise" or "leadership nowhere nearly to take over control of...", at the same time as showing a scene of ARVN soldiers receiving uniforms and trying on helmets. If I were a viewer who were not familiar with or who were against Vietnam War, I would think that the ARVN had never fought in this war before.

    This judgment of the ARVN resurfaced and perpetuated again in Ken Burns' PBS <u>The Vietnam War</u> in 2018... Nevertheless, Galloway did recognize that "the little soldiers" had "fought valiantly," and that the leadership was incompetent. It was true that Lam Sơn 719 commander was corrupt and incompetent, who might not understand or might have despised the war plan developed by the American MACV, **but some cowardly units or a few incompetent leaders should not imply that the ARVN was as such.** If this army was that incapable, it would never have recaptured by itself the cities and national routes lost during the 1972 Easter Offensive, which was fiercer and bloodier than the Tết Offensive of 1968

    The term "Vietnamization" would break, if they were still alive, a great many hearts of those ARVN who had already sacrificed their life, as they might feel the repercussion from their tombs. While the world might not think the RVN was worth helping, the Vietnamese survivors might not even know how destructive and degrading to the warriors that term can be. How would the Vietnamese younger generations, more educated and more politically active, bear such a shame, thinking the world might look at them with disdain as their elders and ancestors were incapable of protecting their own country? How could such proud people with some 4,000 years of struggle and victory over Chinese, French and Japanese invasions, fail to fight Communism and defend its own land? To me, Nixon and Kissinger had gravely insulted the RVN and its people with the term "Vietnamization."

Some did, Some Didn't

I was often asked whether I fought against the U.S. whenever I said I had fought in the Vietnam War. In an earlier paragraph, I had talked about acquaintances and colleagues revealing their lack of knowledge about the war, so it was one of the funniest things that happened to me when someone asked me if I were Hmong because I was Asian and an American ally, and that I had fought in Vietnam. Again, even my colleagues, who were public schools

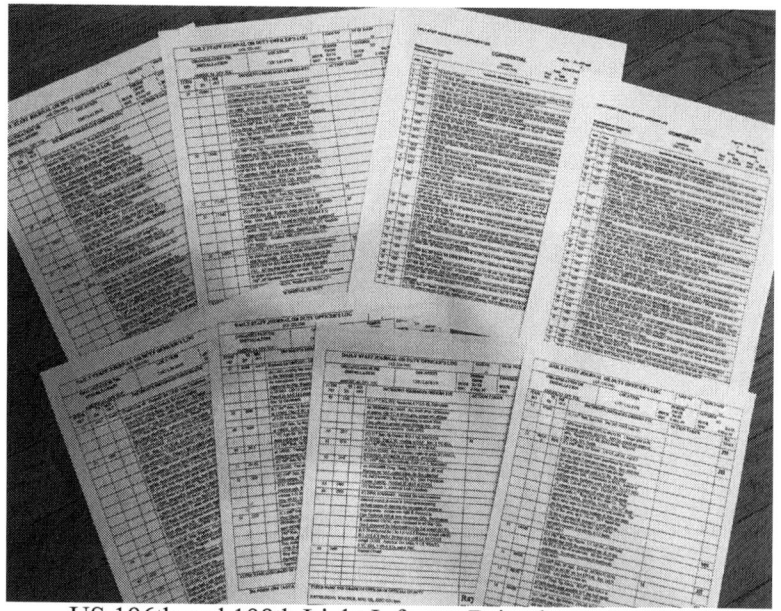

US 196th and 198th Light Infantry Brigades accounts on The Battles of Tam Kỳ and Núi Yon Hill – May 13-14, 1969
Source: https://www.americalfoundation.org/cmsalf/images/DTOC-Logs/6805_DTOC_trans.pdf

principals, thought I was Hmong probably for those same reasons. Even worse, many Vietnam Veterans, who should know more about the Vietnamese having fought in Nam, had asked me the same questions. Most recently, a U.S. Vietnam Veteran wanted to know how "my troops had defeated the Americans." It was so obvious that he, who had actually fought in the war and might have been sent to reinforce us the ARVN, still could not distinguish the communist Vietnamese, who were the enemy who had fought against the Americans, from us, the Vietnamese-Americans who came from South Vietnam, his allies. The key answer was very simple: our armed forces did not fight side-by-side enough. For many reasons that I will discuss later, we fought our own war.

It became more and more upsetting when our Coalition research team[499] tried to look up the 1969 Tam Kỳ battle in the Province of Quảng Tín for information about a personal friend, Hoàng Diên, who was in command of an ARVN company of Regional Forces that had retaken Núi Yon Hill from the NVA. His friend's

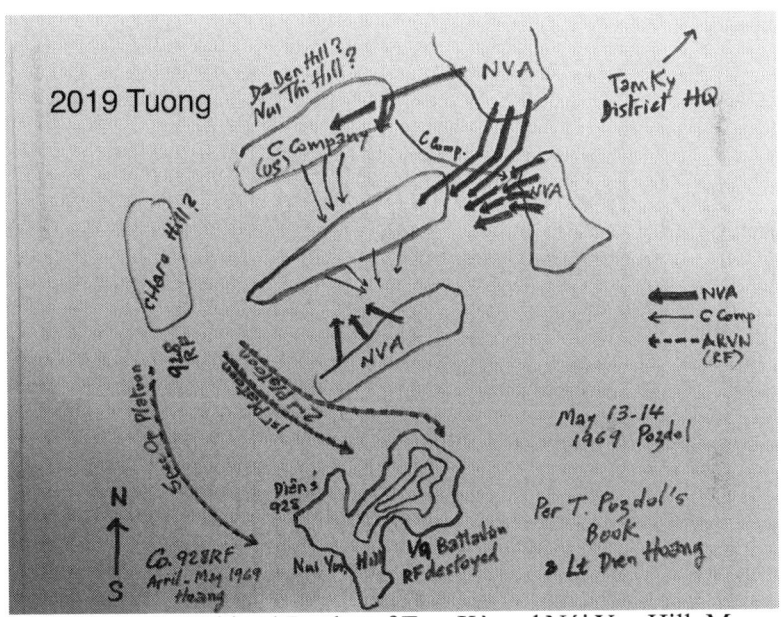

Map of the Combined Battles of Tam Kỳ and Núi Yon Hill, May 1969
Sources: Map from Thomas Pozdol's Book "Tam Kỳ and Núi Yon Hill

---

[499] The Coalition of Allied Vietnam War Veterans, headquartered in the Twin Cities, Minnesota, is developing an educational history website to help clarify the complicated and misunderstood Vietnam War. The website War against Communist Aggression, according to the plan, will be released to the Smithsonian in 2025.

combined battalion 1/9 that stationed on and defended Núi Yon Hill was totally wiped out by the NVA. Internet information in English as well as U.S. Army records, such as America's "Daily Staff Journal or Duty Officer's Logs[500], seemed to report mainly U.S. troops activities with very little mentioning of ARVN even though both campaigns took place side by side in the same region. A book Tam Kỳ: The Battle for Núi Yon Hill mentioned a tiny bit about the ARVN 2nd Infantry Division and some other unknown ARVN; nevertheless, it somehow treated the U.S. units as the main ones engaging the NVA while the ARVN event was left out. This was one of the evidences showing that the war was definitely treated more as a war between the U.S. and Hà-nội, not between the South and the North.

    While the American accounts did not include anything about my friend Lt. Diên's company 928 retaking Núi Yon, Diên's version confirmed there was no substantial American troop around, except for the MACV Capt. he had rescued from the rubbles of a collapsed bunker of the hill fortification. This captain was the only survivor of the 6 advisors to the destroyed ARVN Regional Forces Battalion 1/9. It sounded as if either the Americans or Diên was lying. Most recently, Pres. Trump had awarded a Medal of Honor to Specialist 5 James C McCloughan[501] as a hero of Charlie Company (U.S. 196th Light Infantry Brigade) fighting in the same battle for Núi Yon Hill as my friend Lt. Diên's company 928, during the same campaign in May 1969. After the victorious battle of Tam Kỳ ended, along with two of his soldiers and two dozens of U.S. troops from the United States, Diên was awarded a Bronze Star with V at the Tam Kỳ District headquarters. He was also promoted to full Captain and commander of the newly formed Battalion 131 comprising five Regional Forces companies.

    By comparing the maps in Thomas Pozdol's book to the ones hand drawn by Diên, I realized the two maps did not relate to the same description of battles of Núi Yon in the book <u>Tam Kỳ: The Battle for Núi Yon Hill</u>. Author Pozdol's version was focused more on American accounts of the battle near Tam Kỳ on un-named hills which Diên guessed them to be Núi Thị or Dương Con, these being located between Tam Kỳ District and the actual Núi Yon Hill. Diên's battle site was the actual

---

500   https://americalfoundation.org/cmsalf/images/DTOC-Logs/6710_DTOC_trans.pdf DA FORM 1594
501   https://www.army.mil/medalofhonor/mccloughan/

one where his company 928 recaptured the Núi Yon Hill by itself from the NVA.

Tam Kỳ is not the only case of unmatching information. The same situation is repeated in the reporting of other battles. Ia Drang, Hamburger Hill, Khe Sanh... have been described in many English reports, news and books, as American military campaigns; but probably no one but the ARVN knows that the South Vietnamese Infantry, the "Red Hat Angels" Airborne, Ranger, Marines, Regional Forces... have been there fighting, too, as courageously as the Americans? Whenever and wherever American troops engage the enemy, if not the same units, the ARVN are also there next door engaging the other elements of the same enemy. It is a Vietnam War, for God sake! People must remember that! As the enemy is hiding everywhere, the ARVN has to be there, oftentimes around and away from Allies bases, like ocean water surrounding islands. The ARVN are not just in Sài-gòn, Huế, or big cities busy overthrowing its government as the U.S. High Command[502] thinks. They, the soldiers of the Republic of Vietnam, have been fighting against Communism since the 1950s, and have never stopped fighting, until doomsday in April 1975 when they are ordered by their commander to lay down weapons.

This brings up a serious question: What had gone wrong with communication and coordination between the U.S. and the ARVN militaries? Obviously, a great many American combatants did not know much, if at all, what their counterparts the ARVN were doing next door, and, vice versa for the ARVN! The high commands on both sides might, but not the "little soldiers". No wonder why the American and Vietnamese People had both been blind and deaf about their allies during the war up until today, 50 years later. I have said it before, and I am going to say it again: during the whole time growing up in Việt-nam until I resettled in the U.S., I had never seen documentary film showing our allies fighting in military campaign (note that a 10 min breaking news "Phim Thời Sự" in Vietnamese shown and the beginning in theaters during the war was always part of any movie in any theater in the 1960-70s), except in entertainment movies such as John Wayne's "The Green Berets."

---

502   See section on Wannabes

By the same token, most 1960s-2019 U.S. documentaries by the television news media, as I have seen so far, have failed to show we were dear allies! Government Psychological Warfare has failed to show that we have fought for a good cause: protect lives and preserve Democracy! The American television programs for G.I.s in Sài-gòn (you have heard of Robin Williams' <u>Good Morning, Vietnam!</u>, haven't you?) that I watch on a daily basis to learn English does not reflect any ARVN campaign either.

Obviously, the press media and the RVN government have totally confused the people. Not tens or hundreds of thousands, but millions of innocent nationalist Vietnamese civilians of South Vietnam, the RVN, have been slaughtered. Over a million more of actual South Vietnamese combatants have fought on the same side with Americans and allies in the War against the communists. Yet, people are still ignorant about the war. The same thing is happening with the Vietnamese. Thousands of ally life were sacrificed every week for their democracy, and yet, they vaguely know what the allies were doing for them except spending money ate the clubs.

This has proven that both governments have done a very poor job at explaining the war to educate its citizens during the war (political warfare). After the war has ended, not much is done to debrief them, consequently leading to Vietnam Veterans being treated like trash upon their return home, and the ARVN either mentally struggling in secluded resettlement, if they are lucky enough not to be incarcerated and physically and mentally suffering in communist concentration camps. **Such is the fate of the minnows**[503]!

By the same token, how many Americans know that there was another war fought in Laos and in Kampuchea in the 1960s-70s? Those heroes ought to be recognized and honored, as they truly deserved it. What has been taught in World and U.S. History in high schools? As a high school principal, I have always been concerned whether History teachers are receiving the support and resources they have so much needed, especially when not much of the available materials are conveying the real truth about the war. Text books have the minimal information, so teachers have to scrounge around for supplements on the internet, and what kind of information can be trustworthy when most available documentaries are

---

[503] A follow up from the book <u>Two Minnows</u> by Dr Ha Tuong

misleading? The end results turns out that kids are now (2010s) taught that "Uncle Hồ is a Hero, and that the U.S. and the RVN are evils and aggressors." Obviously, communication breaks up and becomes corrupted when those who have fought hard in the war choose to remain silent (the ARVN, the Royal Lao Army, the Khmer Republic Army), probably out of guilt and shame, while others who have also fought courageously, but actually in a much smaller war (Hmong), are vocal enough to speak up and voice their anger for being used by the U.S.... It's true the squeaky wheels getting oil!

Anyhow, people ought to realize that most of the Vietnam War documentaries available in the news media during the war (including the newest 2018 PBS series The Vietnam War produced by Ken Burns) are usually misleading. The reason is very simple: back then, the rising trend of instant television news has caused networks to compete vigorously against each other for popularity: the winners are the ones that have the most sensational breaking news for the anxious and rebelling television viewers. The end justifies the means! Not like during WW1 (when only newspapers and some rare government news reels are available) or WW2 (newspaper and government-controlled news reels) when the news are still well controlled by the government[504], Vietnam Era news media is a frantically chaotic race of some 400 to 600 television and news networks that use new technology (Telex, teletyping and telegrams) to by-pass any form of censorship, and to deliver the "hottest" information back for evening living room television excitement.

My family in the Twin Cities is actually enjoying unlimited television news now as politics are getting more and more controversial and consequently educative. While following every night CNN, France 24, BBC, Fox News, Russia Today reports[505] on Republican versus Democrat tug war, I cannot stop thinking what it is like during the Vietnam War for viewers to uncover sensational information. **The only difference is that most networks now fact-check and triangulate their sources,** *and use excellent life references (e.g. life interviews) instead of unfounded or subjective ones. The war in Middle East is well coordinated and reflects much more reality as presented World-wide.*

---

504 as I had pointed it out earlier referring to anchorman Don Schelby's report
505 We use multi-media news worldwide to triangulate and verify information.

It is time to make Vietnam War History factual so one can avoid making mistakes in the future, and, most importantly, so that those who have sacrificed for world Freedom and Democracy will be remembered and truly honored. For that sole reason, Honor the True Heroes, along the line of "Who Did versus Who Did Not," there are two main points that have to be corrected:

First of all, watching or controlling the HCM trail, which was located in central RM#3 (controlled by Gen. Daohoeung and Minister Insixiengmay family[506], in Laos low land), was mainly the responsibility of the Royal Lao Army, as verified by the CIA case officers and U.S. Special Forces operating in Laos. Both the Royal Lao and the ARVN veteran groups were totally astonished hearing Vietnam Veterans and Hmong politicians (and Hmong Studies) talk about, as well as seeing and reading fliers and pamphlets from Hmong organization describing Miao Special Guerila Units activities "along the HCM trail," which was so far from their assigned control territories in RM#2 and which was located in Royal Lao Army control. Fact-checking using maps of the 5 regions proved that "the oral history version of the Miao SGU "fighting along the HCM Trail" was totally fabricated, as their area of operation was RM#2 (Northern mountains), not in Royal Lao Army Regions RM#3 and RM#4 (HCM Trail). The Miao tribes were concentrated further North, in the high mountains of the Military Region RM#2, much too far from the HCM trail, which was in Low Laos (see the following maps). The grey shaded areas in the map on the next page were Miao tribal locations in China, North Laos and North Việt Nam – 1960s.

---

506   My friend Col. Khao Insixienmay, a descendant, was an airborne commander there and now lives in Minnesota.

Source: gcanthhmong.com

The Trail was electronically[507] surveilled by devices deposited by both the Royal Lao Special Guerilla Units (verified by Pakse CIA operative T. Briggs[508]) and the ARVN Commandos (verified by the RVN Joint General Staff Bộ Tổng Tham Mưu[509]). The RLA SGUs, not the Hmong SGUs (as they

[507] The Joint General Staffs were located in Gia Định Sector, Northern suburbs of Sài-gòn. Like the U.S. Pentagon, JGS were like the brain of all RVN armed forces, including unconventional warfare units such as the Strategic Technical Directorate that was created in 1960 by the CIA. The book the <u>RVN Commandos clearly accounted for the missions these ARVN SEAL units had performed since 1961 in coordination with the MACV-SOG advisors. Insertions were in North Vietnam, Laos (particularly the U.S. B52 carpet bombing Steel Tiger zones where the Hồ Chí Minh Trail originated from).</u>

[508] ...as Thomas Briggs has described in his comprehensive book on watching the Hồ Chí Minh Trail <u>Cash on Delivery available on Amazon.com. Briggs has been maintaining CIA Archives on their covert activities as well as tactical moves by the Royal Lao Army regulars and irregular SGUs (incl. RLA, montagnard and some independent Hmong)</u>

[509] RVN equivalence of the U.S. Pentagon directly operated the ARVN commandos, consisting of multi-ethnic and multilingual RVN Vietnamese units of

had claimed in many Hmong studies), were mainly responsible for blocking and repelling the NVA advancement into South Vietnam by blocking them and helping guide Steele Tiger B52 bombing right on the trail. Fighting the main battles in Laos[510] was also the responsibility of the RLA in all 5 Régions Militaires (including rescuing the Miao SGUs in their #2 region[511]) under the leadership of a great many commanders. They had also rescued downed American pilots in bombing missions along HCM Trail in Steel Tiger; therefore, the RLA ought be recognized and honored. In Laos RM#2, CIA had funded Gen. Vang Pao to recruit and employ Miao personnel in SGU mercenary battalions[512] to ward off Pathet Lao and to help the CIA and the U.S. Air Force security-guard the radar or supply sites[513] (such as TACAN at U.S.AF-controlled site LS-85, then TSQ-81, and MSQ) and the landing trip in Long Cheng; they had also rescued some dozen downed American pilots (as CIA had reported[514]) in the Barrel Roll area. The Miao, as much as the American and ally special forces in Laos, Cambodia and Việt-nam, ought to be honored because they had done their best with very little support[515] they received!

Each unit had its own tasks and each person his/her own duty! They might succeed, or they might fail; but the main important thing is that they had done their best with the little they were provided. Among all, the only troops that dared jump North "Nhảy Bắc" were the ARVN SEAL commandos trained

---

"Nha Kỹ Thuật" Strategic Technical Directorate / Sea, Air and Land (MACV-SOG frogmen SEAL)

510   The Three Battles by Col Khao Insixiengmay of RM#3.

511   Even in Miao RM#2, RLA were the main forces. Their units consisted of better trained and equipped professional warriors than the Miao that were irregulars and militia-type army by international military standards

512   These fighters received their pay through their Miao commanders (RM#2).

513   The radar site was operated by the U.S. Air Force. The CIA funded the Miao SGUs to provide security to the radar station. The Long Cheng air strip was operated by the U.S. Air Force to supply the base and to move Miao security troops. (CIA records).

514   Many Hmong studies have claimed on PBS TV that records of Miao actions have been lost during the final evacuation. This explains why data are unverifiable. Nevertheless, this claim cannot stand justified as the Miao tribes have no written form of language and most adults and children were pre-literate during the war, until the 1970s: most accounts available to the public have been based on oral history. The only records, if any, have been maintained by the CIA.

515   As examples, the Miao combatants usually received not more than U.S.$5 per month, plus rewards after mission were accomplished. The Vietnamese commandos in Vietnam received U.S.30 to $15 (RVN 15,000 to 25,000 piasters) per month and were pre-awarded death compensation before mission. As a commission 2nd lieutenant, I received 25,000 piasters per month, enough to buy a 100kg bag of rice.

and organized by the U.S. MACV-SOG. Most of the thousands of these ARVN frogmen (OPLAN 34 commandos[516]) were captured during their long-term insertions in the North: some were massacred, others incarcerated in gulags. They ought to be recognized and honored. They were the true secret army that had fought in this war, but they never bragged about it. Even nowadays, they were still maintaining silence, as they had vowed when they were still active, the vow of the "death-loving" (cảm tử quân) commandos that their sacred story would never be unveiled.

These well-trained ARVN SEAL teams of the "Nha Kỹ Thuật" STD (Strategic Technical Directorate) were dressed up like communist local peasants and equipped with AK47 (or Eastern European or Communist Block made weapons) before they were flown North for long-term insertion. They were mainly of Kinh (Việt), Tày (Tai), Mnông (Nùng), Thổ, Thái đen (black Thái), and sometimes Mường (not to be confused with the current Miao-Hmong from Laos[517]) ethnicity who could communicate in the appropriate language with whoever villagers they might run into in the North. Even though created and organized by the MACV-SOG in the early 1960s, they were rarely, if ever, accompanied by their American operatives[518] (MACV-SOG). Normally, the Airborne units were inserted by air (dropped by parachute or heli-transported) to plant electronic listening devices along the HCM Trail areas (Chain of Trường Sơn Mountains in Laos RM#3 and RM#4). Meanwhile, those frogmen ("người nhái") dropped off at sea in the Gulf of Tonkin had the missions of gathering intelligence on NVA troops and equipment movements in North Việt Nam; usually, the latter were moving by train from the Hà-nội areas to Vinh where they crossed the border by foot at the passes of Mụ Giạ and Ban Karai to infiltrate into the Lao Royal Army controlled RM#3. NVA forces were usually "ferociously mighty," and only B52 could break them up. Others' missions were to identify NVA leaders, make attempts to convert to the

---

516   Spies and Commandos: How America Lost the Secret War in North Vietnam by Ken Conboy. Also, Congressional Bills PL-104-201 and PL-105-18 were passed to compensate RVN members of MACV-SOG teams who were captured and imprisoned when infiltrating North Vietnam. Among the recipients who have received this compensation and Presidential Awards are Minnesota residents Trần Quý and Hoàng Thiệu.

517   The Hmong now living in the U.S. were referred to as Miao before 1970s, but were later called by the same Mường by the Vietnamese starting 1975 when this ethnic group decided to change their ethnicity appellation to Hmong, meaning "free people."

518   American faces would be recognized right away…

South, or even assassinate if needed.

The ARVN SEAL sea insertions were launched into North, while other "death-loving" commandos units (e.g. Thunder Tiger and MIKE Forces) were operating across the border in Kampuchea and Laos in missions such as scouting, reconnaissance, sabotaging, and position pin-pointing "đề lô." Most of the time, they had to stay behind (insertions) and vanish into the local population until their mission ended. A great many were also captured and incarcerated in multiple prisons and camps. Many survivors, who were caught and imprisoned long term at the Hỏa Lò prison ("Hà-nội Hilton"), currently lived in California and Minnesota.

The commandos described above had done much in terms of covert operations; nevertheless, they kept them all secret even until today. Survivors after prison incarceration are spread out in the U.S. A few who lived unpretentiously in Minnesota had received Presidential awards. Many Royal Lao Airborne commandos also led a quiet life in Brooklyn Park, as did some Khmer MIKE forces survivors. No one knew of their existence, as they devoted their life mainly to religious practice at Buddhist temples, with the hope their soul would be secured when they were to leave Mother Earth.

The next map from source alphahistory.com shows the two major U.S. carpet bombing operation zones (shaded areas). The area on the top is B52 bombardment zone **Barrel Roll** (High Land sectors A, B and C) where the Pathet Lao Army PLA (communist), some NVA troops that were sent to support the PLA, the CIA and the Miao SGU irregulars (American-backed, commanded by Gen. Vang Pao) engaged each other, to support the Royal Lao main forces. The sectors on the bottom of the map D, E, F and G were the B52 bombardment zone of **Steel Tiger**, where the Hồ Chí Minh Trail was located and where the Royal Lao and the ARVN Strategic Technical Directorate SEAL commandos were operating to keep it under control (sectors D, E, F, and G, with the Hồ Chí Minh trail crossing Lao-Viet border at sectors D, E and F between Vinh and the DMZ). Low Land Laos was populated mainly by tribal Mon-Khmer and Thai Lao[519]. Also, Tchepone, in sector F on the map, was the area where the ARVN had focused their attacks on in 1971 to block the head of the HCM trail at its source (Operation Lam Son 719 - Laos incursion).

519   Indochina Ethno linguistic Groups Map – 1970. UT Texas Online http://www.lib.utexas.edu/maps/middle_east_and_asia/indochina_eth_1970.jpg

The following map (www.u22sr71patches.co.uk.jpg) showed the actual Hồ Chí Minh Trail expanding from Mụ Giạ and Ban Karai Passes, South of Vinh, North Việt Nam, crossing the border into Low Laos (Regions Militaires 3 and 4 controlled by RLA), then expanding from the Kampuchea border (Sihanouk Trail) to the areas facing Sài-gòn. This information would help the young Hmong generation as well as misinformed Americans clarify who did what on the trail.

       Gen. Vang Pao and his Miao SGU were based in RM#2 where U.S. carpet bombing in the Barrel Roll zone took place. Meanwhile, the Royal Lao Army was responsible for RM#3 and 4 cooperating with the CIA, watching and "annoying" the Hồ Chí Minh Trail ("roadwatch") and, if they could, engaging the incoming powerful NVA divisions that crossed the border at the two passes of Mụ Giạ and Ban Karai, between Vinh in North Vietnam and the DMZ. Responsibilities had been clearly defined and we could see clearly that the Miao SGUs had nothing to do with the Hồ Chí Minh Trail[520] or also called the "supply line" (to avoid referring to the term HCM Trail which obviously was not located in the Miao territories) as they had claimed. To the U.S. and South Vietnam, after the Tet Offensive of 1968, the Trail had become the utmost tactical value in the war because **it was the veins that brought blood to the brain of the NVA and VC military machine.**

       Both NVA Army Gen. Võ Nguyên Giáp and Gen. Bùi Tín admitted that Hà-nội had planned pre-emptive attacks on the Miao and Royal Lao positions using the NVA and Pathet Lao forces in Laos, with the sole purpose of distracting U.S. and RVN intelligence from their preparation for the most major offensives in South Vietnam. That would be an application of the strategy of "Dương Đông, Kích Tây" (tackle the West, but attack the East instead):

       Pre-emptive attacks in Laos[521] were designed to distract the Americans (CIA) from the real offensives:

       The 1953 Attack in Laos (Pathet Lao started the 1953 civil war, as Laos was receiving Independence from France Nov. 9, 1953) distracted the French-American forces from the Việt Minh's preparation for the last battle against the French at Điện Biên Phủ,

---

520   YouTube interviews and PBS TPT series Minnesota Remembers Vietnam (2018). Also, Wisconsin Hmong SGU Flier (2018)
521   https://alphahistory.com/vietnamwar/laos-during-vietnam-war/

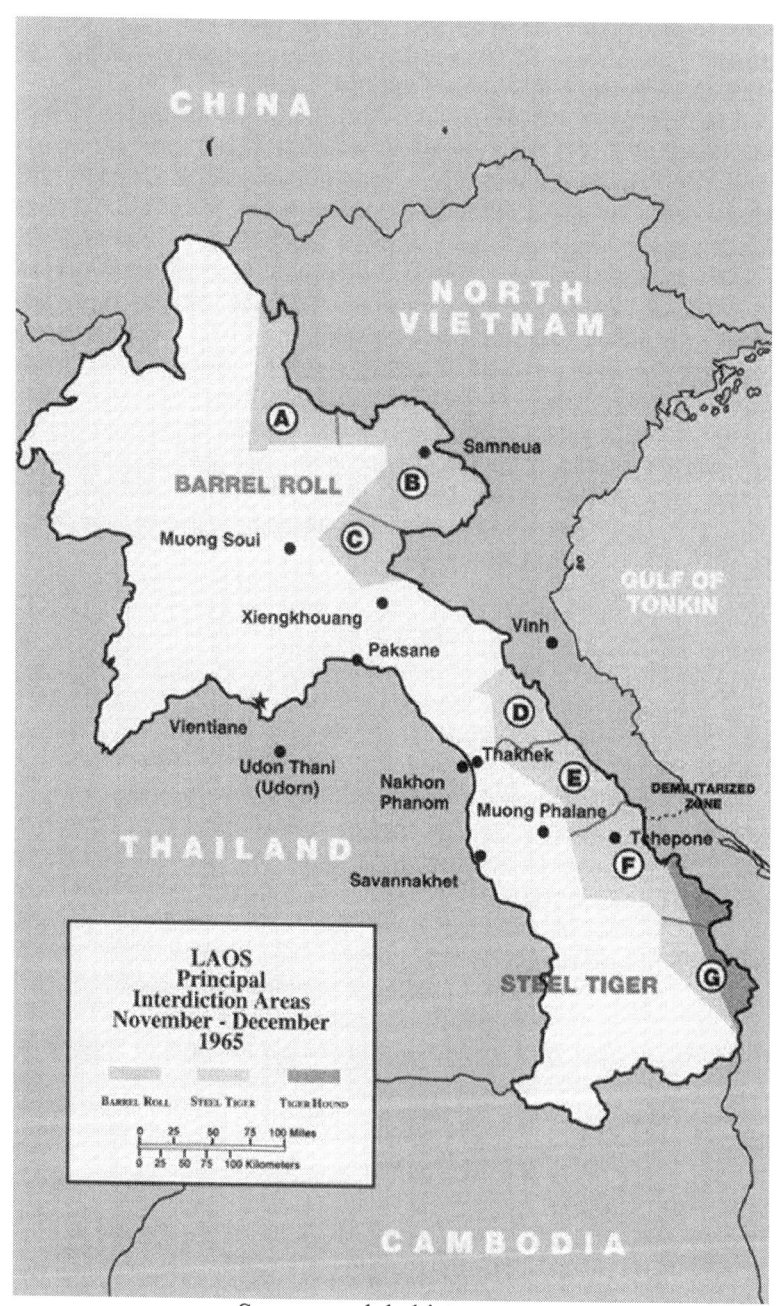

Source: alphahistory.com

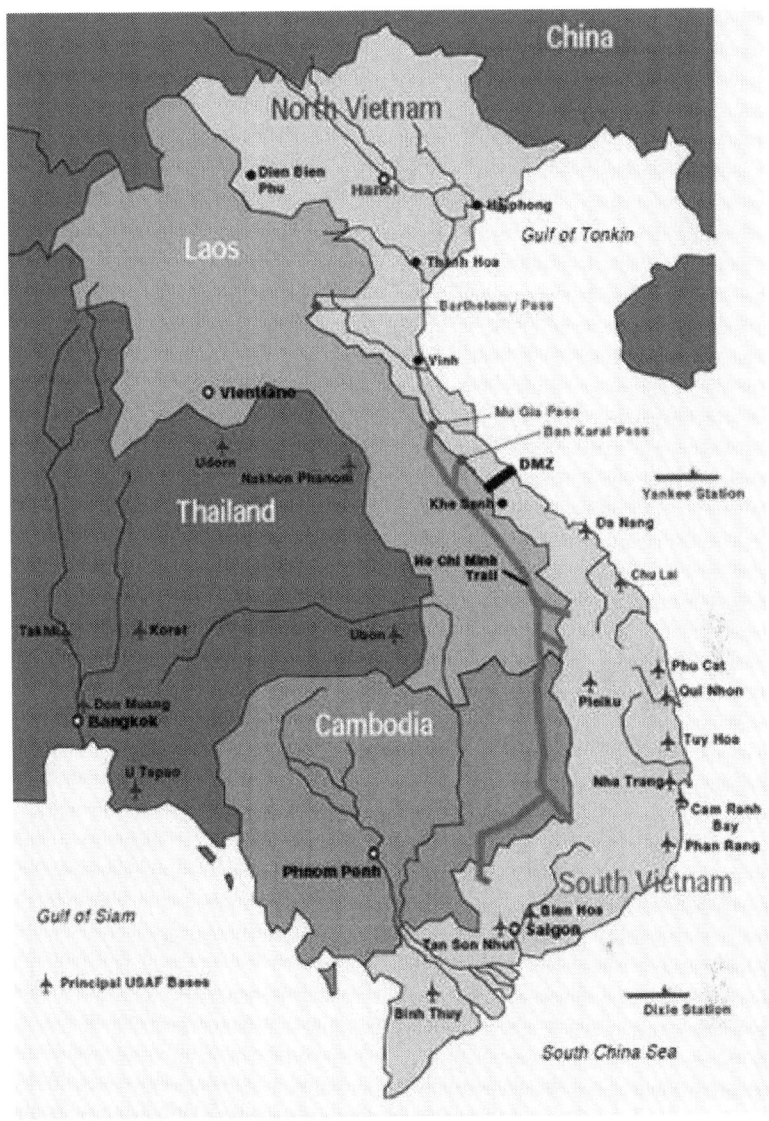

Source: www.u22sr71patches.co.uk.jpg

    The 1967 attack in Laos (allowing much of Northern Laos to come under the control of the Pathet Lao and the NVA) relieved U.S. pressure on HCM Trail allowing troop's movement into RVN for Tết Offensive.

    The 1971 RVN failure during the Lam Son 719 incursion in Laos expanded into 1972 Easter Tide Offensive in South Vietnam. This was also Hà-nội's test to see if the U.S. and Allies were truly leaving the war for good in compliance with the Paris Accords that were being negotiated.

Last but not the least: 1973-1975, as the U.S. had completely stopped bombing the HCM Trail (Paris Accords), and as the Pathet Lao had expanded its control throughout Laos and was taken over the Kingdom of Laos National Government, the NVA had free access to the trail, upgraded it to highway standards to bring in divisions of main forces. In compliance with the Paris Accords newly signed, the U.S. stopped bombing the Trail and, consequently, this trail had become an artery which expanded into a hemorrhagic deluge of NVA and equipment that overwhelmed the supply-starving ARVN.

Sun Tzu-inspired tactics: "Dương Đông, Kích Tây" or "tackle the West, but attack the East instead" (also used by the U.S. and allies during the WW2 June 1944 "Overlord Campaign" of Normandy)

They Might be Humble, but Certainly Not Lazy or Coward

In 2018, the ARVN soldiers who have actually fought in battlefield events (1960s-1975) refuse to participate in oral history interviews to the PBS TPT camera crew for their <u>Minnesota Remembers Vietnam</u> Series. They know the American public has been told in many documentaries, including the most recent Ken Burns' 18 hour series, defaming them for being lazy, and coward or did not want to fight, and that their leadership for being incompetent[522]. They felt guilty thinking the media was insinuating they were the reason why the war was lost. Of course, they got angry to the extent of suffocation because they could not defend themselves or dared not speak up. Nevertheless, those ARVN, even the ones who were interviewed like me[523], would still never have the chance to be heard: the interviews were not included in the series. Once more, one part of the truth had been silenced out by the media.

---

522   The most recent film that describes the ARVN as so is PBS Ken Burns' 18-hour series <u>The Vietnam War.</u>
523   I was interviewed on camera by Minnesota PBS – TPT for this <u>Minnesota Remembers Vietnam series, but no part of this interview was used.</u>

At least, the Minnesotan Vietnam Veterans had a chance to tell their stories that had been kept in for a long time. They absolutely deserved to be heard and honored. Unfortunately, the other main forces that had also bravely fought in the war, Royal Lao, Khmer and South Vietnamese who also deserved it equally, were completely left out by this state-funded program[524]. Once again, they were shocked to find out their counterpart Hmong were the only SE Asian group highly featured in many episodes, ironically with many video clips showing Royal Lao[525] forces fighting instead. Why, they wondered?

In a letter addressed to Minnesota Congress, the Director of the Minnesota Asian Pacific Council recommended the series to focus on the high-land Hmong stories because **they were the most numerous, supposedly 90,000** Hmong in Minnesota. The other groups, the low-land Lao, the Khmer and the South Vietnamese, were ignored. Why? Because there were less of them? The Asian Pacific Council Director told me so in person at the 2017 Lao New Year celebration to justify her judgment. In my opinion, 60,000 low-land Lao, Khmer and Vietnamese should also be considered a large population, too, shouldn't it! Would they dilute the importance of Hmong heroism? Therefore, why should the other groups speak up? It's all about dirty politics!

"I would speak up to tell them we fought courageously, **but I do not know enough. We cannot tell our stories if we are not sure about the details,**" said the 2015-2022 Chairman of the Minnesota Fellowship of the RVN Armed Forces. I was angry, too. I had been wondering why these ARVN would not dare tell what they witnessed, what they had gone through at the battle field, or what they had actually done. Of course, they were honest and would not make up stories, but at least, why not say what they knew, and just skip the parts they were not sure? As honest story-tellers, they would not need to make up things to embellish their accounts as some others had done. Their stories would be true, because they were not illiterate, and that they know their coordinates, they know their precise orders of the day, and they belonged to an organized regular army.

---

524   With the support of the Minnesota Asian Pacific Council community and legislators, TPT received $650,000 to develop a Story Wall and a few television episodes to recognize Minnesotans, both American and SE Asian Refugees, who were involved in the Vietnam War.

525   Low-land Lao people look completely different from high-land Miao.

Once again, their Vietnamese submissive and humble culture that had been blocking them from bypassing their higher ranking officers (who obviously had more authorities to speak up…) was prohibiting them again from speaking up. But, among the higher-ranking veterans, who would be courageous and capable enough to step up and defend their defamed legacies? Of course, the arrogant ones (the wannabes), who had been bullying the more knowledgeable (but lower-ranking) ones, were by now to old and not educated enough to step up and tell their stories about the entire generation of real combatants? Instead, by not doing anything to defend their honor, they ended up proving they were truly incompetent. Even worse, they had also failed to delegate this honorable task to the younger generation. The voiceless continued to be voiceless and powerless, while the ARVN's legacy would disintegrate with time. More Maslow and Reptilian Brain complication to take into consideration!

Consequently, everyone kept their mouth shut, remained in hiding and tried to bury their shame… They had courageously fought and done their best, with insufficient fire power after all allies had disappeared and all military aids had been cut. They had successfully taken back townships and cities captured by the NVA in 1972 Easter Offensive and afterwards. They might not have done well with the advisors, but they fought well within their culture and with their direct commanders. They had fought until the last minutes when they were ordered by their Supreme Commander to lay down all weapons. I was there to witness that, too. I knew that because I had also fought with them. Now, they chose to remain voiceless and ashamed, and absorb all the punches while the world (news media and internet) was publicly putting them down.

That was the reason why practically no one[526] in the Viet-American communities spoke up against the PBS <u>The Vietnam War</u> (2018). That was why they made no claim on their heroism: they felt they did not deserve their legacies. The Royal Lao and the Khmer veterans had no comment about the PBS series either, probably because of their culture requiring them to be humble and not to outspeak the "Westerners". After all, we were raised from the old kingdoms and empire that had major restrictions on their citizens. That was the very simple reason why they had been left out of the film media, in

---

[526] Only one Vietnamese scholar responded against PBS film in a meeting in Vietnamese and a Viet Journalist in English.

the same way the Paris Accords had done to them. This was something the Western culture would never understand. Once subjected, always subjected, until they are truly culturally liberated! The Hmong might not be directly subjected to either this kind of repressive culture, or to the French colonization: they had been Free since they had escaped from China, the free Hmong will always feel and act free[527].

The majority of the Vietnam Veterans I talked with, most recently (Sept. 2019) the 12th Cavalry Regiment at its 14th Annual Reunion in Branson, Missouri, all agreed on Burns' <u>The Vietnam War</u> being a disaster, if not another betrayal. Many activist groups popped up from their resentment and anger: they were getting organized to counter act this series. My way of counter acting this television press media sham was to bring up these PBS and TPT (MN local PBS) series, as well as other 1960s-1970s television press media Vietnam War documentaries, repeatedly in this book. I need to caution factual researchers, professors and school teachers about these materials being detrimental to the cause of Veterans if used in researches and classes. Was it true that some students in the U.S. were taught that "Uncle Hồ was a Hero, and that the U.S. and the RVN were bad guys" or that NVA and VC were courageous combatants as portrayed in the film? Directly or indirectly, purposely (use controversial information to make the lesson sensational) or inadvertently (poor choice of lesson plans and/or resources) they could do so[528] regardless of the true events. I had already heard complaints from Vietnamese-Minnesotan alumni who had to stand up in front of teachers to argue and assert that what s/he was teaching was false! As a matter of fact, we might be facing another phase of misleading and brainwashing the younger generations and dishonoring the Freedom Fighters of the past, as it had been done during the war, only this time by the trusted educators.

In my opinion as an educator and administrator, PBS programs had been outstanding materials for all viewers and had been recommended for use in educational institutions as instructional supplemental resources. Nevertheless, to my astonishment, the 2018 <u>The Vietnam War</u> had seriously

---

527   The Miao came, not from the Royal, but the remote mountain independent tribal culture, embracing Freedom from China.

528   Former advisor and Dean of Hamline University, the one who has actually interviewed defect-NVA Gen. Bùi Tín, Dr. Stephen Young, has reported that many schools are teaching students Uncle Hồ was a hero and that the U.S. and the RVN were "bad."

shocked U.S. and ARVN veterans as if they were hit by lightning. The producers had for some unknown reason (perhaps the Burns family's past antiwar politics?!?), just like the other journalists covering the war in the 1960s-1970s, failed to reflect the importance of deep cultural understanding. According to many Vietnam War researchers[529], Ken Burns seemed to still be influenced by past antiwar sentiments. Most of the Vietnam Veterans I talked with were highly upset that the series appeared to be overwhelmingly filled with admiration for the "champion-that-had-outwitted-and-overpowered-America." The Vietnamese-American communities were outraged the ARVN was treated unfairly by defaming putdowns. Both groups unanimously expected a PBS sequel that would reverse what was portrayed in the "first series." It was all about Social Justice for the millions who had sacrificed; either that this task would be accomplished with PBS initiatives and funds, or that the Veterans would do it themselves. Many groups, such as the Vietnamese-American activists in Massachusetts, a film-making group in California, the Coalition of Allied Vietnam War Veterans (CAVWV) or the 12th Cavalry Regiment[530], had already started producing their own audio-visual or on-line documentaries to unveil the omitted 85 to 90% of Truth...

No one really cares about the RVN Victory or Agony

The Lam Son 719 campaign ARVN defeat in 1971 was followed by another strategic mis-happening during the 1972 Easter Offensive, causing I Corps to almost fall into major disaster, if it had not been rescued by the re-assignment to I Corps of the talented and "clean" IV Corps commander, Gen. Ngô Quang Trưởng[531], in May 1972.

Gen. Trưởng's arrival had completely flipped the I Corps corrupt condition upside down[532]. His excellent skills in strategically coordinating his heli-lifted ground and armored troops with the U.S. air support, ranging from B-52 carpet bombing, U.S. 7th Navy Fleet shelling and air strikes... to

---

529   Researchers believe Ken Burns is influenced by his father who was seriously involved in antiwar leadership in the 1960s.
530   This project is spear-headed by veteran Thomas Crabtree.
531   http://www.historynet.com/the-most-brilliant-commander-ngo-quang-truong.htm Ngô Quang Trưởng "the most brilliant commander"
532   In some situations, he had to execute some subordinates as examples in order to regain law and order.

the operations of U.S. Navy, U.S. Air Force/ Marines fighter-bomber, had made HistoryNet rank him "the most brilliant commander" of the RVN. Both I and II Corps recovered fast from the 1972 Easter Offensive as Pres. Nixon ordered B52-bombing in North Việt Nam ("Christmas Bombing"); consequently, as I had mentioned earlier, most highways were cleared and reopened, and all cities recaptured from NVA/VC. I and II Corps were stabilized. The South was stabilized... until Pres. Thiệu ordered the abandonment of Central Việt Nam in late 1974... Nevertheless, Gen. Trưởng's glorious exploits were not known to the public in the U.S. either, as the "people back home" were really scared of even hearing the word "Nam." **It was so sad that the people in the U.S. seemed to be more interested in bad than good things!**

Ngô Quang Trưởng[533] was born in Cochinchina under the French colonization. Like his counterpart Lt. Gen. Đỗ Cao Trí, he was among the first paratroop officers trained by the French who later transferred to the Vietnamese National Army. He also helped Pres. Diệm control the Buddhist ARVN rebels and Buddhist monks' uprising of 1966, and then recaptured the Citadel of Huế from NVA/VC during the Tết Offensive of 1968. He deserved being given the command of IV Corps in the Mekong Delta. His last assignment was I Corps before the end of the war. A "clean" and straight forward commander, he had successfully used law and order to stabilize his units and zone of command, and pacify the confused and terrified civilians. A truly loyal military man and a fervent Confucian (as discussed in an earlier chapter <u>Confucianism and its politicized infusion</u>, he was placed in a difficult position towards the end of the War where he had no choice but to follow Pres. Thiệu's controversial[534] order to abandon Central (I Corps) and reposition troops in "defendable" territories. This strategic withdrawal had unfortunately caused morale to seriously drop among the armed forces and the people, most of the latter being either full fledge anti-communist or neutral "side-with-nobody." Gen. Trưởng and his family were split apart during confusion of the Fall of Sài-gòn, but they were later successfully reunited in Virginia, U.S.A, where he lived with his family until he passed away in 2007.

---

533   https://en.wikipedia.org/wiki/Ngô_Quang_Trưởng Gen. Ngô Quang Trưởng's biography

534   Vietnamese paramilitary magazine <u>KBC, Issue #5, Garden Grove, California; Article about Gen. Trưởng's meeting with President Thiệu</u>

CHAPTER 15

Maslow Revisited

  This chapter is educational, not just for Vietnamese, but for most veterans, immigrants or people who are caught estranged in a new society or environment during their acculturation or re-socialization. It is usually applied in educational psychology to help shape teenagers trying to cope with school peer and culture. When working with immigrant adults and students, I find it extremely useful in rehabilitating people who resettle in a new, possibly hostile, socio-economical environment. It becomes my preferred tool for working with the at-risk as well as under-privileged students (e.g. students of color).

  Maslow's Hierarchy of Needs can explain why immigrants adjust differently in their new countries. Some are successful, some not. It can be used to explain how soldiers in foreign wars adjust to the new environment and how veterans are handling their PTS and war residues upon their return home. Nevertheless, with the ARVN veterans, it is difficult to tell how they are coping with their new life in the U.S. (or other countries). They usually smile and act normal in public, but most are unhappy inside: when they don't have to maintain their public face[535], they can easily become crabby and argumentative. Are they suffering from PTS? I strongly believe they are, and that they are very good at masking it: they are like dormant volcanoes: some might erupt, some might never.

  Even now, various Vietnamese nationalist groups (of the former Republic of Vietnam) in exile in the world **still do not agree with each other on many socio-economic and political issues** related to nationalist matters. As seen reflected in their daily nation and world-wide blogs and emails to each other (in Vietnamese), each individual email communicator or blogger tries to vent their frustration, or even anger, oftentimes based on unfounded gossip or distorted information ("tin vịt," news from the ducks meaning fake news). As cyber-interaction is indirect (not face to face), they are not aware that those emails can hurt others (defamation of character) mainly

---

535 They mask their true feelings when they are with outsiders and, particularly, the "foreigners."

because they cannot see the reaction of the people they are interacting with. In most cases, they do not care about what the others feel. Being self-centered is the essential mental status they all have, as part of Maslow first phases "adapt to the environment."

Their email destination oftentimes reaches from 50 to 80 addresses, and the text can be a few pages long of blabbing redundant culturally incoherent discussion they call "biện luận or phản luận" rebuking someone else's statement. Discussions are usually based on exchange of gossips, and information from unverifiable sources (they still don't understand the fact-check process). Psychologically-speaking, their arguments have a major chance of clashing each other mainly because everyone has been acculturated differently, depending on where and whom they hang out with, where they work, or their level of education and the intellectuality of their social domain. Basically, they might not trust anyone. It is that fear of the unknown that prevents them from opening up and conduct a normal relationship, a fear of being outwitted by others. They have been victims or witnesses of falsification and betrayal in the course of war, surviving and migration. It is hard that, on the one hand, they want to hang out with their cultural group which they might not trust, and, on the other hand, they are afraid of the general population which they think they are incompatible with. They might have to hang out by themselves!

I bring up this mental issue here, not because I intend to disrespect the Vietnamese community, but because I want to sensitize the more educated and acculturated younger generation to their elders' calamity, so this group can grow and rise up for better in the society. **Out of the mud rises the serene lotus.**

One thing for sure is that the Vietnamese community, old or young, is becoming more and more stagnant, and more or less refuses to reach out to the community at large (English-speaking mainstream). Why so? As they are divided, people are becoming insecure, and being insecure, they certainly cannot feel at home in a culturally strange society where they think they can be intimidated by their own ignorance, particularly when they might have low self-esteem believing they are the misfit. Culturally, Family Honor is utmost important. It is the mental health of the misfits that makes them experience complex difficulties at the level of "sense of belonging" and "self-esteem" (based on Maslow's pyramid of Hierarchy of Needs): failure to fit in. It will absolutely affect their relationship with others, and especially with their loved ones; nevertheless, they will never admit it as it is a shame to the family if people think there is something wrong with their mind.

Rarely have I seen the Vietnamese first, or even second generation immigrants serve as volunteers in the community at large (English-speaking), except for those who are highly educated and who can relate well with the "Westerners." I have been trying to justify to the general public why the Vietnamese immigrants cannot get involved in community service[536], which is a very American "thing," or participate in networking with the Vietnam Veterans. This second choice should have been easy as they all come "from the same boat," the Vietnam War... It might be very difficult for the first generation Vietnamese immigrants to understand what mainstreaming is, even though their ancestors always teach them "nhập gia, tuỳ tục," which means "abide by the host's customs when entering the house." Instead, they behave as if they are still living in Việt-nam!

The younger[537] as well as the second and third[538] generations, especially those who grow up sheltered in that closed cultural environment, need to be aware that the American culture highly values community services and mingling. This culture perpetuates cross-cultural teamwork and leadership, in the idealistic case, and especially nurtures kindness towards other races. Americans in general try to avoid repeating the same mistakes their elders (older generation) have made causing segregation. The younger Vietnamese Americans might be more successful at mainstreaming into the society because they grow up more American than Vietnamese. Nevertheless, it might still be less likely to happen if there is a grandparent in the family who always assists in rearing children: they usually try to encourage, or rather do their best to make these children hang onto their root cultural values; this style of raising children may contribute to them being unable to develop assertive leadership skills in their youth when interacting with others.

---

[536] Many are very involved in their own monolingual community, but not in the community at large. Also, not many groups of former ARVN are involved with the Vietnam Veterans either.
[537] Born in Vietnam and coming to the U.S. as a young child.
[538] Born in the U.S..

Traditionally, in "well-raised" families, Vietnamese children are trained to strictly follow directions from parents, or from other parental designees such as nannies, aunts/ uncles and grand-parents, when they are between 1 and 18, or even older. In many cases, family hierarchy counts and might give authorities to higher-relationship ranking regardless of age. This means a younger uncle has more authority than a nephew.

"Children can be seen but are not to be heard" when adults are around and have to "đặt đâu, ngồi đó» (remain where they are sat). They are shaped onto Confucian requirements (as I have explained in an earlier chapter on Confucianism). Failure to shape children usually brings blame on the parents (especially the wife) and, by extension, shame to the grandparents and ancestors. This rule will bend them to be submissive to the higher hierarchy when grown up, so that they can later serve their supervisor or commander properly. Being bold, assertive, argumentative, or outspoken usually means to the society that the parents have failed in educating their child who is consequently considered as "con nhà mất dạy" (child uneducated by the family, a very serious shame to the family name). Children growing up submissive become better servants who later on experience hardship when leadership is expected. Do I have proofs? My two siblings and me! Thousands of high school students I work with for a decade!

This traditional expectations, as I observe Vietnamese students perform in group and team work in the Minneapolis schools, might possibly still exist. I notice that many Vietnamese children nowadays are in general slower in interaction and much less assertive with others, sometimes subdued and apathetic, and, by extension, incompetent by American standards. Hesitation can be interpreted by American standards as timid, if not undaring or even cowardice. **This might be the reason why the press media has characterized some ARVN units as lazy and cowardly, and leadership as incompetent (e.g. History Channel <u>Vietnam in HD – A Changing War</u> (1960-70), <u>Peace with Honor</u> (1971-71), PBS Ken Burns' series <u>The Vietnam War</u> - 2018).** *For sure, they are not lazy or coward. Cross-culturally thinking, they are just polite and awaiting permission from those in the higher hierarchy, and dare not violate the deeply-engrained culture. In the U.S., they are culturally confused and cannot* handle the new expectations! After many years making efforts in motivating them to be more socially and politically active with the English speaking activists and politicians in Minnesota, I conclude that **the ARVN, who are now elders (between 65 and 80 of age), are culturally incompetent.** They "freeze and fly," as Goleman has described in his "reptilian brain" theory whenever they realize they have to participate in events that involve "foreigners", even though the latter are their allies. Do these activities activate their "French colonials" snapshots that might be intimidating to them? This behavior might be caused by some unresolved core issues or a lack of educational experience.

     Personally, as a child, I had been shaped into a "con ngoan", so it has taken major efforts for me to "undo" it in order to become assertive and do well as an American public school principal. Not until I started teaching math to English-speaking students that I can relax from and bend the traditional Vietnamese cultural tensions: this is a turning point for me as I realize I have better connection with these "mainstream" kids than with those in my own culture, urging me to focus my doctoral research on African Americans, instead of researching the Asians as my boss, an associate superintendent, has suggested. My Doctorate degree and humanistic (multi-faceted) education and training have given me a high enough status to be self-confident to work against my Vietnamese and family culture. Principals must be able to teamwork with other highly-educated "Westerner" colleagues and handle challenging parents and students. My colleagues believe that I thrive in doing so!

     "Knowing our Foes…" is needed as we want to know the opponent's strengths and weaknesses; however, neglecting or ignoring "Knowing Ourselves" will doom us to disgrace and failure, as we cannot measure up to the opposition's strength. We might be good in some areas, but there are plenty of others better than us. When are my fellow misfit Vietnamese immigrants going to realize how self-destructive they can be, being self-absorbed and mutually-divided (because of acculturational differences)? Even if they are able to realize so, will they be able to achieve the higher levels of self-realization and self-determination to catch up with the mainstream? Not until they can heal from their colonial mentality (when they do their best to crush others down so they can elevate themselves) and start their self-awareness and self-determination process, will they be able to reach a higher Maslow level and **their plight for true Democracy and Freedom will not be attained.** They will continue to hide in their comfortable mono-lingual and mono-cultural ghetto, and mask their fears with their smiling face in the presence of "American-born." Hiding behind expensive cars, jewelry and appearance show offs[539] to conceal their cultural uneasiness works only among their own, but not among the mainstream culture where they feel awkward knowing they have no substance to help pump them up.

Along this same line of psychological rational thinking, the other first generation immigrants (the elders) from the Khmer Republic and Kingdom of Laos (low, midland Lao Theung and highland Hmong) might be experiencing the same blockage at the Maslow Sense of Belonging level that leads them, one way or another, to social and political mutual-segregation.

From "Lost in the New Independence" to "Lost in Freedom"

We have seen how lost the newly emancipated Vietnamese are when emerging from the French Colonization in 1954. The new Independence has caught many pseudo leaders and wannabes off guard and caused them to clash each other, while the common "little people[540]" continue to be submissive. This time, when Freedom arrives with a totally new Democracy, they are caught off guard again. As the pseudo leaders clash with each other (e.g. coup d'état, rebellions), the common people, who have little to no ambition, continue to mind their own business and serve passively.

The younger resettled immigrants are trying to negotiate the next levels of needs, such as security and belongings, e.g. connection, out-of-the-bubble friendship,

---

539   Expensive alcohol, lavish sea food and fancy party, huge real diamond jewelry, expensive new sports cars (Ferari, Porshe, Mercedes Benz,) .. The question is: how can they afford them?

540   In the context of this book, "little people" refers to the common citizens who, more or less, have no connection with the ruling class, whether or not they are in the "network" or they are connected to the "bribery system." The "little people" are part of the bottom socio-economic class, the majority of the population.

relationship with neighbors and diversity[541], finding one's place in the new society, as described in Maslow's Hierarchy Scale[542]. They are actually navigating interchangeably between the old culture and the host one, and, in most cases, having to develop two completely opposite personalities (e.g. a new behavior at school versus a traditional one at home). Meanwhile, most of the adult immigrants are pre-occupied with managing some cultural social connection with each other (Vietnamese), while trying to cope with the new environment (physiological needs e.g. food, housing, rest, job). The issues of having a big super ego "cái tôi lớn" (in French "le grand moi"), and being stuck at the belonging and self-esteem levels (Maslow) result in the development of a self-preserving narcissism. This condition will keep them from achieving the higher levels of self-esteem and self-actualization as they are pre-occupied with "self-embellishment", or even "self-revering." It is an understandable way of saving face and preserving the family name. By elevating self, they usually think of other Vietnamese as inferior. Nevertheless, the more they try to boost their super-ego to increase self-esteem, the more they might realize their efforts do not necessarily bring more respect from others. No one likes to be pushed down! Respect has to be earned, but not by putting others down. Consequently, they found themselves alienated in their own bubble.

We have to remember that, in order to survive and succeed in a new socio-political environment, ambitious people with a history of being colonized or dominated by another race tend to crush others in order to self-elevate[543]. By doing

---

541   Not like in the mono cultural Vietnam, America is very diverse which can cause major culture shock for many.

542   http://www.simplypsychology.org/maslow.html Physiological, security and relationship levels are the lowest levels of psychological maturity on the Maslow Scale. Growth in the context of Democracy and Freedom at the esteem and self-actualization levels does not happen until the individual knows how to exercise their rights in a responsible way so that others' rights are not infringed. Many people in free countries, the U.S. included, still did not achieve the latter levels when they abuse their rights and affect negatively by imposing their personal egocentric values when others'. http://www.edpsycinteractive.org/topics/conation/maslow.html Educational Psychology Interactive expanded this concepts even deeper by splitting the Self-actualization level into 4 stages, from low to high: Need-to-know-and-understand, Aesthetic needs, Self-actualization, and on the top Transcendence. When an adult thinks he knows all already, and refuse to open up and learn more, he/she is definitely stuck at the lower levels and will continue to behave like misfits, and believes all the others are misfits (low self-esteem, bullying,... being a low level of mental health concerns).

543   As referenced in Dr. Ha Tuong's dissertation <u>Discourse of Respect among African American High School Students: Jawanza Kunjufu's Raising Black Boys, and Developing Positive Self-Images Discipline in Black Children and Paulo Freire's Pedagogy of the Oppressed.</u>

so, they believe they can stand out and "sell" themselves, hopefully to the colonizers who might notice them out of a crowd and give them a power position in the society.

Times have changed, and the new question is: why do they still feel a need to stump on others, now that there is no more colonization? In fact, they have Democracy in their hands! For those who are currently in America (or other civilized countries), where everyone is equal and respected, the Vietnamese are still "weird!" They are so "weird" many of the more cognizant ones have written blogs and books to make their own people aware "how ridiculous Vietnamese behave." Blogs such as Sự thật khó nói mà phải nói! (the truth that is difficult to present, but must be presented!) have been circulated between email groups most recently (2019). The most popular book Thói Hư Tật Xấu Của Người Mình (The Rotten Habits and The Despicable Behavior of Our People) has been circulated on-line since 2011 based on the milder ones Chõu lòu de Zhong Gúo Rèn (The Ugly Chinese) written by Bo Yang two decades ago and the Ugly American published by William J. Lederer and Eugene Burdick in 1958.

As civilized people, should they respect themselves? Who do they want to "shine" with? After all these years living in America, they continue stumbling onto these new mysterious obstacles, Freedom and Equality, which they do not know how to handle without abusing them.

Self-preservation urges them to speak up against others to show off their "worthwhileness." In their efforts to elevate themselves above others by diminishing their fellow immigrants, they end up making the most serious mistakes when believing that, with their new democratic power and freedom, they can say or do anything they want at the expense of others' democratic rights. The most common violation of others' rights is defamation of character. They oftentimes find themselves out of place when expressing their thoughts that turn out culturally ridiculous.

A concrete example of abused Democracy was displayed in inappropriate comments posted by Vietnamese individuals in "Facebook" published in the local popular Vietnamese weekly magazine Sài-gòn Nhỏ (Little Saigon), issue #603 of June 3, 2016. One post, probably heavily influenced by his traditional Vietnamese values, criticized sitting Pres. Obama for his *"seriously inappropriate behavior unbecoming for a president of a powerful country"* when the latter was visiting Việt Nam in May 2016. *The post described Pres. Obama sitting on a "vulgar plastic" chair and eating in a common low class restaurant, or leisurely feeding Koi fish (instead of fast feeding them) at "Uncle Hồ's" bungalow, or walking in the rain "instead of having an umbrella"... Obviously, those "Facebookers" were either new to the U.S., or immigrant elders who refused to grow out of their old restrictive culture of the 1930s, or who were ignorant about the American and modern culture*[544]*. They were showing, not just their total lack of cultural sensitivity and civility, but also their utmost bullying behavior "unbecoming of a civilized person." Psychologically-speaking, their self-esteem might have been quite low, showing signs they* **are stuck as a misfit in Maslow self-esteem and belonging levels.** *Also, obviously the editor who had chosen to post such an article might as well belong to the same Maslow category.*

One must realize that people nowadays, either American-born or well acculturated immigrants, love a powerful leader who is modest and open-minded, and who can bring him or herself to the same level of the people. It is all about being socially updated and having adequate general knowledge of current and cross-cultural issues, which people, who are either stuck in their past or maladapted to the new society, cannot cope with. Nevertheless, these people might argue that my viewpoint is distorted, that I have sold out, resulting in me not being able to reason in the "appropriate" way a respectful Vietnamese ought to…

---

[544] Definitely, the "Facebooker" who commented that President Obama had infringed into the peace of Vietnam by being there was a pro-Communist! The former nationalists as well as the non-political people of Vietnam were all enthusiastic about the visit, as shown by how heartfeltly they had welcome President Obama. Nevertheless, it is a post by a Vietnamese in the U.S., as Little Sài-gòn paper is a U.S. paper.

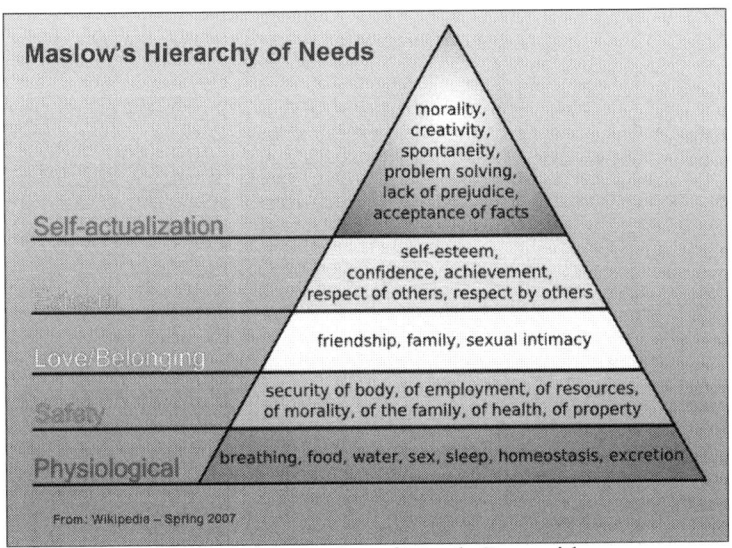

Maslow Hierarchy of Needs Pyramid
Source: Wikipedia 2007
(Refer to an earlier section Psychology: Maslow and the Hierarchy of Needs)

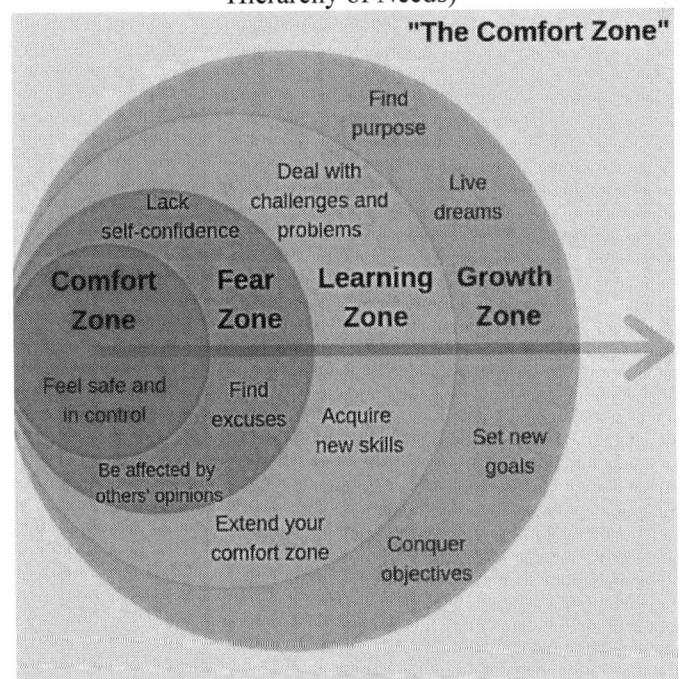

Comfort Zone, another view parallel with Maslow Hierarchy of Needs
Source: Unify.org

Frustration grows even more when those, who tend to live more in the past[545] where they feel is their **comfort zone**, realize they are intellectually and professionally so behind, and so socio-culturally out of place they feel they desperately have nothing worth "selling" in order to earn respect from others. This dead-end might make them feel high anxiety, if not fear. The Unify chart on the previous page explains well visually in a different way what the Hierarchy of Needs does, nevertheless, they both show that the individual, whenever s/he is in conflict with the environment that is causing fear, will revert back to her/his Comfort Zone or Maslow's Love/Belonging, or even lower level Safety.

When applied to the Vietnam Veterans, that comfort zone can be Việt-nam they remember and can visualize. Many Veterans have experienced serious grievances upon their return home being mistreated by fellow citizens and family to the extent they believe their Comfort Zone is actually wartime Việt-nam. They feel they are much more respected there than at home. African American Veterans always say "they never called us N….r! **Việt-nam, a place of Fear then is now Home, while Sweet Home in the U.S. has actually become Fear.**

In my opinion, new immigrants who come from an old tradition and strict culture need to be retrained, rehabilitated and re-socialized so they can start a new fruitful life with each other in a truly democratic and free society. Unfortunately, many socio-scientists strongly believe it is nearly, if not totally impossible, to re-socialize people or animals[546]. Sadly, as social scientists have discovered, the older those immigrants are, the more impossible it will be for them to "re-socialize."

---

545 First generation Vietnamese whose educational level is low (with no university degree) are usually use the past as reference; consequently, they tend to replicate past achievement or lifestyle as if it is comfort food, instead of progressing into the future and embracing updated out of the box knowledge, especially when the latter seems to be contradicting what they have experienced in the past. For this reason, a great many immigrants still use archaic practices in the new environment, causing serious conflict with safety or legal regulations. As examples, mothers after delivering babies, requesting hospitals to use charcoal burner to keep area under the bed warm, or dancers performing steps that dated French colonial époque 1920-1940s, women being treated as inferior "Chồng chúa, Vợ tôi" (Husbands are lords, wives slaves)…

546 Such as the re-socialization of Tarzan… or the domestication of a ferocious animal…

That is why we don't see many middle age (in the 40's) to elders (60-80's) Vietnamese participate in the mainstream! If they cannot outgrow from their roots and move on, they will end up either adopting the submissive fetal position (Goleman's fright) or fighting back as in resistance (fight). They will not be able to move up from the Maslow's Self-esteem and Belonging levels, or, if we refer to the Comfort Zone theory, move out of Fear into Learn and Grow Zones. Consequently, they will either have to hide in their **cultural and linguistic Comfort**, otherwise they will develop mental health issues and eventually become social misfits or rebels in the mainstream!

This might be the case nowadays of the mass shooters who have gone out of control in American society because they feel they cannot fit in. We can only escape from our ignorance and weaknesses by opening up our mind, learning and growing. However opening up our mind will encounter obstacles when we are too attached to our past or our comfort zone (where one belongs well) but the environment refuses to accept us as the way we are...

Justifying "their" absence in community-at-large activities

What really causes "them" to be unsuccessful in those two levels of Belonging and Self-esteem? Fear is the cause[547]. Fear of the unknown, fear of confrontation, fear of interaction with others, fear of failure... These might be the main reasons why many former ARVN (they are mostly in the upper 60s or 70s now, many of whom without an education background higher than high school diploma) and their family (the men are the head of the docile family) very rarely interact in a natural setting with the community at large ("cộng đồng Mỹ, American community" or "Ông Tây, Bà Đầm, the Westerns"). This may also be the reason why the former ARVN members seldom look for occasions to meet up and to participate jointly with the Vietnam Veteran groups, except for some rare occasions when they can appear together in larger numbers in their uniform, such as the memorial commemoration of Black April in St.

---

[547] As I have observed when working with people of various socio-economic, culture, linguistics, race, age and gender diversity in Minneapolis public schools and communities.

Paul, Minnesota[548] or Vietnam Veterans Day in Forest Lake... "Strength in numbers" actually works.

    This behavior is actually the same as the Vietnam Veterans who choose to go into hiding: they don't feel they really belong to the community at large or perhaps because of a low self-esteem. Such are the dilemma faced by the Vietnamese ARVN, the maltreated Vietnam Veterans, the Republic of Khmer veterans as well as the Low Land Laotians, whether they are cognizant of it or not!

    While working for a couple of decades with the Hmong community in Minneapolis Public Schools, I have noticed that, even though the elders feel culturally totally alienated from the society, these new immigrants end up being more active participant than the other SE Asian immigrants. Their community has actually been empowered by a handful of welcoming resourceful "Americans." Right from the beginning of their resettlement in the U.S., the Lao Family (Hmong Association) leaders have made the right decision to hire "local resourceful native-born staff" to work for and to advance Hmong Community organizations. Their Maslow transition has been more or less smoother than the one experienced by the Vietnamese, and maybe the Lao and Khmer. As a result, the Hmong community has successfully established a better social and political connection with the community at large and the mainstream society.

    The second generation of these Hmong immigrants whom I have worked with has more or less successfully integrated with the society as they have grown older[549]. One can observe that the second Hmong generation speaks as fluently, acts and makes decision as effectively as any other American-born mainstream kids. These younger ones, in particular, seem to flourish well as they have been catapulted through layers of education in all fields and into elected political positions, mainly thanks to their well-focused goal-setting and, probably, because of the well propagated Hmong Oral History of heroism[550] of their elders. On top of that, the

---

548   ARVN and U.S. veterans organize events jointly to commemorate yearly memorials and remembrances.

549   I have been following up with former Hmong students on Facebook

550   As verified by documentaries (news media and CIA, tactical map and cross-referencing sources, some of the Oral History contents were found falsified. By the nature of Oral History, stories counted by story-tellers whose education levels might be elementary or even less (none), it is inevitable that things are made up to make the stories sensational. Hmong culture has been based on Oral History, which

Hmong leaders' involvement in political lobbying activities has opened up many opportunities towards political strength[551], social assertiveness and economic benefits. With that momentum, the other SE Asian groups believe it is the Oral History that has helped promote the Hmong sacrifice during the war, to the extent it is believed and verified that many details of their involvement in the war have been exaggerated, or even made up, in another word "fabricated," so the history of the war in Laos and of the Hmong participation has been murky and misleading…

Concerning documentaries that were misleading, the Royal Lao and ARVN veterans wanted to make it very clear that the Miao Special Guerrilla Units SGU Bataillon (as television press media referred to as the CIA <u>America's secret army</u>) might have fought in Laos alongside with the RLA against the PLA (Pathet Lao Army, communist or leftist, Hà-nội -backed), and some NVA units (reinforcing the PLA), or helped rescue U.S. pilots[552], but that happened strictly in RM#2[553] which is under Gen. Vang Pao's control. **They have never, ever fought in any part of Việt-nam**[554]…*Some Hmong*[555] *first generation elders, bloggers*[556]*, Hmong SGUs or even Hmong Studies researchers had openly claimed that*

they excel, as these tribes did not have a written language until mid-1950s. https://omniglot.com/writing/hmong.htm

551   The Hmong community has succeeded in launching five state representatives and one senator into Minnesota Congress as of 2017, as well as many others in city councils and school boards in the Twin Cities.

552   http://www.jefflindsay.com/hmong.shtml Why are the Hmong in America?

553   Hamilton-Meritt, Jane (1993,1999). <u>Tragic Mountains: The Hmong, the Americans, and the Secret Wars for Laos, 1942-1992. The Miao SGU were strictly operating in their assigned RM#2.</u>

554   Members of the ARVN often wonder why, as shown in website photos, current Hmong veterans were wearing RVN medals that were issued only by the government of the RVN. Hmong armed forces, whether they were "the irregular" ethnic clandestine (CIA guards in RM#2) or official "regular" (as part of Royal Lao Army), had never fought in the war to support the RVN, so they should not wear them as they were never rewarded those medals. Only the armed forces from Taiwan, New Zealand, Australia, S. Korea, Philippines, Thailand,…besides the U.S. Army, had actually fought in country, therefore they would be awarded RVN Medals.

555   Note that the terms Miao and Hmong are used interchangeably depending on whether the events happen in the past during the war or in current timeline after 1975.

556   Paige Ashford and Desiree Acker's report on Prezi site https://prezi.com/yvf4wzvq9jkd/ho-chi-minh-trail/ and many adult Hmong former "secret Lao army" members whom I had run into at workshops had reported that the Hmong army was recruited by CIA to destroy the HCM trail and bridges, and, if necessary, to engage the NVA.

the secret army had "fought along the trail" or helped to "block, or attack the Hồ Chí Minh trail" (YouTube and PBS TPT interview documentaries), and also to "engage the NVA divisions that are moving on the trail if needed" (Wisconsin SGU militia organization promotion pamphlet). Those accounts are all made-up... Some Minnesota Hmong legislators have even confirmed in front of congressional committees some false claims made by the Miao SGUs and oral history and use them to support their political agenda... (Lao, ARVN and Khmer veterans and I were there to testify, so we heard their testimonies)...

SE Asian communities and U.S. Veterans realize that those claims have been exaggerated and fabricated for a long time for the purpose of gaining and strengthening the Hmong political advantage in the state. They have successfully helped push the Hmong politicians' agenda onto the table in order to achieve even more power... **Again, I cannot overstate that the truth of the matter is that Gen. Vang Pao Miao SGU Battalions (with majority Miao ethnic with some Lao-Theung) have never been near the HCM Trail that actually starts from the center of RM#3, but much further from their RM#2**. *Are the ethno-centric oral history confused Miao SGU with RLA SGU? The Lao SGUs in RM#3 and RM#4 actually have some Miao members, but they are fighting as Lao soldiers under the RLA command. Col. Insixiengmay (former commander in RM#3) confirmed that the RLA SGUs usually have 15 to 20% ethnic (speaking mainly Hmong, Mon-Khmer, Thai languages among the 140 ethnic groups[557]) in their formations. Also, the T-28 fighters that they claim are Hmong planes are in reality part of the RLA Air Force...*

This brings up some new topic for Maslow conversation: A core issue that some of the Hmong veterans now encounter at the Maslow self-esteem and belonging levels (as discussed earlier in <u>Psychology Playbook</u>, Chapter 5; also, see Bibliography <u>American Sniper: The True Story</u>) has probably caused them to try to attain societal recognition to match up with the claims and illusions of heroism, by illegally wearing the RVN and U.S. Purple Heart medals! This core issue of wearing un-awarded medals also exists among some Royal Lao, ARVN and Khmer Republic veterans, but not to the level as with the Hmong. To enhance and give them even

---

557   https://en.wikipedia.org/wiki/Demographics_of_Laos#CIA_World_Factbook_demographic_statistics

more meaning wearing the medals, many do not hesitate spending thousands of American dollars and buy ranks (from U.S.$3,000 for Colonel to U.S.$5,000-$8,000 for General) so they can parade with stars in their grand ceremony uniforms. **Psychologically-speaking, this behavior create a new Comfort Zone as it boosts their self-esteem and gives them the impression they can be respected and honored when hiding behind the ill-gotten medals they wear on their uniforms along with their purchased military ranks.** This might be a way of raising funds for some; but, for others (the victims), it is a coping mechanism that actively reshapes the environment around them without the individuals having to passively shape themselves. Anyway, it has become a "win-win" situation, hasn't it? No matter how wrong it can be being imposters, the feel of the uniform, the stars and the medals must give them a sense of purpose, whether it's real or not. Meanwhile, the former ARVN reshape the self-esteem by guiding their children into pursuing a military career and becoming real Colonels and Generals in the U.S. Armed Forces[558]...

Oftentimes, Vietnamese wish they can someday catch up with the Hmong community. Low land Lao leaders (Royal Lao, Lao Theung) sometimes complain their people prefer more to work with the Hmong than with their own. Both groups are from Laos, but, in Laos during the war, thanks to having a written language in their culture, the Lao had a better socio- economic head start over the tribal mountain- dwelling Miao. The Vietnamese self-help groups sometimes have to rely on the Hmong community for assistance to fund their cultural activities. This happened when I was still working with Vietnamese associations in the Twin Cities in 2008-2010.

Recently, younger Viet-American children have been recruited to promote Hmong organizations during the 2019 State Capitol Asians Meet Legislature Day. It seems that everyone is looking up at the Hmong veterans as the "official" Asian Vietnam War participants. After having participated in legislative lobbying efforts (2019), I can say that the majority of MN legislators are confused by the ethnic details: they seem to think there is no longer ARVN, Royal Lao or Republic of Khmer affiliation in the MN community, and believe that the Hmong activists are representing all SE Asians.

---

[558] As of 2019, many Vietnamese-American younger generation Hậu Duệs have been awarded the ranks of Colonel (15 in 2016) and Generals (1Major and 2 Brigadier Generals, and 2 Rear Admirals) in the U.S. Armed Forces.

The ARVN believed that many of the medals worn by these Hmong SGU veterans in the U.S. are either Republic of Vietnam (Vietnam Campaign, Injury, Gallantry…) or U.S. Army (Purple Heart) issues. The wearers never earned them because they never fought for the RVN and the U.S.A. These photos can be found on the internet

Many politically active SE Asian Minnesotans believe that the Asian Pacific Council, an official state organization created to serve all Asians in Minnesota[559], acts as if it is representing only the Hmong population[560].

It is interesting to see how differently people from the SE Asia War survive in the new environment: Some give

---

[559]   I was one of the Asian community representatives who witnessed the governor signing the 1982 resolution.
[560]   It supports the PBS TPT series <u>Minnesota Remembers Vietnam focus on the Hmong, besides the VN Veterans, "because they are the largest SE Asian population (Ms. Sia Her: 90,000 compared to 60,000)</u>

up as they cannot cope with the environment, while others can successfully adapt by reshaping it. Meanwhile, the Viet, Lao and Khmer younger generations refuse to pursue their elder generation's legacy. Particularly with the Vietnamese, the generation gap is severe and the ARVN legacy is in great danger of being forgotten within the next 10 or 15 years.

We can see clearly here an example of successful psychological adaptation: the Hmong have been able to reshape the environment into a new Comfort Zone by elevating their history and culture, meanwhile the other three groups, like a snail or turtle sensing danger, pulled back into the cultural refuge. As a result, the former can move up and dwell well in the Learn and Grow Zones, whether they are young children born in this country or young children migrating from the mountains of Laos, with the idea and public recognition that their elders and creeds[561] are heroes.

Nevertheless, the younger Hmong are suffering a great identity loss because of their immersion into the American culture and language, especially when, thanks to their better education, they can fact-check Hmong oral history and realize some of the important information has been exaggerated. They are also very aware that their elders are highly politically divided, and some do not want to have anything to do with their community internal politics. In the meantime, the elders who have been involved in the war are probably combating PTS and mental health problem when trying to live up to their **ethno-centric**[562] **Oral History** heroic expanded claims. This explains why they adore uniforms, stars and medals. As a result, these veterans have actually created a new world of Hmong heroism[563] so they can fit into its Maslow level of Belonging by **swapping it with the Sense of Purpose**: it helps them come up with their **newly adapted Comfort Zone**. People respect them and they do not need to fit in it."

---

561   The Hmong resettled in the U.S. came from many different tribes that might see themselves as very distinct from each other. Blue (Green) and White Hmong live in the U.S., Red in Vietnam… https://henry.mpls.k12.mn.us/uploads/hmong.pdf

562   …if worse, as it's elevating one culture while ignoring its own allies. My involvement with the Hmong immigrants and younger generation showed me that the majority is friendly and receptive, however, the political leaders are abrasively defensive and tend to block other allies out and focus more, if not totally, on their own clan advancement.

563   The internet is now full of photos of Hmong Veterans wearing South Vietnamese medals and U.S. Purple Heart, parading with military stars… This is based on a true admittance of a Hmong SGU commander as well as Hmong researchers.

Ethno-centrism is actually the symptom of uncomfortable societal adaptation. Their image of heroism, as perceived and adopted by the population at large, has created a new reality that helps the Hmong individuals accommodate to the society: to change the way the others view them by making people around them believe that they are Heroes. "Vietnam War refugees are the best of all immigrants," as many local Minnesotans have told me, and find out later that I am not Hmong. Obviously, Minnesotans love and show a high level of respect and admiration for the Hmong. Meanwhile, other Vietnam Veterans ask each other on Facebook: "Why are they treated better than us, Vietnam Veterans? We are homeless. We are still poor. We are dying of Agent Orange…"

How does Maslow's theory apply to Việt-nam during the war and now?

We should remember that, during the over-a-century long colonization, the fertile South was well-guarded and "cherished" by the French compared to Central (An Nam) and North Vietnam (Tonkin). Competition between the colonized Annamites for potential employment by the colonizers was harsh among Cochinchinois collaborators; this being the possible and more plausible reason why the South Vietnamese had become politically divided while adapting to changes under the French rules (Maslow's Safety, Esteem and Belonging stages of needs). Meanwhile, Tonkin and An Nam, where people were stubborn and land was economically less productive, did not attract the French much; this consequently had allowed the people there to stay united, with more or less the unshaken and unbroken Self-Realization and Self-determination stage to fight back because of the old anti-foreigner mentality and culture.

After the French were defeated by the Việt Minh in 1954, and after the occupiers had left Indochina, as the socio-political climate totally changed from subjugation to independence, the new State of Vietnam was also reset to the Maslow's basic need stages: the physiological (feed the hungry in the North and Central, rebuild economy), existential (make a living to survive and advance) and belonging (to be unified

North-South under new ideologies of Independence, Freedom, Democracy) levels. Adaptation had more or less affected the rebirth of this new nation whose people had at last partially recovered its long lost Independence from foreign power and who were trying to adapt to a new Democracy.

Socio-politically speaking, the three former Indochinese parts of Việt Nam were undoubtedly going through a completely different metamorphosis because of the profound effects of the lingering "divide and conquer" strategy the French had used during occupation. The North found itself still struggling with a devastating famine, and the Central facing the conflicting Imperial versus Republic ideologies, besides surviving the same famine as in the North. Meanwhile, the fertile South was trying to appease divisive political factions while being torn between American materialistic influence and the old versus the new socio-political identities. This fractured Việt-nam was still very chaotic and unpredictable, and the people in the three parts were still discriminating against each other.

While the nationalists were unknowingly (or in denial by egocentric pride) experiencing crisis[564] when embracing a new confusing Democracy, how was the budding Communist Party coping with the new State of Vietnam, each interpreting Independence and Democracy differently? Would Hồ Chí Minh be able to lead through the stages of self-realization and self-determination after he had declared Tonkin (North Vietnamese Indochina) independent from the French and the Japanese (August Revolution - 1945 Independence Declaration)?

In my opinion, North and South Vietnam had little opportunity for a potential unification because, as soon as the Việt Minh had defeated the French in 1954, its membership's mind and strong will were instantly taken over by and exposed to one way of thinking: "Uncle Hồ's" way. It was prescribed by his "Red Little Bible[565]" which every family had to memorize and strictly adhere to, as if it was the Christian Ten Commandments or Bible, or the Buddha's teaching. As a matter of fact, the Northerners had to adopt the Red Book and

---

[564] They did not realize how Democracy had affected their behavior especially when they were subjective and had not been exposed to modern psychology

[565] Similar to "Uncle Mao's" little red book *Quotations from Chairman Mao Tse-tung* he had widely distributed after his Cultural Revolution (1964-1976). Edited by People's Liberation Army Daily and translated by Central Compilation and Translation Bureau. https://en.wikipedia.org/wiki/Quotations_from_Chairman_Mao_Tse-tung

"Uncle Hồ's" teaching as the new religion, with their children being raised separately by the state[566], thus setting up a totally new societal system. They had to be "tamed" and set on "autopilot" since the beginning of their induction.

Even today, as shown on the current 2016 VTV Việtnam Television programming, they are still living in the same dream that has been implanted in their genetic traces DNA by their "Uncle Hồ" since the 1945 Proclamation. The dream, which has been ascertained to be the only truth, has to be maintained pure and unequivocally righteous by the top-down directions of the Party. The people are born and raised obedient and loyal as their ancestors have been, but, nowadays, to one and only one cause, via a one-way brainwashing, so they can turn into faithful believers. "Uncle Hồ's" way expects "no religion, no patria (motherland), and no family!" At one point, children are raised by the government away from their parents who have become their comrades. "Religion is Opium of the people," meaning forbidden. The Party is above all! Resistance is futile (resistance results in being sent in exile by one's own children).

Current VTV programs still preach what has been preached since the 1950s! "Uncle Hồ's" Youth (like Nazi Hitler or Chinese communist Mao's Youth) are still active in their uniforms and proudly show off their exemplary-award, the red scarf (VTV4 2016). **Children story books, as proudly presented on VTV4 (May 2016), still preach "Uncle Hồ's" words glorifying anti-American slogans.** National television Programs still show skits of NVA anti-aircraft batteries being championed as the ones that have shot down the most American B52 bombers (Nov. 2018). Pictures of "Uncle Hồ" are still given out as awards to children role models, and his bust is still high on the family and offices altar. Images and legends of party role models from the old past are still paraded on the television screen every day along with their heroic "annihilate the Americans" accomplishments. The winning government election candidates have been pre-determined even before the actual elections occurred (2016 Assembly election, as reported on social media). The voters accuse many candidates for having Chinese citizenship, alleging that Hà-nội's officials are colluding with China to subjugate Việt Nam to this rising world power, again, and, as a result, to transform

---

566 … to help their parents focus better on dutifully demanding work and labor… the parents and children called each other "comrades when showing love… 1950s-1970s

the country into a southern Chinese autonomic province.

Ironically, anti-capitalist propaganda is imprinted in most gathering Agenda, while, interestingly, capitalism is being practiced throughout all socio-economic daily activities among the ruling echelons and their network. As of Nov. 2018, Sun World and Vin Cartel-like industries, Vin-schools, Vin-mall, Vin-stores, Vin-automotive, Vin-technology... are growing everywhere in major cities, most probably having shares belonging to "officials." Meanwhile, the common people are still dirt poor, four decades after they have been liberated. The flourishing picture seen by tourists in present day Vietnam is misleading them into thinking that Vietnamese people are happy and well-off. In reality, they are still swimming in Maslow's basic need levels scrounging for their daily family scraps of food and money for scooter gasoline while a small minority is enjoying the self-actualization level and its benefits.

The Communist Party has painted a beautiful picture of Democracy and Freedom for the People, from the indoctrinated cadre on the top to the common people on the bottom, as something completely different from what the rest of the World knows and hears about. Thanks to modern technology such as high tech tablets, smart phones, the internet and social media postings from the "real democracies," (in spite of government oppressive cyber security laws ((censorship)) passed on the 1st of January 2019), the truth is still unveiled and rings loud and clear. The people dream and wish they can experience real Democracy and Freedom abroad in person[567]. In the meantime, they can at least witness it via internet, social media and counterfeit film media news.

Except for some activists who are clairvoyant and educated who can use their rational thinking skills to find the truth (they are usually considered by the government as dissident "phản động"), most are still as obedient and submissive as before, or remain subdued and silent: they have no choice but accept their fate "an phận," and live a day by day life to avoid trouble. These people have no choice but to be content with their status quo, as their ancestors have traditionally been, as long as they can run their daily business and have enough to live by. With the communist tight control,

---

[567] Resourceful and daring people line up every day at embassies (e.g. China, Cambodia...) to apply for work visa with the hope they can find a better job abroad. Oftentimes, they fell into traps of sex traffickers, or even become victims of organ "robbers" who end their life away from their beloved ancestral land.

revolution is futile, and protests are obliterated at its infantile stage. The only thing they can do to exhaust their inner pain is to discreetly complain with the tourists they think they can trust, and smile amicably at the others. These people might have realized they cannot function well in the current political-social environment which they feel they do not belong to. The can find an escape only in their imagination, by discretely dreaming with expatriates so that their mind can at least achieve an acceptable self-esteem level and self-actualization. Such is their Comfort and Safe Zone, somewhere they can be in their next life!

CHAPTER 16

A Completely New Environment to Cope With

A few days before Sài-gòn fell, RVN Gen. Dương văn Minh or "Big Minh" assumed by default the interim Presidency[568] from the resigning interim Pres. Trần văn Hương. The latter was the 2nd Republic official Vice President in Pres. Nguyễn văn Thiệu's team, replacing VP Nguyễn Cao Kỳ in the second term. Around noon on April 30, 1975, escorted by NVA officers to the Sài-gòn radio station, Big Minh went on the air and, in the name of the last president of the RVN, ordered the ARVN to lay down all weapons and surrender. The highest ranking NVA officer, Col. Bùi Tín, a Báo Nhân Dân People's news press media cadre, accepted Minh's surrender.

The Way of the Warriors: Either chose a Heroic Death or Obey Supreme Commander and Lay down Weapons

Of course, the ARVN had to comply with this Presidential Order, especially as most of the RVN commanders had already been evacuated to safe harbor by the U.S. That was the only choice left as there were practically no usable reserve weapons and ammunition left as all of U.S. military aids had been cut off two years prior. The ARVN felt they were "bức tử" (pushed into a Death trap). Many military and paramilitary commanders who had refused U.S. Embassy assistance to be evacuated chose to commit suicide with their family to preserve their Dignity and Honor. After all, they envisioned that to survive with a meaningless existence under the new regime would be worse for both their family and themselves.

Among these heroes were Maj. Gen. Nguyễn Khoa Nam (career officer since the 1950s French-sponsored Vietnam National Army) and his deputy officer Brig. Gen. Lê văn Hưng (one of the first officers of Thủ Đức Military Academy 5th class of 1955, after the French left). They shot themselves dead, as did Maj. Gen. Phạm văn Phú (one of the first career

---

568   http://www.history.com/this-day-in-history/south-vietnam-surrenders

officers from Đà Lạt Military Academy Class of 1954 who fought in the 5th Airborne Vietnam National Army Battalion in support of the French in Điện Biên Phủ), Brig. Gen. Lê Nguyên Vỹ (graduate of the French-Vietnamese Regional Military School in Huế in 1951, commander of the ARVN 5th Infantry Division), and Brig. Gen. Trần văn Hai (Đà Lạt Military Academy Class of 1951, commander of the ARVN 7th Infantry Division). Col. Hồ Ngọc Cẩn, Chief of Chương Thiện District, continued to fight in spite of Sài-gòn order to surrender.

Col. Cẩn was captured and executed by firing squad, he bravely requested not to be blind-folded and said before he was shot: "If I had won this battle, I would not convict you the way you did to me, I would not dishonor you the way you did to me, I would not interrogate you the way you did to me. I fought for the people's Freedom. I had made contributions to the people and am not guilty of anything. No one can convict me, will History judge you or me as Rebels? If you want to kill me, just kill me. Do not blind-fold me. Down with Communism, Long Live the Republic of Việt Nam."

Other officers, seeing no way out (not being able to cope with any of the five Maslow's needs under Communism), had also chosen the Bushido-type of Heroism and committed suicide, along with their whole family, rather than remaining alive to be dishonored by the incoming conquerors. Such was the case of ARVN Maj. Đặng Sĩ Vinh[569]. Two hours after RVN interim Pres. Big Minh ordered all ARVN units to surrender, ARVN Maj. Vinh had the whole family, including seven children and his wife, eat their last dinner together then take an overdose of sleeping pills. He used his side arm Colt 45 to end everyone's life while they were asleep, and then shot himself, leaving behind a letter and the rest of his meager[570] salary for burial.

A National Police Lt. Col. Nguyễn văn Long had committed suicide by self-inflicted gunshot in front of the National House of Congress on Tự Do Street just a few hours before Mười and I walked by on April 30th when we were looking for our way out. Many others had done the same thing. I had a similar plan, Plan C (in case Plan A, escape by sea, or Plan B, join the resistance, failed) to use either my tiny pistol I was guarding for my boss or a grenade which I had carefully

---

[569] http://www.vietmemorial.org/myweb/the%20sucides.htm
[570] He was among the "clean" commanders.

hidden in the roof of our Gia-Định home outhouse. Many thousands other military and paramilitary had also decided to end their life instead of surrendering as they were ordered. Such was the Vietnamese Way of the Warrior, "Tinh Thần Võ Sĩ Đạo," that had elevated the honorable Confucians to the Dignity of a Fighter, and to a level of Martyrdom which they would never receive. They would not even be guaranteed their tombs would be respected as most tombs and memorials might eventually be desecrated. Such was the fate of the National Military Cemetery "Nghĩa Trang Quân-đội Biên Hoà" where tens of thousands of ARVN were entered.

While some had chosen to die a Heroic Death, many others had continued to preserve their Confucian values (serve the leader) by obeying to the last commander in chief, Gen. Minh's order and surrender to the new regime. It would also be the Vietnamese Way of the Warrior to obey the orders given by their commanders. Most high-ranking officers and civil servants who stayed behind were taken far away to re-education camps in North Central or even North Việt Nam to endure rigorous treatment to redeem **"mistakes they had committed against the People of Việt Nam."** Everything they did while serving in the RVN was considered crime against the People. Depending on their rank in the system, and the level of importance of their assignment in the Republic, they might be incarcerated from 3 years (common soldiers) to fifteen (colonel). Many never came back because of poor health or "sudden death or inexplicable accident." In my case, I would be incarcerated for at least eight years, and my wife Mười seriously dishonored.

Sweet Re-Education or Death Camps

My friend's father, a Military Police Major, was sent to a camp in North Việt Nam where he "mysteriously" fell down the mountain while he was sent there to pick twigs for stove fire. His remains were never to be found. "Rebels" were executed on the spot without due process, either openly as a warning to the other inmates, or discretely as in the case of the MP Major. Another of my friends' boot camp comrade was accused for having assembled a transistor radio to "communicate" with the outside world: he was sent to the firing squad without investigation. Not just RVN military personnel and civil servants had to be "re-educated." Protesters against the new establishment or potential rebels, even some former VC members of the NLF, as identified by the commune or People's Court as traitors or dissidents, were also sent to concentration camps, as they had dared confront the government, in contempt of the "People's Court." My brother-in-law Anh Tư was one of the victims.

Eventually, many ARVN fallen soldier's graves were dug up, bulldozed or desecrated throughout South Vietnam by the new local victors. History had shown that the worst blood-thirsty ones were not the NVA, but the local VC: they were out to take revenge, vent their fury or show off their power. The symbolic "Lamenting Soldier" monument at the National Military Cemetery of Biên Hoà was pulled down and many graves were "touched up." Many other cemeteries, such as the Mạc Đỉnh Chi historic cemetery in North of Sài-gòn, were flattened and transformed into public parks. Mạc Đỉnh Chi was renowned as many public figures since the French occupation, such as late Pres. Ngô Đình Diệm and his brother Nhu, were laid to rest there. It became Lê văn Tám Park[571].

Unfortunately, the ARVN soldiers who had sacrificed their life during the war were not let to rest in peace. A systemic retaliation campaign against their tombs was launched as one of the ways the government gave out strict warnings to potential rebels: "see what will happen to you (be executed) and to your final resting place (be desecrated) if you fight against the People!" In the meantime, everywhere in the country, new monuments were raised to honor and martyrize the VC and NVA soldiers who had died. These places were not necessarily cemeteries with graves… but just monuments to commemorate the martyrs of the People's Army.

For the living ARVN, hundreds of gigantic so-called re-education camps were opened all over the reunified country to "re-educate" the children who had betrayed the People. Reassuring promises made to the surrendering ARVN were widely propagated to lure them into registering for re-education camps: "Don't worry, Comrades, you won't stay too long in these re-education camps… After just a few weeks, you will return home to be with your loved ones." They actually reserved startling surprises. "Not too long" turned out to be

---

[571] When a teenager, I had visited this cemetery many times with my parents. Now, each time I travel back to Sài-gòn, I usually stay at Evergreen Hotel across the street, so I'm very familiar with this area.

from 4 to 20 years or forever (dead during re-education).

What the South had feared started happening slowly. Việt Nam was again subjugated, this time, not by foreign invaders, but by its own children, the liberators. Like the "đấu tố" persecutions by the People's Court during the Land Reform Era of 1940-60s, the Vietnamese started purging and vindicating their own citizens, their own relatives and family members.! Friends against friends! Children against parents! Neighbors against neighbors! Comrades against comrades!

"Mười phần, chết bảy, còn ba, chết hai, còn một, mới nên thái bình"

The old prophecy is coming back, again[572]: when, out of Ten, Seven would die, then Two more would die, leaving One surviving, then there would be Peace… Die here does not really mean pass away. It relates to how the Vietnamese will change when they turn against each other…

Neighbors were expected to report to the hamlet committee if they heard someone listen to the radio. Indeed, no one was allowed to have a radio. True news and current issues were forbidden fruit. The government strictly forbade its citizens getting updates on other countries' democracy. Re-education camp qualification was automatically issued to those who violated the laws. Listening to forbidden stations such as Voice of America or British Broadcasting Company meant the perpetrator was planning to stage a revolution against the government. An unauthorized or unreported gathering of more than 4 people would result in the same consequence: Re-education camp. All families with relatives serving in the RVN government or military had to volunteer[573] for "Thanh-niên Xung-phong" Labor Camp in the North or be subject to "punitive consequences…"

Share or Repent

… All families had to account for missing members to the local hamlet. Families with members serving in the ARVN would suffer consequences unless they sent one member to volunteer in the "Thanh-niên Xung-phong" in a faraway labor

---
572   See section Effects of Karma on Politics (Chapter 2)
573   …bring your own tools, food and supplies…

camp for three years, at their own expenses as the government had no funds for such operation. The latter also promised the volunteer would be exempted from future military service. My brother Hải was only 17 back then. He was the only one in the family who could have enough strength to endure the hardship of camp. He volunteered, so that my aging father would be spared. His group was stationed in Central where they had to dig canals for irrigation. Diligently, he worked for three years at the family expenses for his food, tools, medical supplies, clothes... After three years, he was discharged with honor "lao động vinh quang" (the glory of serving in labor) and returned home, with the joy that he would be exempted from the military draft, as the government had promised. Most importantly, he had saved our Father [574] from a major calamity.

The next twisty part was that government was then sending its main regular troops, the "bộ đội," to Kampuchea to fight the China-backed genocidal Pol Pot's Khmer Rouge. Pol Pot had already massacred 2.7 to 3 Million of his own people, what he called the "corrupt and educated." He had also been beheading Vietnamese immigrants in Kampuchea in retaliation for this race for having invaded Champa (the same genocidal Cáp Duồn that had been going on against all Vietnamese living in Cambodia). In spite of the government promised not to draft "Thanh-niên Xung-phong," Hải was drafted anyway. He was sent to Kampuchea.

His suffering did not stop there. After three years of risking his life fighting the Khmer Rouge in Krong Siem Reap (where the Wonder of the World Angkor Watt is located), when he was discharged from the People's Army and returned home, he received absolutely no assistance in rehabilitating and finding a job. The reason was that Hải had a brother who had not just served in the ARVN, but who had also defected to the "Đế Quốc Mỹ" Imperialist American. That brother was me, "the People's traitor,"! Hải was still jobless as of now, 45 years later. To earn some pocket money to support his family of 4, he had been giving rides to anyone who needed them on his Huber-like motorcycle, taxiing people around for a meager living. He refused to bribe in order to get a job in private companies[575], this becoming a regular labor tradition of the

---

574   The government did not care about women..

575   Most of the companies are in the "network" owned by the privileged comrades who are connected to the government bribery-system. Applicants has to bribe the guard so he lets you in to submit the job application, bribe the secretary so

South, especially nowadays when economy was booming...

Families living in a large house[576] had to share it with other families or move out into a smaller one. Property no longer belonged to their owners, but to the commune (probably based on the Hồ Chí Minh's 1946 Land Reform to correspond to the ideology of Commune Belonging of Communism). A great many families that had either some members (or the whole family) flee abroad for freedom, thus leaving the properties unoccupied, fell into this category. The space was commandeered by the people's hamlet or city council Hội-đồng Nhân-dân and allowed new families, mostly the comrades "cán bộ" newly arrived from the North or from the country-side, to come and share the space as if they were members of the family.

... My mother-in-law was in her 70s. As Mười and I fled abroad, she was then the only one living in her beloved spacious three-bedroom "villa" in Gia- Định" (Northern suburbs of Sài-gòn) where she had shared so many fond memories with her late husband and the eight children. It had a huge landscaped front yard with many fruit trees. It also had a second unoccupied semi-attached servant's quarters that could lodge another large family. Eventually, the government would jump right onto the occasion as if it was a pot of gold, by having two huge families of high-ranking NVA officers move in to occupy. They took over the upper main house and another family of four the servants' quarters. The liberation continued. All of them came from the North.

My mother-in-law somewhat appreciated it, though: all of them called her "Mẹ" (mom) and themselves "chúng con" (we children), and were very polite and helpful. She needed someone there all the time as she had medical conditions. I thought some of these Northerners were well raised and probably well trained to show they were not cruel conquerors, but that they had come in to really liberate the poor South.

A few years went by, Sister Chị Tám, a pharmacist who had been living in France, was able to sponsor my mother-in-law to migrate and join her France in the late 1970s, then Sister

---
she would not bury your papers on the bottom, bribe the decision makers to get the job, and bribe them again annually to keep it... Bribery originates from the Chinese tradition of gift giving to show respect and gratitude. An Nam (Đại Việt) has been paying tribute to China for thousands of years since the 26[th Century BC...]

576   A couple can have only one room

Chị Chín to Wisconsin U.S.A a few years later. Her house was torn up into pieces and two tall apartment complexes were erected in place of the villa. When I came back to visit in 2001, I could barely recognize it. Two multi-story buildings stood tall on the property lot now. It had become a prime property with a large front yard. Maybe some influential NVA officers had made a big profit out of it…

… Brother Anh Tư was not that lucky. His modern "villa" in Sài-gòn, not too far from the Independence Presidential Palace (now War Remnants Museum), was "shared" with another high-ranking NVA officer and his family. As his children had defected to the U.S. before the Fall of Sài-gòn, his house became "half-empty," thus urging the government to commandeer two bedrooms and the rest of the house in "appropriate proportion." Either out of anger or thinking that the new democratic system would be fair and would listen to his plight for justice, he went to a public hearing to voice his concerns with the People's City Council of Hồ Chí Minh City "Hội-đồng Nhân-dân Thành-phố Hồ Chí Minh." "I never served in the 'rebels' government or military' ("Chính-phủ hay Quân-đội Ngụy") and I bought my house with honestly-earned money. I am a dentist and have done nothing wrong against the Revolutionary Party. I should be allowed to live in my own house…" The council, which also served as the People's Court, slapped him with contempt of court and sent him to re-education camp for a light sentence of two years for behaving as a traitor-inspired dissident.

The camp was located where the ARVN III Corps (Camp Long Giao, Long Khánh Province) used to be located. His wife decided to abandon their house. They tried to escape a year later, but were caught in a government trap. He was sent even further for re-education in Central Vietnam. After a year, they successfully escaped by boat and came to the Minneapolis, Minnesota to join their children…

The new regime tried to avoid the purge that the South was feared of. The latter had expected that millions would be massacred as it had happened in Kampuchea, China, the Soviet Union… The new government knew that any sign of harsh punishment would make a much greater number of Southerners leave the country out of fear of retaliation. That was why the millions of Northerners had left the North in 1954 during the 10 months of Passage to Freedom. Between the carnage caused by the Bolsheviks Red Guards during the up rise instilled by Soviet Socialist Lenin/Stalin that had massacred millions of "debauched people" (including the family of Tsar Nicholas II), the one by communist China Mao Tze-tung that had rid of the so-called corrupted class, and the genocidal one committed by China-backed Pol Pot Khmer Rouge annihilating from 2.2 to 3 Million intellectual and "spoiled" Khmer people, one would not have to think long to decide on escaping Communism. How many "decadent" Vietnamese traitors would be "purged" by the Communist Party? The horrendous history of Communism's take-over had pre-warned the Vietnamese freedom lovers of a bleak fate!

    During the war, the Vietnamese communists had already proven their capability of committing comparable massacres of the innocent, such as during the 1968 Tết Offensive, the 1972 Eastern Tide Offensive, the recent 1975 Hồ Chí Minh Campaign, as much as the spreading People's Court Land Reform persecutions "đấu tố," assassinations, bombardment of residential areas with mortars, rockets and artillery, or booby trapping public transpiration and entertainment events… Nevertheless, at the end of the war in 1975, instead of instantly killing them as it was done in Kampuchea, in order to purge the remnants of the RVN as a whole, the People's Socialist Republic of Việt-nam had chosen this time a "less intimidating and more humane" solution, so it can catch the whole herd. The "Trại Học Tập Cải Tạo" re-education camp plan was implemented based on Chinese re-education camp models. All ARVN troops would follow RVN Pres. Big Minh's order, and, just like lambs, let their fate guide them to designated stations to register for "re-education camps"…

    It was estimated that over 1 to 1.2 Million former RVN officers, religious leaders, intellectuals, merchants, civil servants, and even some communists (identified by the Party as dissidents) and other plain civilians who had shown signs of objections like my brother-in-law Anh Tư,… were incarcerated[577] for up to 17 years. Most were repeatedly

---

[577] Thevietnamwar.info <u>Vietnamese Re-education Camps – The Vietnam War : Re-education camps in Vietnam</u>

interrogated, tiger-caged[578], Conex-boxed[579], abused, tortured, starved... Hundreds of thousands[580] died of diseases, suicide or "unknown accidents", or were simply executed as examples. Slowed-down death was by any means preferred by the regime as it would be much more painful and terrifying[581] than an instant one. If compared to the case of death by hanging, beheading or execution by firing squad, this way would help the Party avoid being blamed by the Human Rights Watch for having treated prisoners with intense cruelty.

... Brother Anh Tư said inmates in his camps would pay "high price" by bartering whatever their family had given them in terms of dry food in exchange for a unknown rare plant that could be found growing on the outskirt of the fields they were taken to for their daytime agricultural projects. The plant had an unpleasant taste but it would help people who could not endure any longer the pain of being "re-educated" find a quick and painless Death. After all, they had died long time ago and those dry fish supplied by their loved ones during family visits would not make them any happier... Not at all...

The Other Solution: Flight or Fight

While most stayed behind and obeyed their Supreme Commander Big Minh as good Confucians, many had chosen to leave. These were either people who were well-informed of current political issues (they stayed updated democracy-centered current issues) or just happened to be hanging out with the informed ones (the educated ones who followed the news) or the ones who had the means of transportation (e.g. fishermen). All available means of air and sea transportation

---

[578]   This cage was made of barbed wire big enough to fit one inmate in sitting or lying down position. The victim, hard-headed dissident, being confined to this position would not be able to stretch or even turn around. He would be exposed to harsh sun and the elements.

[579]   Approx. 6'Wx8'Lx8'H steel or wooden containers used for storing U.S. military suppliers and ammunition were used as in-the-dark solitary confinement cell for hard-headed inmates. As these Conex boxes are exposure to the sun in Vietnam, it can heat up to a temperature of 150 degrees F. Food and water were fed through a small opening.

[580]   https://en.wikipedia.org/wiki/Reeducation_camp   Wikipedia estimated over 160,000 inmates perished during their camp incarceration. Also, https://ongvove.wordpress.com/2010/05/09/ 2010/05/09/trai-cải-tạo-của-csvn-sau-nam-1975 estimated 800,000 RVN personnel were incarcerated in camps

[581]   ...starting with ankle shackles being increasingly clamped until the inmates lose his feet...

had been taken away during the Frequent Winds evacuation[582]. A large majority of RVN commanders and political leaders had already been evacuated by the Americans helicopters a few days before the Fall of Sài-gòn. Others who could not access designated evacuation pick up points chose to venture their way and escape on their own across the ocean, using any operable means of transportation they could find. I chose this solution instead of reporting to re-education registration sites, joining the underground resistance or committing suicide.

Many others, who strongly believed that, someday, help would come, had joined the resistance to reconstitute a clandestine army and fight back. This was Mười's and my Plan B if we could not escape successfully by sea: join the underground forces and fight back. Many who came to the U.S. had also started underground resistance, such as Black April "Tháng Tư Đen," or "Kháng Chiến Quân" Resistance, and later on the more concrete verifiable Hoàng Cơ Minh movement[583] near the Viet-Thai border.

In Minnesota, the Vietnamese self-helpers headed by late Mr. Phạm văn Vy had organized "Liên Minh Việt Nam Phục Quốc" (League for the Restoration of Vietnam) which Mười and I had joined. Even though most movements had failed because of complication in funding or abuse of financing (e.g. Mafia-type of extortion of businesses to fund the cause of the organization[584], and, as alleged, terrorism and assassination[585]), the efforts themselves showed that the spirit of the RVN and its Way of the Warrior were still very strong. The hope and wishes of the Vietnamese immigrants were founded on the belief that the U.S. would never give up on its allies. The latter had returned and liberated France, the

---

582   Frequent Winds is the operation facilitated by the U.S. Embassy in Sài-gòn to evacuate the RVN military and civilian personnel whose life might be in danger if they stay behind. At the same time, individuals who are able to find means of transportation, such as commandeer helicopters, boats…, will also be rescued by the U.S. Navy 7th Fleet awaiting in the South China Sea, if there are able to escape successfully. https://worldhistoryproject.org/1975/4/29/operation-frequent-wind

583   Hoàng Cơ Minh movement was later on identified as the abusive and corrupted Việt Tân (the New Vietnamese) that was inhumane towards its membership; however, other groups, mainly some Vietnam Veteran, looked up at this movement as courageous.

584   PBS TPT Report on Terror In Little Saigon has made the Vietnamese Community in California upset as it has insinuated that extortion and terroristic activities are commonly practiced by many revolutionary ARVN liberation groups that are searching for a way to undermine or even overthrow the current Communist Regime in Vietnam. https://www.tpt.org/frontline/video/frontline-terror-little-saigon/

585   TPT Terror in Little Saigon

Philippines, China and the Republic of Korea from invaders at the end of WW2. This too would be a temporary retreat, just like Dunkirk or Bataan[586]. "Westmoreland shall return," as the Vietnamese in exile all hoped. They trusted the United States did not just abandon an ally like that! This super human power had even provided protection to its former foe Japan. Why not the RVN? If one keeps grinding metal, some day it would become a needle. "Có công mài sắt, có ngày nên kim."

> *"...The Day of True Liberation will come.*
> *It's just a matter of patiently wait.*
> *At the end, everybody will be liberated,*
> *Such is our Destiny..."*

Along the same line of thought, the underground spirit also exists in the SE Asian communities, such as the Lao having an underground group searching for a way to take Laos back, or the Hmong groups (Gen. Vang Pao's alleged connection with underground in Thailand, the current patriotic sentiments of the organizations such as Hmong Homeland, Neo Hom,...) seemingly seeking for ways to recapture their Lost Homeland[587], or the Republic of Khmer group against Hun Seng's regime or the Vietnamese Hoàng Cơ Minh's group actually launching underground efforts in Thailand and Việt-nam in the 1980s… Sometimes, these organizations might resort to violent inclinations or even terrorist means, mainly to help create a sense of purpose and accomplishment which are extremely important for self-esteem and belonging (Maslow).

Time to Wake Up

Việt Nam was reunified. There was no more North, Central or South, but just one Việt-nam[588]. Families that had been separated now could find each other. Equality prevailed. Resources from the South were then shared with the North.

---

[586] https://bataandm.weebly.com/general-douglas-macarthur.html Gen McArthur promised "I shall return" when he left Corregidor in March 1942. Two years later, he returned in Oct 1944 and liberated the Philippines from Japan.

[587] http://content.time.com/time/specials/packages/article/0,28804,2101745_2102136_2102247,00.html Gen. Vang Pao litigation in plan to overthrow the Democratic Republic of Laos in 2007 (Operation Tarnished Eagle)

[588] Not so true for those outside of Vietnam!

Northerners now could catch up: motorcycles, crops, television sets, music players, gas stoves… NVA Bộ Đội troops taking a hike downtown Sài-gòn, now renamed after "Uncle Hồ's", looking up high skyscrapers ("so high they could have twisted their neck" as the Sài-gònese laughed at them), browsing shop windows, eating fancy foods, "buying" fashionable clothes,… "Very truly, Comrades, we (NVA) don't know why you needed us to liberate? You have much more than we do, and your life is so much happier than we were told in the North!" said the NVA soldiers.

Time went by and the Southerners would not have Free Enterprise until 1986. Before that, everything belonged to the commune (from this word, Communism was born): Rice communes would share rice with industrial communes, for example. From what one made, one could keep 10% or much less, as one had to share ones crop with the comrades. My sister could join the seamstresses' commune, but not my black-listed brother: she was a woman and women were not worth being black-listed or paid attention to. My brother hurried up and got married: his wife would not be black-listed[589] and the family would finally have an income to raise the kids, with the hope they would do better.

Time went by, and the people were still poor. They had to bribe at all levels in order to get a job, and continued to bribe to maintain their job. My brother refused to bribe, so, still, he would not have a job. With the hope that the next generation will not be black-listed, he tutored his children so they would … qualify to be tutored by their teachers. Sacrificing for children meant he would have to bribe the teachers so they qualify for tutoring in school. Without it, his children would fail, no matter how smart they were. Teachers needed money too… to bribe and maintain their job… So, with time going by, the liberated started to feel they needed to be liberated. Democracy was not real Democracy. Liberty did not mean Liberty. Everything meant bribery. Bribery at all levels, if one wanted to do ones business! The rich upper class started emerging from the poor, and it was all related to the ruling class.

---

589   As a result, he could not participate in the U.S. Family Reunification program later on.

1973 – POW Exchange

"We only want to live in Free South Việt Nam, ""we'd rather die than returning to the Communist Party," "returning to Communism is to commit suicide," "if you return us to Communism, we will all commit mass suicide," "we are determined to stay in the South…" were the banners deployed by some 200 NVA and VC prisoners of war the RVN was releasing and repatriating to the Socialist Republic of Việt-nam (North Vietnam) in 1973 following the Paris Accords[590]. To the members of the Peace Keeping and Monitoring team's astonishment, they observed the 200 line up in squatting position **with banners and hollering that they refused to be freed.** Some representatives met with both RVN and NVA officials under international guards to reaffirm their free wills. Of course, the NVA ones were upset, while the RVN content. A point was clearly made: Freedom in the South meant much more than in the North. It was about time someone had opened their mind on the truth of Democracy and Freedom, two years before the end of the war… A few years later, among the successfully escaping boat people were former NVA and VC POWs. They had found all means to escape for their life because their comrades upon returned to the North after they were released by the RVN did not really find freedom: they were all executed! They were liberated, indeed.

1975 - An NVA Colonel's Confession

Just a few months after Sài-gòn was "liberated" (my father was still alive), Col. "Thượng tá" Nguyễn Tr. appeared at his door with another man. They could be in the 40s. They both looked inquisitive, and seemed to be surprised to see how dilapidated my father's house was. The colonel, who was higher ranking then his younger brother, an NVA Captain Đại Uý[591], seemed to be the spokesperson. He introduced himself

---

[590] 1973 documentary film of the NVA and VC POW release and repatriation by the RVN with Paris Accords Peace Keeping observers #MF20 – 5741 (PF#30040HD. U.S. Maj. Gen. Lloyd B. Ramsey, Provost Marshal General: "These events were filmed as they took place in South Việt Nam. The Release and repatriation of enemy prisoners of war were photographed as a matter of record with permission of the International Commission of Control and Supervision and the Four Party Joint Military Commission and assistance was provided throughout by the Army of Republic of Việt Nam.

[591] above Thượng Uý Senior Lieutenant, Trung úy Lieutenant, and Thiếu úy Junior Lieutenant

and explained they were my father's sons from the North. My parents were surprised but father realized right away the colonel and the captain were indeed his children he had left behind with his first wife when migrating to Sài-gòn in the 1940s. Even though everyone in my family was more or less intimidated by the high-ranking uniform and wondering whether they might be in trouble, they were actually very curious about these "children" bearing a different last name (this really infuriated Father[592]), and, especially, about father's "other" family that he never talked about during the whole time we were together.

Col. Nguyễn went through a long story to clarify things that had happened since my father came to the South. Besides the eldest son who did not come to visit, a fourth child, a girl, for some reason passed away probably because of the serious famine that struck the North in 1945-1946[593]. Father had already left for the South: he was a member of the Việt Minh in the North, an anti-French and anti-Japanese patriotic organization. The government of Hà-nội assumed he had defected to the South to serve the new State of Việt-nam Prime Minister Ngô Đình Diệm, and, in order to punish his first family, had deprived the eldest son, who was then trying to join and serve Việt Minh, of any citizen's privilege or advancement in Hà-nội government. The family decided to change their last name to Nguyễn so the remaining children could have a way to succeed in the society and serve the Party. They seemed to be patriotic and all wanted to serve Việt-nam. The colonel implored:

"We have served our Ancestral Land for so many years. We have defied the Trường Sơn Mountain and confronted the hardship of Hồ Chí Minh Trail to come and liberate the South. During the past 15 years, we have endured the pain from battle to battle against the "imperialist Americans and their nationalist servants", and brought about victories after victories. Now that we have liberated you, we can return to normal life, but we have nothing. The government has only a thank you send off

---

592   In Vietnamese families, the father's last name is utmost important and is to be highly respected and revered.

593   http://www.amchamvietnam.com/3121/1945-famine-in-vietnam, https://en.wikipedia.org/wiki/Vietnamese_Famine_of_1945 and http://forum.axishistory.com/viewtopic.php?f=6&t=86968 During the Japanese occupation, because of the war, crops not being able to be transported from the rich South to the impoverished North and available rice being reserved for both the French and Japanese armies had created one of the most serious famine in the North, killing over 1 million Vietnamese, most of them were children and elders.

for us, and we are left empty-handed. Yes, indeed, nothing to live with. We have no house, no belonging, not even a sewing machine for our family to make a living with."

My mother, who by default would be the two men's step mother, felt sorry for them. She bought each a sewing machine as gift to help them support their family. That was all our family could afford as they themselves were impoverished by the war with only my sister and our sister-in-law, my younger brother Hải's wife, working to support the whole family of four adults and two children. Hải believed the two officers were disappointed, thinking that South Vietnamese, with all American aids, should be richer than that. That was the first and last time my family saw the two step-brothers. No further follow-up connection was attempted by either family.

1981 – Father kept uttering in his coma

"Run, run, Son, otherwise they would catch you!" For many years, father had been pedaling his cyclo[594] for a living since the end of the war to help my sister, the main family bread-winner, support four adults. His frail skeletonish body had to bend and push on the pedals taking people around for a meager fee. It was much harder during the monsoon season that usually lasted from April through October. He had had a very harsh life since the mid-1960s when my god father returned to France.

Towards the late 1970s, he developed a small nodule in his throat that no one knew whether it was benign or not. My sister took him to the hospital for surgery, and because they could not afford to bribe for antibiotics or bandages, his throat could not heal: it kept swelling and became seriously infected. The family had to tear up old clothes into strips and boiled them in water to be used as bandages. Had they bribed someone, perhaps he could have survived the simple operation. His throat kept swelling until he could no longer swallow food. The bandage made of boiled pieces of clothes could not help reduce the swelling. How could you heal an infection without antibiotics and proper care?

---

594   A type of rickshaw used in Vietnam and Cambodia to take customers around like a taxi. The operator sits behind the customers and pedal it to power forward.

He could swallow only rice juice. Milk was too expensive and my family could not afford anything else. He fell into a coma after nearly a month of starvation. My sister told me that, during his last delirious moments in bed, he kept moaning: "Run, Run, Son! They will get you! Run, run!" He died tragically, but peacefully, the same way he had been living, a peaceful man who always did his best to sacrifice for his children. During the last five years of the war, he and my mother worked as servants to find money to fund my education, this allowing me to achieve my higher education. I came to Minnesota with the Bachelor's Degree their financial sacrifices had provided.

Even though he had served in Việt Minh and might have been a more or less silent passive communist sympathizer (he listened to NVA radio) during the war, he must have found the truth after the Fall of Sài-gòn, and was very happy I had fled to the U.S. Father did not flee the North in the 1954 Passage to Freedom (Geneva Conference allowances), but he did defect to the South nearly a decade before. Now it was his eldest son's turn abandoning Ancestral Land in 1975… Just as the song "In 1954, Father left his birthplace, in 1975, Children abandoned their homeland…"

…Một ngày bảy lăm, con bỏ nước ra đi
Hai mươi năm là hai lần ta biệt xứ
Giờ cha lưu đày ở ngay trên đất ta
Và giờ con lưu đày ở đây trên xứ lạ!
Một ngày năm bốn, cha lùi quê hương
Lánh Bắc vô Nam, cha muốn xa bạo cường
Một ngày bảy lăm, đứng ở cuối đường
Loài quỷ dữ xua con ra đại dương!...

> One day in 75, children abandoned their homeland
> In twenty years, we had abandoned our land twice
> Then you were wondering on your land
> And now I wonder in foreign land!
> One day in 54, you forfeited from your childhood place
> Avoiding the North you ran to the South, as you wanted to stay away from the fierce and cruel
> One day in 75, standing at the end of the road
> Ferocious monsters were chasing me into the ocean!...

Translated by Dr. H. Tuong

Again, Khánh Ly's voice resonated Phạm Duy's second part of the moving song "in 1954, Father abandoned his birth place…in 1975, children left behind their homeland" reflecting the second major exodus of the South Vietnamese "boat people" during their journey looking for Freedom.

1990 September - NVA Gen. Bùi Tín defecting from Việt-nam[595]

Gen. Bùi Tín, the NVA Colonel (1975) who accepted the unconditional surrender of the RVN from its by-default interim President Big Minh, decided to defect to Freedom and took his exile in France after his participation in French communist paper L'Humanité annual conference in Paris. The general went from mildly criticizing the People's Democratic Republic of Việt-nam leadership that, in his opinion, was deviating away from its original ideology, to viably attacking the socialist and communist stances and demanding an immediate **real Freedom and Democracy** for Việt Nam. As a truly socio-political analyst and news media writer, he had seen the real truth and hoped his blog website would serve as an awakening to the truth, with the guidance from the viewpoint of a former NVA political cadre, addressing others who were still lost in the communist delusion.

In his Aug 4, 2010 blog in Vietnamese, Gen. Bùi Tín shared his "dream" about a true Democracy that embraced multiple parties, as he believed competition would urge the latter to make efforts towards strengthening and "cleaning up" their organization, thus truly serving the constituents, bringing about success and benefits to the people, combating wasteful expenses and corruption, and using the free press as a counterbalancing reporting system. If the Laws were seriously upheld and applied fairly, and the officials were adequately compensated, there would be no temptation and consequently no corruption. Nevertheless, he strongly believed that Việt Nam was not functioning in that mode, that the people's secret

---

[595] http://www.c-span.org/video/?c4511560/senator-john-kerry-defector-bui-tin , https://en.wikipedia.org/wiki/Bùi_Tín About Bùi Tín in English, http://www.nytimes.com/1990/12/29/world/ex-follower-of-ho-chi-minh-scolds-vietnam-in-broadcasts.html NY Times Dec. 1990 report

Defect NVA Gen. Bùi Tín meeting with U.S. Senator John McCain
Sauce: www.militarytimes.com

service was harshly oppressing the people while preaching the opposite, that they were themselves "pushing drugs" while campaigning against it and executing the competitors, insatiably laundering funds whenever they could for their own gain. Việt Nam should learn and move closer to and befriend the true 46 democracies, such as the U.S., Canada, the United European Countries, or Australia, Japan... instead of belonging to the worst 33 militaristic dictatorships under democratic disguise these including China, Cuba, North Korea, Myanmar, Congo, Somalia...

Speaking about Việt-nam, the general said: "In our Việt-nam, because of the lack of Democracy, there is no serious and fair law. No Democracy means citizens' rights are trampled on. No Democracy means there is no Freedom of Speech or Freedom of the press, these being the soul of a democracy. No Democracy results in the government not being determined in its drastic anti-corruption measures as it has promised... Social injustice has reached its utmost limits, with people losing their land, home, and jobs expanding everywhere. Some having too much while others have none. Civil servants and workers have meager wages, when officials have power to gather wealth, and launder money from public funds... Rulers' ethics have dried up to the bottom when top provincial officials become "pimps" who are raping young and naïve school girls, transforming them into "sexual slaves." If this had happened in a civilized democracy, such cases would become scandalous crimes of the society – however, in our country, the Party just gave a reminder so it can collect bribes, and the worst thing that could happen is a simple transfer into a different position! No! It cannot be like that. This country does not belong to some individuals. We cannot condone such a level of injustice, rotten ethics, social existence that has degraded to the bottom we have never seen before...

... This single-party totalitarian regime has exposed its whole incompetence, decadence, and destruction. We do not need to argue or rebut...

(Gen. Bùi Tín's personal blog – VOA Viet, Aug 4, 2010 Translation from Vietnamese to English by Dr. Ha Tuong)

Of course, the Democratic Republic of Việt Nam did not like such an unmasking statement. It had the Việt Nam Embassy in Paris disavowal of the general's behavior, and other Vietnamese commentators from both left and right sides, still heavily divided within self because of the ill-treated Democracy, had confusing opinions about the defector[596]. Nevertheless, Gen. Bùi Tín was not the only high-profile politician defecting.

2013 - Another defecting Official, Mr. Đặng Xương Hùng, Consul to Genève, Switzerland (2008-2012)

Mr. Hùng had decided to leave the Communist Party and defect to Freedom at the end of his term as Consul of Việt Nam to Switzerland in Genève. As a political refugee, he wrote a letter to the Vietnamese People, especially those refugees and migrants living out of Việt Nam but who are still relating to and cooperating with the Việt Nam Government. His address might be aiming more at the Vietnamese immigrants in Switzerland[597]

---

[596] https://vi.wikipedia.org/wiki/Bùi_Tín Wikipedia about in Bùi_Tín Vietnamese

[597] We have to note that a great many, if not majority, of Vietnamese expatriates living in Switzerland are sympathizers and pro-Communists. This is the country

and other East-European countries who, ironically, had been supporting the current regime in Việt Nam but would prefer to live in the Freedom of the host countries instead of returning to Việt Nam. In his own words, he said;

> *"We had run away from Communism in search for Freedom and Civilized Knowledge and had been welcome by our new countries just for those reasons. Why are we still maintaining relations, collaborating and supporting the communist regime back there? Was it because we think it has changed and no longer is what it used to be? I myself had that thought in mind, hoping Việt Nam would someday change from its original form[598]. Today, after the Party 12th Convention, we can assert it has not and will never change. The Party thinks it will just build up a one party ruling firmer and stronger, but not to bring the country up to powerful nation. Such is my sole conclusion that I would like to share with you. Don't even think your contributions to the government will (truly) be contributing to your Country, as the Party always claims in their propaganda. Your efforts are just to uphold their Party. Such efforts do nothing but harm and cause difficulties to the People's Movement (Việt-nam), as well as indirectly divide the communities abroad. Your investment in your own property and housing in Việt Nam is related to the People there losing daily theirs to urban development (commandeered). The propaganda "People's land" truly means the "Party's Land."*

Background information: While the yellow Republic of Việt-nam flags proudly fly and soar in the skies of most countries where nationalist immigrants and refugees had resettled[599], the People's Democratic Republic red flag with yellow star is the Vietnamese community pride in Switzerland. The latter represents the communist inclination of that Vietnamese colony there, and, to the nationalists abroad, symbolizes that Swiss Vietnamese community is a satellite planet of the DRV.

---

where one can find most Facebook friends displaying the Red Flag with yellow star in the center. They are not just passive, but active political activists.

598 The 1970s repressive communist regime
599 U.S.A, Australia, New Zealand, France, Canada,...

1992-1995 – Buddhist Monks protesting against Government oppression

To my greatest astonishment, major protests by the Buddhist that took place against the People Democratic Republic of Việt-nam in 1992-1995 did not appear on the Website, including YouTube, especially on Yahoo.com, unless researchers specifically looked up uprising by year: 1993. Interestingly, the Unified Buddhist Temples (Giáo-hội Phật-giáo Thống-nhất Việt Nam) the same one (but perhaps under new leadership) that had been constantly protesting against Pres. Ngô Đình Diệm, then initiating the successful Fall of the 1st RVN in 1963 and also against Pres. Nguyễn văn Thiệu, attempting the Fall of the 2nd RVN in 1970s… was now protesting the communists who they used to directly or indirectly work with, support and/or harbor in their temples.

For some reason, the most significant uprising was in Huế in 1993; however, it was well hidden from the internet. As many as 40,000 Buddhist monks and followers had participated in the demonstration and hundreds had been arrested and incarcerated away from large conglomerations, probably so that the people would not know about. A very few videos were able to be posted on YouTube (no web blog though) clandestinely, some ended up being viral on Vietnamese nationalists social media FaceBook posts. These were not just small insignificant events. They involved self-immolation similar to the one by Rev. Monk Thích Quảng Đức in 1963, car incinerating and a great many Buddhist venerable monks and followers being beaten up, taken away and incarcerated in remote Central and Northern re-education camps[600]. It all happened in Huế at the end of May, 1993. One Buddhist female follower set herself on fire near Linh Mụ Pagoda, Huế. The local government claimed she had terminal disease and had drunk until nearly unconscious, then committed suicide by self-immolation. Her remains were quickly removed by the authorities that refused to let her relatives retrieve: a total oblivion as per magic. The Committee for Protection of Human Rights in Việt Nam ("Uỷ-ban Bảo-vệ Quyền Làm Người Việt Nam") had published on Facebook a list of venerable senior monks Thượng-toạ, Đại-

---

[600] Facebook YouTube in Vietnamese: Phật Giáo Huế Biểu Tình 1993 Huế Buddhists' Protest in 1993 Parts 1, 2, and 3. https://www.hrw.org/reports/1995/Việt Nam.htm Vietnam: The Suppression of Unified Buddhist Church as Human Rights Watch had reported in 1995, http://articles.courant.com/1993-04-07/news/0000103649_1_vietnamese-buddhists-binh-gia-pham-vietnamese-Communist-government-s-persecution Buddhist protester burned self to Death

đức, such as Thích Huyền Quang and Thích Quảng Độ, and followers who were sentenced and incarcerated from 4 years to 20 years or even Life sentences (until they die, not like in the U.S.) in re-education camps Ba Sao (North), gulags Z30A (South) and A20 (Central) for having fought for Freedom of Religion, Human Rights and Democracy. Most received 18 to 20 years, and 3 Life sentences. Articles published by the People's Democratic Republic denied there were any incarcerations (Nhân Dân Newspaper The People), or real self-cremation (Tuổi Trẻ Newspaper Youth). Instead, they claimed the Buddhists had taken advantage of Freedom of Religion to violate laws and disturbed order and security, thus going against Buddhist teaching (as presented by the 3 parts of YouTube reports by the Committee on Human Rights).

2009 –Taxi Driver in Hồ Chí Minh City

Tuấn had been driving me around during my visit in Sài-gòn in 2009. The North Vietnamese language he spoke told me obviously he was a new comer, after 1975. It was distinctively more assertive with an intonation harsher than the one spoken by North Vietnamese who had been living in the South for a long time. That was how I knew he was from the North. Not just anyone from the North, but definitely a NVA commissioned officer by the commanding but respectful sound of his voice.

By the way I looked, broader and more over-weight than any Vietnamese local there, he knew I was a Việt-kiều[601]. By the clothes I was wearing, he knew I was from the U.S. We might have been in opposing units at the front line during the war. He might have committed war atrocities (such as directly or indirectly killing civilians as many VC or NVA had done). He was now a taxi driver and I an American tourist. It was so interesting how our destiny had placed us in this strange position.

Tuấn did not hide his past, and I did not mine either. We connected right away as he committed full compliance to my needs of transportation. He told me he was a former NVA Thượng Uý Senior Lieutenant who had fought in many different provinces in the South. Tuấn could not resist "dishing"

---
601   Vietnamese expatriate

the Communist Party that had betrayed its membership after their mission was accomplished. He was truly pissed off, and I could be more open, trusting he was not just saying things to dupe or entrap me (of course, he could have done that to turn me in for some rewards). Anyway, we did not open up until the third time I called him up for a ride.

No matter how much one had accomplished or how many victories one had achieved, one would not have any benefit if one did not belong to the ruling "corrupt cadre" (in his own words). After he had realized he was used up like a wedge of lime, being "squeezed up to the last drop of juice and thrown away" (Vietnamese analogy "vắt chanh bỏ vỏ," meaning being take advantage of and discarded after one could be of no more use), Tuấn became a rebel and turned against his own party. He declined his promotion, trashed his "worthless People's hero medals and recognition certificates," resigned from the People's army and moved South to restart his life. He was now perfectly happy driving his rental taxi and enjoying his freedom from the bribery system.

1976-2016 – Catholic Priest Thadeus Nguyễn văn Lý[602]...

...had been considered by the People's Democratic Republic of Việt-nam as dissident since the Fall of Sài-gòn. A recipient of a Czech Republic award, the Homo Humini Award (People in Need), for fighting for Human Rights which he shared with Buddhist Venerable Monks Thích Huyền Quang and Thích Quảng Đội in 2002, he had been persecuted by the Việt-nam government and served many terms in prison as well as house arrest during several decades. In spite of the government politburo's harassment and repression, he continued expanding his movement for Freedom of Speech and Human Rights that resulted in his arrest at the Huế Archdiocese in February 2007. Photos and video of his 2007 court session, which the court had allowed the press to attend to show that the "democratic" government was embracing democracy, had gone viral showing Father Lý trying to advocate for himself by yelling out "Down with Communism" while the secret service

---

[602] https://en.wikipedia.org/wiki/Thadeus_Nguyễn_Văn_LýBD Wikipedia in English, http://www.nbcnews.com/news/world/vietnam-frees-dissident-priest-father-nguyen-van-ly931 NBC News: Father Lý was released just before President Obama's visit to Vietnam in May 2016.

and uniformed court guard were muffling up his mouth and taking him away. There was neither due process nor attorney representation, of course. He was sentenced to 8 years in prison for committing "very serious crimes that harmed national security."

Catholic Priest Thadeus Nguyễn văn Lý being muffled at Court trying to defend himself during his trial due process
Source: luongtamconggiao.wordpress.com

Coincidentally, he was released just before Pres. Obama visited Việt Nam in May 2016.

Father Lý's and his team's efforts in teaching via their underground "Block 8406", <u>Manifesto on Freedom and Democracy for Việt Nam</u> 2006 as well as their online blogging had been conveying to the People of Việt-nam the real Truth about Democracy which had been hidden from them. While the Catholics were fighting their own war for Freedom and Democracy, the "non-governmental" Buddhist organizations[603] were also struggling and demanding for their People's Rights

---

603   The government-certified Buddhist sects have propagated more and more across the country to support the DRV government and to mask the Free World from seeing the non-certified ones (the genuine Buddhist organizations) being repressed. Monks in the former organizations could be seen in leaked out social media with behavior unbecoming for Buddhist teachers (gambling, sexual conduct, parading with girls, et. al.). The government-certified sects have the most monumental and newer temples that are major tourist attractions. Some are said to be built like amusement parks

to be respected. More than ever, the common people all over the country, from big cities to small townships, dared come out on the streets to demand the DRV to stop brutalities against the people, commandeering the people's houses and properties, muffling up their voice. In spite of the new cyber security laws passed on January 1, 2019, viral photos and videos could still be seen by the Free World on YouTube, FaceBook, and emails, and no matter how hard the commissaries tried to disguise or hide, their face and actions could still be viewed in testimony of civil and human rights oppression. Nevertheless, people's outcries for justice oftentimes were unheard and unwitnessed because they were mostly in Vietnamese and contained locally.

May 2016 – President Barrack Obama

Pres. Barrack Obama, "a Beacon for their Hope," when visiting the People's Democratic Republic of Việt-nam, had caused a major frenzy time among the population in the two cities of Hà-nội and Sài-gòn (Hồ Chí Minh City). Weeks before his arrival, a multitude of videos and news clips had already gone viral: his two Cadillac One limousines, his Air Force One liner, his staff, his two Navy Seal squads, 24/7 update news, his female driver, her training, the hotel, his rooms, his secret service, the K-9 team of 3 led by one that bore the rank of Captain,… Then, tons more of photos and blogs invaded YouTube, FaceBook and the internet blogs and emails. Anything about Obama, whether in the U.S. or abroad, had become a major interest for the Vietnamese everywhere, thanks to the new cyber technology! I would say nobody else, including Pope Francis, had stirred up so much interest and welcoming feelings. Lines of people lined up on both sides of the streets to greet the Obama's, with banners, Obama photos, U.S. flags, and tons of smart phones taking photos and videos… It was a total frenzy of joy and hope! They were hoping the U.S. would return and make changes to the Democracy they currently had!

Việt-nam welcoming Pres. Obama
Source: English.vov.vn

This phenomenon had happened neither to Pres. Bill Clinton[604] in 2000, nor to Pres. George W. Bush in 2006. People took advantage of this visit to display their anger and frustration at the Vietnamese regime oppressing human right movements, and, especially the Taiwan-owned steel company Formosa in Hà Tĩnh Province, Central Việt Nam, that had deliberately dumped significant toxic waste into the East Sea (SE Asian name for South China Sea) killing millions of fish, shrimps, lobsters, clams, oysters... and consequently harming and killing people and Việt Nam economy. The Vietnamese were really looking up to the U.S. to intervene against the China invasion of Trường-sa[605] (Spratly archipelago, at the same latitude as Sài-gòn) and Hoàng-sa (Paracel archipelago, at the same latitude as Đà Nẵng), these two islands being Đại Nam's establishments since the XVIIth Century under the Nguyễn's Dynasty. They were also claimed by some other SE Asian countries. The XX-XXIth Century dispute was probably caused by the area having petroleum reservoir potentials.

---

604   President Clinton was the American President who had opposed to Vietnam War and avoided military service three times. In 2000, he (and First Lady Hillary and daughter Chelsea) was the first U.S. president visiting the Socialist Republic of Vietnam, 25 years after the U.S. troops and allies withdrew from the war.

605   1974 Battle of Paracel Islands between the RVN and the PRC People's Republic of China Navies. https://en.wikipedia.org/wiki/Battle_of_the_Paracel_Islands , Wikipedia on Paracel https://en.wikipedia.org/wiki/Paracel_Islands and on Spratly https://en.wikipedia.org/wiki/Spratly_Islands

Vietnamese human rights groups that were now realizing that the Vietnamese people used to have rights, but that these rights had been taken away when the communists took over the South and unified Việt-nam, were hoping that Obama visits would serve as a beacon to a true liberation. Many activists held in confinement, such as Catholic priest Lý (earlier paragraph), were released; however, human right groups were barred from meeting with Pres. Obama. Nevertheless, the latter was able to sow his seeds of democracy and economic advancement just by being present in Việt-nam for a short time. Whether we want to admit it or not, the overwhelming welcoming feelings for Pres. Obama had proven one important fact: the people had recognized that the U.S., the true Samaritan, had a real Democracy message to teach them, this had been taken for granted during the war. **Freedom and Democracy: One never appreciated what we had until we lost it!** They had probably hoped it would still not be too late, as they were taking major risks of losing even more of their meager Freedom if they were arrested and incarcerated.

What really matters?

At this point, we can see clearer the liberated people in the South were actually not liberated. The liberators did not really liberate anyone, but, on the contrary, their efforts were not recognized and appreciated by the "liberated ones". They have been definitely confused when transitioning from colonization to independence in the 1950-60s. Furthermore, the wannabe-leaders have confused them even more in the 1970s. We have to admit it is tremendously overwhelming to find out suddenly we have all the freedom. It feels as if the once-upon-a-time colonized people have been codependent on the guidance of the French. Whether it is repressive and restrictive or not, it is still helpful and supportive for the common pre-literate people, as **it sets a framework for the system, similar to the restrictive system of the Annamite Empire.** It is like dysfunctional children, or even adults, needing a structure that helps stabilize their life. That is how the colonial structure, no matter how undesirable and horrific it can be, has actually provided a sense of structure and purpose to the Annamites when it leads their way to modernization through a century.

My experience working in public high schools has

shown me that students transferring from a structured (e.g. private or parochial school) into a public inner-city school usually are "lost" for a period of time. They are lost in Freedom! In many cases, wild animals raised in captivity, when released, do not really know what to do or where to go right away. It is not easy to be freed.

That is why the nationalists have turned against each other as they were wandering in the loose, while the communists stay unified as they are bound within a strict communist system (sanction, purge, elimination...). The former have been released from a totalitarian regime into a confusing Democracy they have never experienced before. In the meantime, the latter are united because they are well under control, switching from one totalitarian system, the colonizers, to another similar one that is even harsher, the Communist Party. During the war, peasants being treated inhumanely or even massacred by the totalitarians usually dare not protest and choose to be subjugated, but they resist and fight back when mistreated by the more lenient nationalist government.

Anyway, that confusion can be remedied only when one has experienced both sides: the totalitarian governments versus the democratic ones. Not until then, can they see clearly and think rationally. The road cannot be determined by someone (self-determination) unless he/she has reached that self-actualization stage (Maslow). This is how Sen. John F. Kennedy has foreseen and determined that the U.S. cannot intervene in the sovereignty of countries that have just been liberated. The newly freed people cannot think analytically and decide which road to take, and which ones to avoid until they have outgrown into their own united[606] self-actualization and self-determination (Maslow top level of <u>Hierarchy of Needs</u>).

---

606   It will still be a dilemma even if the people are in self-determination stage but cannot think normally as united. Being divided will lead the country to chaotic lawlessness.

CHAPTER 17

The Awkward Truth about Winning and Losing!
(The key words are still capitalized)

Reality is always Reality. How it is perceived is crucial to our existence. Real Freedom as perpetuated by a True Democracy is always REAL, as we live it and feel it in the air we breathe; nevertheless, nobody has the same interpretation of Freedom or Democracy. How one defines it and how one achieves it have a myriad of possibilities. In a fight for Freedom, whether it is from the perspectives of the invaders, oppressors or of the defenders, the outcome has to be Democracy, which in this case, both the communist and the nationalist are fighting for their version of Democracy. The principal question is: Which Democracy would the People of Vietnam prefer? The answer to this question will definitely determine who wins and who loses. **The true winner has obviously to be the one who acquires that Real Freedom in a Real Democracy.**

Unfortunately, once Freedom and Democracy was supposedly acquired in April 1975, the People of South Việt-nam gradually realized they could not agree with each other any longer, especially with what the ruling class had claimed in the name of the unified People's Democratic Republic of Vietnam or "Việt-nam Dân Chủ Cộng-hoà", and the sacred word "Dân Chủ" (Democracy). Years went by, and what they were experiencing was not what they had been expecting. It was unbelievable to conceive that a small political illiterate group was able to impose its interpretation of Democracy, as in "Democratic Republic" with "Liberty and Equality," on the others who were the majority group, the People of South Vietnam. The saddest thing was that the "little people," as they started calling themselves "dân oan or the poor and innocent," were now no longer free.

The Democracy they had been promised by the Communist Party, as little as it was under the Republic during the war, no longer existed. No matter how minimal Democracy and Freedom had been before the April 1975 Liberation, it was still Democracy, the **Real** one, as the Freedom under it still allowed the "little people" to own property, to speak up and protest in the name of Democracy and Human Rights. It was not really Freedom defined in a civilized free world yet, but it was still the **Real** Freedom.

Now, for that reason, we have to re-examine the definition of "win", "victory" or "victors." What makes meaning to the people when one say…

Win or Lose?

Following this line of thoughts and combining it with the attitude of the diverse group of "little people" towards the new Việt-nam, one will have to accept that the war for Freedom and Democracy in Việt-nam has not been won by the "victors," but, if we open our mind to our inner-soul level of self-realization (think rationally and mindfully, with purpose), we will see clearly that the "losers" might have actually won it, because those who have been militarily liberated have actually lost their Freedom and Democracy. It is just a matter of common sense, as the Vietnamese have been longing for them for a few thousand years.

Let me put it in a different way. The victorious Communist Party, even though it has won the war militarily in 1975, has not won the people's heart, according to how the Vietnamese people in Việt-nam are currently reacting against the regime nowadays. The people in South Vietnam, purportedly liberated in April 1975, can no longer feel and breathe that air of Freedom and Democracy as they used to during the Republic and when the allies are still there in the 1960s and 1970s. Who wins then? Each time I visit Việt-nam, I can feel and hear people's opinions and emotions gravitating, on the one hand, towards despair, anger, and regret when they look at where they are now, and, on the other hand, towards wishes, hope, and prayers when they think about a future where things that are happening now can be undone. The Big Question is: How to undo it? Who will help them undo it? Everything is under tight control. The "little people" get "shaped up" immediately as soon as they gather and try to make their voice heard. Protests are repressed as soon as they start budding. Of course, these people, who have been raised and trained to be submissive, will blame everything on their fate, and, like a docile child, bite their lips and accept their beatings and incarceration. The only thing they can do is to turn to China and protest against its abusive involvement in Việt-nam which is more or less allowed by the collaborating Việt-nam officials (bribery?). It is actually a clever peaceful invasion using economy and demographic expansion[607].

Never before had I felt it when visiting Việt-nam. My last stay there was in 2019. It was the culminating point when anger could not be held in any longer. As for the Southerners, especially those living in the Mekong Delta where a great number used to be non-political, if not sympathizers, it became obvious to me that the "little people," which represented the large under-privileged majority, were openly rebelling. They could see with their own eyes discrepancies happening and the unfair treatment and had gradually realized what they had been told half a century ago was nothing but point blank lies. Who were magnanimous, who were cruel, who needed to be liberated, who were the real friends and who were the villains... all became clear now. They could see they were still poor, meanwhile a small percent at the top were cracking rich. They vividly witnessed corruption at the maximum at all government levels in plain public. They could witness others losing their properties in government land grabs, and the confiscated lots being sold to the Chinese or used for building high-rises for the "network" profits. They could see innocent being clubbed and imprisoned for speaking up in open daylight. They could see they had no security against crime while criminals, many being in the "network," escalating in their action with impunity. Furthermore, their Ancestral Land was in danger again because of the governmental colluding with the eternal invader, China, and repressing the citizens. How ironic! They had no more fear and would not hesitate to swear and curse out their government, out loud in public...

It is clear that the victors of the 1975 Liberation have actually become the losers. The people hate their version of Democracy and Freedom! It has surely taken a long time, 44 years, for the liberated ones to reach the self-actualization and self-realization phase.

---

607   Migrants from China have been moving into Vietnam to buy land/properties and start family. Heavy industry factories from China are also moved into Vietnam causing major demographic conflict and excessive pollution.

Some resisters are even daring enough to openly honor the memories of the Army of the RVN and the shadows of the yellow flag. They are mourning on Facebook the South Vietnamese and the allies who were lost in battles, and their comrades' bulldozed mass graves. Yes, they are doing so in spite of the new Cyber Security Laws recently passed in Việt-nam[608]. Memories of the RVN veterans are now resurfacing, being brought back in risky public show of chagrin by the new snapshots of desecrated tombs and cemeteries. Isn't that more evidence that the losers in the military war are the true winners? The "little people" who have chosen to stay behind in Việt-nam after the end of the war know now they are actually the victims: they are uprising in various ways and demanding the Freedom and Democracy they fear may never come back. Thanks to the world-wide freedom of speech they have discovered on the internet, they can see and understand more. Even in their repression and despair, they are slowly achieving self-actualization and self-determination levels and gradually working toward the top of Maslow's pyramid! They have grown up in deprivation but are beginning to emerge from its shackles.

During their visit to Việt-nam, U.S. Presidents Obama and Trump have been welcome as saviors and champions, with a high hope they, as the defenders of human rights, will use their power to eventually liberate them, not knowing that such power is actually in the voting hands of the Americans. Occasional news of U.S. Navy patrolling the China Sea is seen as miracles. They wish that their destiny can be redirected so that their nightmares of the doomsdays will soon come to an end.

If war has been won militarily, but Freedom and Democracy are still absent, then the war for Freedom[609] is actually not over at all. It is like going deer hunting: if the hunter finds and shoots a ten-pointer, but it runs away... Is it winning or losing? As we have not actually got the catch, this being the goal of hunting, it is not a win. Ending the hunt means losing. Winning a war has to be winning the people; however, if the people grow more and more against you

---

608   These laws do not shut down the internet as China laws do to the Chinese people. Agents and informants are actually spying on the internet and social media to identify violators and arrest them

609   The DRV claimed they were supporting the National Liberation Front NLF liberate the South from American invasion. The war for Freedom had started in the 1940s for Independence from the French.

because what you have been promising them is just a hoax, you have not won the war. You have lost people's heart! With that in mind, one can now see who has won the Vietnam War, and who has lost it. **It is an awkward truth, isn't it!**

It is strange to say that the losers, the nationalists, the Americans and Allies, are the winners, while the "winners," the Communist Party, have actually lost it. Rationally-speaking, this is the very Truth. If the party has really won, there will be no protesters for human rights, for Democracy and Freedom, and there will be no demonstrations against the government, whether it is in the North, Central or South, in the big cities or small townships. Definitely, there will be no people openly cursing out the government or longing for a foreign power, particularly a "former-invader like the U.S.," to come back and re-liberate them[610].

Heroes or Villains?

We have dissected the "Vietnam War style" awkward terms of Winning and Losing. We have identified the winners and the losers. We now need to determine who the Heroes, Villains, or in between "Pseudo-Villains" are. Here is another awkward term, indeed, "pseudo-villain!" I am going to briefly look into the villains first, as we have already kind of discussed and seen who they are, then we will honor the Freedom fighter Heroes, based on the same viewpoints and logics of those awkward terms Win-Lose.

I define villains those who, accidentally or purposely, directly or indirectly, have caused deadly harms to the people who are upholding, protecting or seeking Freedom and Democracy for personal and group belief and ideology. This category can include from simple factory workers who have made weapons and ammunitions to the "big shot" tyrants such as Mao Tze-tung, Adolph Hitler or Sadam Hussein. Individuals or groups are villains if they are cognizant that their act will hurt or cause harm, whether it is being performed to unnecessarily[611] oppress, cause physical and mental

---

610   When President Obama visited Vietnam in 2016, the Vietnamese were praying that the U.S. will come back and help. When President Trump was elected President in 2017, protesters were parading with banners saying "President Trump help us."

611   When there is no national security purpose.

injury, disable or take out others, especially the innocent or defenseless people, or, for the purpose of self-interest, fame, power, popularity, business, political gain, religious belief, or just plainly out of egocentricity. These villains have also chosen to use tricks, perjuries, or dirty maneuverings to persuade others to support directly or indirectly their cause, to directly or indirectly render innocent people victims of oppression and to use them as a mean to achieve their end goals. The more defenseless the victims are, the more cruel the villains.

As shown throughout the wars in Việt-nam, Laos and Kampuchea, the villains involved in the war (those who deliberately kill and massacre) as well as in the U.S. have been directly or indirectly responsible for the death of millions of victims in the 1960s-70s and for the mental and or physical discomfort of millions others thereafter. Of course, those who "would indiscriminately kill the 'wrong' people (note that the communist never admit they have killed these innocent people) rather than release the wrong ones" would definitely be the true villains as they had committed their malicious acts out of purposeful irrational insanity. Nevertheless, by antagonistic argumentation from this definition, those who have no choice but to kill or hurt the aggressors to defend self or innocent others, in my opinion, might not be villains, but can actually be heroes.

As categorized above, we can identify so many villains we have no way of adding all of them up. In my accounts of "who has done what" in previous chapters, I think the readers can easily guess who the heroes and who the villains are. Surely, Pres. Nixon, Sec. Kissinger, as well as other politicians from all parties, some members of diverse agencies (including the CIA, Pentagon, etc…), public figures, or celebrities who have relentlessly taken advantage of their power, influentially or popularity to directly or indirectly mislead or instigate others to induce physical and or mental harms, death, et. al. **with no remorse or redemption,** would fit in this group. Even if they have superficially apologized, it is not enough. They should do something more wholeheartedly sincere and concrete, such as sponsoring or funding Vietnam Veterans events, using the media and movies to undo what they have done and said in the past. They really needed to walk their talk, otherwise millions of apologies would be like "nước chảy lá môn" (rain drops sliding down the taro leaves = do not matter). **PTS is not the injury that Veterans suffer from. It is the whole country**

that is suffering it and the whole nation is responsible for healing it.

Moreover, the political populist characters that have desecrated heroes and snuffed out the souls of the powerless or disempowered ones will be part of this group. Those news media professionals who, for the sake of competition or opportunistic career advancement, have chosen to instigate and mislead their innocent and vulnerable followers (most of whom happen to be in need of news relating to their in-country family members), instead of presenting to them the whole truth to help them sort things out and make peace with themselves, and, finally, the religious and spiritual groups that have taken advantage of disciples' trust to manipulate their own agendas (e.g. the Buddhist factions that are sympathizers or pro-terrorists in Việt-nam)… are all part of the villain world.

This gang of villains, who happens to be Hà-nội's supporters, is huge, indeed. It is unbelievable that Hà-nội has so many supporters against their own Motherland. I am wondering how many of them would have learned the Truth after the war has ended, and, if they have, is there any within this group who has a sense of guilt and redemption!?! They have more or less betrayed their own soldiers (their protectors), whom their country has sent into harm's way to sacrifice their life and well-being **for the protection of Humankind, Freedom and Democracy they so much enjoy.** They are somehow subsequently or indirectly responsible for the millions of innocent lives that have been untimely extinguished. Many have pretended, consoled and convinced themselves they have done nothing that is meant to harm others (compassion is kind) . In this way, they can continue minding their own normal business and make a living. Some may have publicly apologized; however, apologies that are not accompanied by concrete acts of sincere redemption mean nothing but hypocrisy. It is too late to undo the bodily and mental damages, but, at least, something can still be done to release, appease and help heal the sufferings the victims have been bearing for over four decades.

Meanwhile, the Lucifers and Satans (or "Quỷ Dạ Xoa,") are the meanest among the mean ones who have not hesitated to personally end innocent begging lives: they are nearly a million heartless communist executioners. A great many of them have passed away during the war and, yet, have earned the honorable titles of heroes and have the label "martyr" on their embellished and well-taken care of graves.

Some of the luckier Vietnamese (communist), more opportunistic surviving comrades are the ones who, thanks to their vow of total allegiance to and connection to the Party, are now occupying high socio-political level of power. They are now sowing the seeds of their success down to their next generations and sharing benefits with those news members who seek to join the "network" via bribery. All are now harvesting from the "little people" the crops of corruption, repression and elimination, and enjoying their cumulated material wealth and political power through a system that is nourished by "their mutually-supporting connections[612]." Many of these have been sending their children to the U.S. (preferred choice) or Australia (2nd choice), not just for schooling, but primarily as stepping stones: to buy mansions, businesses and expensive cars[613], to open savings account[614] to which they can transfer corrupt money and prepare for a quick escape (in case the system collapses). Additionally, they develop plans for their children to marry U.S. citizens[615] in order to become naturalized American citizens, and to wait for their parents[616] to unite with them in retirement in the U.S.[617]. The ultimate goal is always traditional Vietnamese multi-generational family-centered: to enjoy together their collected benefits looted from the "bần cố nông" (the powerless resourceless poor)…

---

612   The government cartel.
613   Lamborghini, Ferrari if upscale, otherwise Mercedes, Porches sports cars if lower scale.
614   High school age children with millions of U.S. dollars in savings.
615   for their own citizenship
616   Upper ranking government officials and their network can get a green card VISA if they bring in a large amount of money, e.g. U.S.$500,000 in 2017, U.S.$1.2 million by official U.S. Immigration Standards in 2019. Their huge savings are earned from a corrupted network system. In 2017, U.S.$3 billion worth of housing was purchased in the U.S.. They can also legally immigrate and reunite with their children who are now U.S. citizens.
617   The current U.S. immigration laws allow migrants who brought in U.S.$500,000 (2018 requirements) to automatically qualify for a green card. This status gives the resident all rights and privileges citizens have, except the right to vote in political elections. This is a very easy way for corrupt Vietnam officials to attain permanent residence in the U.S. as most of them have at least U.S.$1,000,000 in the savings. My colleague, St. Paul school principal, was wondering why an 18 year old exchange student from Hà-nội had U S $8,000,000 in her personal savings account. She was spending ravishingly money, such as giving out as gifts brand new $400 iPods so classmates would make friends with her…Other qualifying status are: own a business that can employ 3 American citizens (1980s requirement), or be a special and rare specialist such as surgery specialist, software specialists, scientists, rare religious leaders,…

In a similar way in the 1960s-70s, some American activists are able to culminate by triumphantly gravitating to a higher level of popularity and glory being voted top lady or elected to political power. Some others propagate their legacy to the younger generation by expanding their version of social-justice and humanity legacies. Meanwhile, in the world of U.S. news media, the misleading documentaries produced back in the 1960-70s are still being exploited and are further expanded into even more misleading media (2018) showing U.S. Soldiers fighting pitifully against an army of heroes and martyrs, and accusing the ARVN[618] for being lazy and cowardly, guided by incompetent leadership. Meanwhile, the films glorify the communist NVA and VC as patriotic heroes. The director has hired 3 Vietnamese co-producers in the DRV[619] and involves fewer former RVN consultants (as the latter complained in behind the scene interviews). This arrangement has created an ideological imbalance in favor of the "other side." Either through ignorance of cultural knowledge, or through the efforts of ignoring rational and factual truth, and the stubborn and fanatic 1960s anti-Vietnam War pride, and perhaps the sympathy for, if not collaboration with communist groups, it consequently leads to more serious journalism fallacy in spite of the truth reporting obligation "**in the name of Social Justice and Democracy.**" The veterans I have talked with cannot believe such a talented producer can end up creating such a deplorable documentary for a publicly-funded "beloved" network[620].

Villains should pay their debts, as those in WW2 have! Would an international court someday make a judgment against them, as the Nuremberg one has against WW2 genocidal murderers, before they all pass away? Would it sanction them by at least freezing their assets? Would it jail or execute them accordingly? Or **Justice goes only after the losers (Hitler's people)? Meanwhile, it seems that only the losers, in this case the poor, voiceless and powerless peasants and ARVN of South Vietnam, as well as the veterans in the U.S., are punished while the devils of Hà-nội can go free and**

---

[618] Vietnam Veterans were shocked when they heard such a put down against the ARVN. They asserted that their RVN counterpart have demonstrated utmost level of courage and efficiency.

[619] I have watched the accompanying PBS report on behind the scene the making of The Vietnam War 2018 series. A member of the project advisory committee has even made a comment about the imbalance between communist (more members) and nationalist (less members) components of the project participation

[620] PBS is my favorite network, indeed, in spite of The Vietnam War!

enjoy their loot and "patriotic heroism." Such court, in my opinion, would find a good excuse for not having any concrete evidence (survivors cannot stand on trial in court) even when the perpetrators have been openly and directly bragging they have committed genocide. Of course, it is impossible to prove that the "Conscience of America" has indirectly caused a few millions more Vietnamese, Laotian (Hmong included) and Khmer lives to be wasted! May be the victims are not worth the effort of bringing about Justice. They are just voiceless yellow people.

Unfortunately, those who are paying heavily are the oppressed and voiceless veterans and war victims. By any means, many of these defeated freedom fighters have been keeping their tears silent and invisible, and they will take their sorrow and pain to their eternal untouchable world.

When reflecting on the "**Conscience of America**," one needs to realize that the three million Americans who have rotated in their tour in country during the war have actually seen Death or been close to it: either killed in action, killed by a stray projectile or shrapnel, killed by injury, slowly killed by Agent Orange or PTS, obliterated by friendly fire. The rest at home, 200 million[621], have no idea what death or the threat of being killed by "gooks[622]" is. Most of these American people have no idea how ruthless and inhumane communist soldiers can be. Regardless of their age, most of the Communists soldiers, particularly those coming for North Vietnam (NVA)[623], think they are patriotic heroes trying to protect their Ancestral Land. Fueled by snapshots from the television news media, most of the Peace-loving and compassion-filled American souls expect one has to fight this war of attrition with the heart of a Saint… against a blood-thirsty enemy. Like fighting against the Zombies, there will be no way to win if we have empathy and compassion for an enemy that has no soul.

---

621 According to the 1970 U.S. Census, there were 203+ Million in the U.S.. https://en.wikipedia.org/wiki/1970_United_States_Census

622 When commenting on the racist term "gooks" used by a U.S. Marine veteran John Musgrave in PBS Ken Burns' series The Vietnam War, Blogger Deborah Peterson Small was highly "pissed off" at the level of inhumanity the GIs were at. https://www.alternet.org/2017/10/vietnam-and-war-drugs-what-we-forget-we-repeat/ Journalist Joseph Galloway believed honorable men were compelled to dehumanize the enemies so they could kill them.

623 As a matter of fact, a great many of the NVA were patriotic young men and women who believed they had the duty of protecting their Ancestral Land against and liberating their brothers and sisters from foreign invaders, as their trusted leader "Uncle Hồ" was able to convince them.

Would the U.S. have lost WW2 against Germany, Korea, and Japan if the people back home (US) had known and seen on television back then that some 50 Million, a great many being civilians and children, were slaughtered during WW2 by, not just the "bad side," but also by the allies (American included)? There was minimal antiwar movement or protest[624]. It was all about the Conscience of an Emotional America of the modern democratic era.

Readers might realize that the Villains and Heroes as I have defined are among those I have discussed earlier. Nevertheless, it is more difficult to identify those in the middle: the pseudo-villains. In order to make it more understandable and concrete, I will be referring to Daniel Goleman's Emotional Intelligence brain interpretation when analyzing these categories. Daniel Goleman believes that the Reptilian Brain (located at the base of the main brain) causes people to react even before they could deeply think emotionally and rationally. This will help them decide what course of action to take: **Flight** (and in some occasions, Freeze or Fright) versus **Fight**.

Quite a few pseudo-villains[625] have been misled either by kind words used by the communist propaganda machines or via the one-sided press media. These can be the North Vietnamese youth who have been tricked into volunteering to cross the "Dãy Trường Sơn Mountain on the Hồ Chí Minh trail heading South to "liberate the poor victims from the American invasion." These young and eager souls have been led by the Reptilian Brain instigated by the party indoctrination[626] that causes anger and that controls their mind (Fight category). Pseudo-villains can also be the naïve American civilians or the upset returning medal-trashing veterans[627], such as Lt. John Kerry (Fight category), or the conscience-objectors who

---

624 According to well-documented programs produced by Ken Burns on PBS and other documentaries.
625 such as my two NVA Colonel step-brothers who have paid visit to my parents upon Fall of Sài-gòn
626 The propaganda was that the U.S. intervenes in South Vietnam to start an invasion so it can take over the whole Indochina from the French Colonials.
627 2016 <u>Vietnam in HD: On the Frontline TV documentary by HistoryHD: the Vietnam Veterans Against War organized a 5-day protest at the U.S in April 1971. Capitol to express their anger against the Vietnam War. Many veterans took turns throwing away their earned medals. On this occasion, Lt. John Kerry made a powerful testimony at the U.S. Congress against the war. All had escalated from Nixon Kissinger's incursion and carpet bombardment into Cambodia, the Kent University incident that involved 4 students being shot dead.</u>

have made decision based on philosophical belief that humans are sacred and should not be harmed (mild Fight category). Whether they know what consequences their actions might result in, they have more or less, in the name of Humanity, joined the rank of antiwar movement to incapacitate the own government, stab in the back their own soldiers and help the enemy defeat the "invasion of Việt-nam." The least serious pseudo-villains are the frightened ones who have escaped to Canada to avoid their citizens' duties[628] and/ or getting involved in the Nam carnage (Fright or Flight categories), the conscience-objectors and those who can afford to make up reasons to continuously defer the draft to avoid serving. After all, as I have mentioned earlier, rationally-thinking, they might be right when thinking there is no reason Americans should be fighting in Nam, a country which the U.S. has no connection with or business being there! As for those who have chosen to allow chemicals (drugs, psychedelic substances) and the new lifestyles of "living fast and ignoring the present reality 'sống vội' " (Fright category) to take control over their conscience so they can enjoy instant gratification (Woodstock festival, drugs, insatiable sex, speak out your mind, Freedom, Freedom, Freedom[629]...) as well as take advantage of the "last moments" of safety and "freedom from draft", they have a lot to think about their behavior. Nevertheless, in a way, they all are also war victims, because they have been bombarded with wrong information they have no way of fact-checking to verify the truth as we can nowadays. They have been coerced to behave and act in the name of compassion and humanity.

It is important that the villains openly redeem by actively supporting the mistreated veterans. Doing so will lighten up the heavy sentiments of guilt they might be carrying, if they have a real conscience. Vietnam Veterans, in my opinion, can forgive as past mistakes can and should be learned (so that they won't be repeated) and erased for their better healing from their PTS, as well as for a better America.

---

628   Military draft.
629   Freedom was one of the Woodstock song.

Like Eagles, they watch their Preys

This reasoning algorithm leads us to a clear focus on the politicians and the news media at both fronts in South Vietnam and in the United States. They are definitely the BIG villains. They are the main brain of a mechanism that has brought down the war, thus bringing the three countries Cambodia, Laos and South Vietnam and their surviving people into a total subjugation to a totalitarian regime, the SE Asian Communist Party which is primarily headed by Hà-nội, which may reign for a long time. It also makes us think about how well the Sun Tzu wisdom "Know our Foes, Know ourselves, One Hundred Battles, One Hundred Victories" has been well used by the Vietnamese Communist Party. It also confirms that its marionette handlers, the Soviet Union and China, are, not just faithful allies, but also good at picking their servant in that proxy war. It is not the classical espionage and intelligence games that modern warfare uses to learn about the enemy and prepare for war. It is all about the overt and covert psychological mind control and manipulation that take advantage of the American Humanity and Conscience to fight the war, not just at the battlefield in Việt-nam, but also in the heart of America. As both NVA Generals Võ Nguyên Giáp and Bùi Tín have so proudly stated "**North Vietnam won the war, not militarily thanks to its powerful might, but politically thanks to the strong support of the People of America and Hà-nội's efforts outwitting our enemy the imperialist America.**" After the war ended, NVA Generals Võ Nguyên Giáp and Bùi Tín admitted that Hà-nội "would have surrendered if the Pentagon had continued bombing for two more days" during the last Xmas Raid over their Capital in December 1972.

"Know our Foes, Know Ourselves…." efforts have been relentlessly used by Hà-nội on both the people in the U.S. by infiltrating directly and indirectly into the antiwar movement and via the popular living room television news media, "playing" with these two popular masses simultaneously. These two "rear[630] battlefields" reinforced each other and **are monitored (9am news - U.S. time) closely by Hà-nội on a daily basis.** Like an eagle, Hà-nội was watching its preys, the press media feeding into antiwar movement against their Government, fighting each other until exhaustion before eating them all.

---

630 In comparison to front battlefields

Probably, the biggest mistake Gen. Westmoreland and previous U.S. military commanders had made was to allow the hundreds or even thousands of press media networks to tag along with the combating troops (embedded teams) at the battlefields. At the Minnesota Concordia College Round Table discussion in April 2019, former anchorman Don Shelby compared war press coverage between WW2 and Vietnam War campaigns. He said that "the first thing Gen. Eisenhower did when engaging with the press before June 1944 D-Day was to meet with all press teams and lay down the ground rules about what they could say, and what they could not." He had a funny smirk on his face when he continued "on the contrary, the anchor-teams were allowed to go wherever they wanted, even into dangerous areas, sometimes without security escort, and came back to the hotel to send back (or teletype) to the U.S. reports for breaking news… without having to go through any censoring…"

Meanwhile, the South Vietnamese news media Networks had to endure a major censorship implemented by the RVN government to prevent Freedom of Speech from poisoning people's mind. Probably, that same censorship had not allowed Western news network to collect information about ARVN war plan and campaign strategies. As I had said earlier, if the Americans could see in the news their men suffering and dying during WW2, they could have seriously protested against their government conducting the war of liberation in Europe. Out of the staggering total military death toll on all sides of 80 Million[631], 419,400 Americans died tragically, including the 407,300 military that were killed of all causes. Nearly 7 times more American GIs died in WW2 than in Việt-nam (within the 5 years of WW2 compared to the 10 years in Nam), and the public in the U.S. withstood it without major protests, thanks to the true cooperation of the news media. Did that mean that American people had cared less about their men and women well-being during the WW2 than Vietnam War? Absolutely not! They would have raised hell if **the news media had posted all miseries and atrocities as they had done with Việt-nam!**

Compared to approximately 3.8 Million Vietnamese civilian of both North and South Vietnam[632], 62,000 Laotian

---
631   https://en.wikipedia.org/wiki/World_War_II_casualties
632   https://en.wikipedia.org/wiki/Vietnam_War

and Miao civilians and 310,000 Khmer civilians were killed directly by war events, plus nearly 2.7 Million Khmer Rouge genocide[633], WW2 culminated with 50 Million total civilians dead, either directly or indirectly war related in WW2. Death toll was at least 10 fold more during WW2 than the Vietnam War. Again, American people would have raised major hell if they had known or read about what was going on in the daily newspapers as they could during Nam. Luckily, they were kept uninformed; therefore there was no significant antiwar protest in the 1940s. It was not because they care less about Humanity or did not have compassion or "Conscience of America" during WW2. **The news media were not allowed to antagonize people's Reptilian brain with front line snapshot images and stories.** If they were, protests back home could have caused the U.S. and allies to lose that war, too! Then, two atomic bombs were used by the U.S. government in Japan to instantly massacre hundreds of thousands of mostly civilians and cause slow death to many, many more, without any disobedient protest in the U.S.... Did you get the picture? news media war...

New Art of War: Win the invincible by attacking their Conscience

Snapshots published on television or in the newspapers control our reptilian and emotional brain and reinforce its Fright or Fight status, thus making people, especially those who have loved ones fighting in the war, pump up their fear leading to rebellion. The U.S. news media therefore plays an utmost important role in shaping the Conscience of America, and, consequently, redirects its senses causing uncertain reactions.

It was that "Conscience of America" which was growing steadfast in the U.S. that the enemy had started targeting in their "rear war" against the U.S. government in its own "backyard." The U.S. "Achilles heel" had been identified and so well used by Hồ Chí Minh in his eternal resistance plot. The communist propaganda machine realized that the American people's humaneness and compassion could be used against themselves: "lấy gậy ông đập lưng ông" ("use

---

633 Consequential reasons can be escape as refugees, concentration camps, injuries. In addition, 2.5 to 3 Million Khmer civilians were executed by the Khmer Rouge in post war.

their stick to beat their own back"). According to NVA Gen. Bùi Tín, Hà-nội, like a tiger waiting for the proper moment to jump onto its prey, scrutinized the news in the U.S. and used the palpitation of the antiwar movement reaction to guide its war decision making. They had made the right move during the critical point when the Paris negotiations were in their crucial decisive stage: apply even more pressure on Kissinger even when Hà-nội was on the brink of collapsing because of the Xmas Bombing in December 1972. It was a close call, but the right move on the part of Hà-nội had turned all the odds upside down, thanks to "Biết Người, Biết Ta, Trăm Trận, Trăm Thắng - Know our Foes, Know Ourselves, One Hundred Battles, One hundred victories." Even nowadays (2019), their VTV3 and VTV4 channels were still bragging how they had won the Christmas Bombings! **"Our antiaircraft batteries have overwhelmed the imperialists' B52s…!" "Our victory has dragged the American invaders back to the Paris negotiation table to sign the surrender!"**

Oftentimes, the VCs had nicknamed the American forces **"Con Cọp Giấy," meaning "Paper Tiger:"** For them, the showing of claws and fangs to intimidate were just fake menaces. The Pentagon and the U.S. high commands were more concerned about complying with the Geneva Conference rules of engagement to avoid the antiwar movement and the world criticizing it, while the Nixon was more haunted by the thought he might lose the second term election. Anyway, it would be a win-win situation for both Hà-nội and Nixon Kissinger team. "Uncle Hồ's" team did not surrender and Nixon and Kissinger got what they had wanted, the former was elected for his second term, and the latter received a Peace Nobel (a tie with his DRV counterpart Lê Đức Thọ) as well as possibilities of advancing his career even more.

After Paris "victory," Hà-nội knew it was the proper time to launch the final Hồ Chí Minh Campaign, when it watched triumphantly the U.S. and allies continue to withdraw troops and aid to the RVN being cut off. The U.S. and allies not returning in 1972 in spite of its Easter Tide Offensive (bloodier than the 1968 Tet Offensive), occupied territories remaining occupied (Paris Accords) and the ARVN morale being low because of lack of supply and support were all favorable signs according to the three factors "Thiên Thời, Địa Lợi, Nhân Hoà:" Timeliness, Location and Humanity. The fruit was indeed ripe and ready to be harvested! While discontentment continued to brew in the Hearts of Americans and the protests were dominating the political theater in the U.S., the communist implanted activists in the antiwar movement were stirring up the turmoil brew even more to finalize the internal liberation. Among the protesters, there were Buddhist monks of the "small vehicle" in saffron and brown robes (grey in black and white internet photos) protesting in the U.S. as well as in Vietnam cities. I strongly believe that, based on the photos[634] and documentary films I had watched on the internet and public TV, especially the ones showing Vietnamese monks demonstrating in Việt-nam and the Asian Buddhist[635] monks in protest with crowds in the U.S., that a well-coordinated effort to sabotage the American Government's war planning was in part orchestrated by the Hà-nội government and executed by the local communist infiltration. The Sài-gòn regime had "allegedly" linked them, either to antiwar movement while studying or preaching in the U.S., or to the VC when leading Buddhist protests (e.g. Phật-giáo Ấn Quang) against the RVN re government. *More and more Facebook postings by Vietnamese activists are now confirming this allegation.*

     Unlike the European news media (such as Paris Match, Pathe Journal…) that had been more or less partially covering communist atrocities, the U.S. ones, on the contrary, usually omitted most of such news. Were they trying to prevent American viewers from reacting against communist cruelty, thus possibly harming the antiwar movement efforts? Probably, they realize that news could be more sensationalized only when they could satisfy the needs of a majority to "squash" the war, even at the risk of hurting the American troops in Việt-nam or betraying the government. Selected snapshots had been picked out to fuel and antagonize even more the people's compassion for the enemy and instigate resentment against the war machine; consequently, condemning their own soldiers for committing the unacceptable acts had become the most effective weapon in their efforts to win in this press war for popularity! An example of this was the popular History HistHD television channel showing documentaries filmed in the early

---

634   Monks can be seen in photos demonstrating with antiwar movement in the 1960s. These photos must have been removed from internet posting in the late 1990s.

635   The "big vehicle" monks belong to the Northern sect (China, Japan, Korea…). They are vegetarian and wear grey robe. The small vehicle monks belong to the original sect from India/Ceylon, receive food from the people (begging monks) and wear saffron/brown robe. The latter had participated in multiple antigovernment protests in the RVN, and were

1970s and recompiled in 2016 entitled <u>Vietnam in HD</u> that was portraying American G.I.s having downed morale and smoking weed (at least 50% of drafters admit to trying marijuana, opium or heroin) while the ARVN despicably failing in taking over the war from the U.S.[636] If I were an American citizen then, I would not support the war either!

Many opportunistic news reporters (including people's beloved and trusted Walter Cronkite) and networks blocked viewers' from the incriminating images of the massacres and executions committed by the communists (such as the Huế massacres of 8,000 innocent civilians in 1968). Consequently, the press had prevented people from understanding that these atrocities had been designed to subjugate Vietnamese peasants to bend to the demands of VC locals and the NVA main force "bộ đội's:" they had no choice but to fully cooperate with and comply to the enemy's needs, e.g. supply food and medicine and gather intelligence about the Allied troops and ARVN. Hiding communist atrocities from the American public helped keep communist action more humane and their cause justified. Exposing atrocities committed by the U.S. and ARVN troops heightened and overwhelmed the people's sense of guilt and, as a result, antagonized the culture of compassion and humanity back home; thus, setting the misinformed people up against their own government and army. It was very simple! This game actually forged among antiwar activists a profound hatred against their own government and soldiers who were serving their citizen's duties in Việt-nam.

Vietnam War researcher Nam Pham has stated in a May 2019 blog on Facebook that, "during the war, many American reporters had made their names and careers thanks to their work in Việt-nam. Such were Neal Sheehan and David Halberstam. Most of us did not know that they heavily relied on a communist spy who worked for the Time Magazine throughout the war, Phạm Xuân Ẩn. Basically, a communist spy had traded information with American reporters, and, obviously, this information was transferred to the VC high commands so they could use it against the American forces. After the war, the spy was promoted to the rank of NVA/VC general. According to the Communist Party commendation issued to Phạm Xuân Ẩn, he had significantly contributed to many battlefield victories. In short, he had received military

---

[636] Vietnamization was defined to the public as training the ARVN to take over the war

information from the American reporters who had the privileges of attending military briefings. These intelligence leaks had helped kill American troops and South Vietnamese soldiers.

There were also stories that David Halberstam and others wrote their war reports while sitting inside the famous Caravelle Hotel in downtown Sài-gòn, instead of from the front line as they had led their readers to believe. This means that many of their 'hot' accounts, which might be revealing information that promotes communist cause or undermining the nationalist (American and allies are included) efforts, might be promoted."

Even though the Vietnamese Communist Party has won the war militarily, the voiceless soldiers, Americans as well as Vietnamese, have actually won the political one, if one judges socio-psychologically by the fact that the people in Việt Nam are now changing their mind to embrace and wish for the return of the Freedom and Democracy the Republic of Vietnam has tried to defend in the 1960s-1970s. People with a decent common sense and some critical thinking skills can reassure these combatants they are actually the **Heroes**, because they have fought on the right side, for the right cause: the "True" Freedom and Democracy. The "little" people in Việt-nam are now admitting and voicing this sentiment openly. YES, absolutely YES! The nationalists have definitely won the political war against communist aggression!

They have done their best then in or out of country, in spite of the little back up they have received, followed by an intense betrayal, and horrendous losses… Again, their very own back up, the U.S. Government, has failed them in a war it has started. It is indeed with discontentment we are able to find out four or five decades later, as preceded with the leak of the declassified Pentagon Papers that, right from the start of the war in the 1960s, the Pentagon and the White House intention has always been, through many presidencies, **not to help a friend, South Vietnam, win the war against Communism, but just to save face… "70% – To avoid a humiliating U.S. defeat**[637]**," but really "not to help a friend…" The U.S. has never been defeated ever before, and it refuses to be defeated**

---

637 https://en.wikipedia.org/wiki/Pentagon_Papers#Internal_affairs_of_Vietnam , http://www.history.com/topics/vietnam-war/pentagon-papers and http://www.thevietnamwar.info/pentagon-papers The Pentagon Papers, officially declassified in 2011

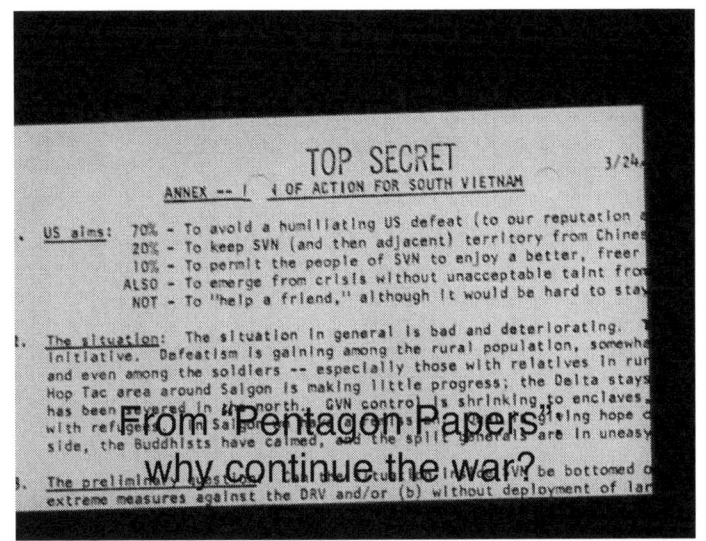

Source: https://vietnamwar.blognook.com/2017/09/21/99000070-to-avoid-humiliation/
70%: to avoid a humiliating U.S. Defeat
Not to help a friend

by the Vietnamese (as discussed in earlier Chapter 13).All predictions and evaluations are pointed out this one will be a defeat. As Cronkite has "determined," the only way out is to negotiate! As "Uncle Hồ" has preached, all patriots' efforts will lead to victory! That is the primary reason I have entitled my first book Two Minnows: we are all minnows in this proxy war! Long after the war has ended, we are still minnows, as Vietnam Veterans are still suffering and ARVN voiceless!

Meanwhile, more Losers come on stage, such as the docile and naïve pro-communist, or the collaborating/sympathizing peasants who do not have enough grey substance during the war to see and think rationally on their own in order to find out they have been duped. The latter, especially the Vietnamese war non-participants, are now the victims of the communists they have "closed their eyes on[638]" during the war. While living in Peace, they actually feel oppressed and exploited in all aspects of life by practically all levels of officials in the government. Whether it is in a big city like Sài-gòn (III Corps) or Huế (Imperial City, I Corps), or small like

---

[638] Let them do whatever they want so they will let us mind our own business.

Vĩnh Long (Mekong Delta, IV Corps) or Hội An (II Corps), the "little people" are now so upset they begin to speak up and voice their resentment in public. But, as mentioned earlier, in front of tourists, everybody smiles as if they are happy. They've got their Peace and foreigners think that this is what they have longed for. Nevertheless, even if they know that the "little people" are suffering and being oppressed, no Westerners will dare touch Nam any longer. They are all minnows!

CHAPTER 18

We are Home, America!

Towards the end of the war, after most U.S. military personnel had been withdrawn from Việt Nam in 1973, I eventually heard rumors that the GIs were encountering major problems when "returning home" in the U.S. At that time, the ARVN and many civilians of the RVN who were following the news had thought the former had been rejected by their own "people" and family because of their drug addiction or mental health problems, this being portrayed in some Sài-gòn transfers of news from foreign media stations.

After my resettlement in the U.S., I discovered that many documentaries on major networks, the same ones used by the VC in their propaganda, had been portraying the GIs in such a devastated mental health condition they had to resort to weed. I strongly believe the VC had pushed marijuana into U.S. Bases using children as suppliers. I strongly believe what my brother Anh Bảy (member of the RVN Presidential Cabinet) had said[639] about NVA efforts buying opium from the Miao peasants on the border of Laos and North Vietnam and using it to corrupt GIs (as discussed in an earlier chapter). The NVA had probably realized the GIs preferred weed to opium because it was the least expensive drug[640]. Some documentaries even showed GIs using the barrel of their rifle as bongs to pass on pot smoke to each other[641]. This snapshot had absolutely been presenting to television viewers an unfavorable image of the returning soldiers. Now we might have a clue this drug problem could have added more negativity among Americans back home; nevertheless, drug was not the major problem that I could think of. It was the television news media that killed the returning veterans. The country needed a scapegoat, for sure, and the poor victims, the Vietnam Veterans, became more victimized.

More truth came out later to educate and inform me and my ignorance about the plight of Vietnam Veterans. It was not a matter of family loyalty. It was not because of drug addiction.

---

639   ARVN intelligence collected from captured NVA prisoner
640   For more ecstasy, some might lace weed with heroin or opium.
641   Vietnam in HD: On the Frontline film by History Channel HistHD 2016, beginning of the Changing War section.

It was not because the people of the U.S. wanted to punish GIs for being defeated by the Vietnamese (strange that people in the U.S. still could not make a difference between their enemies Vietnamese communists and their allies, the nationalist Vietnamese!), to be conquered by a tiny country… As a matter of fact, most GIs themselves felt that losing the war was the main reason (circa 1970s) that had led them to this insane mistreatment. So what was so despising and degrading about the GIs? All GIs from other wars had been welcome back with great parades, confetti, and American flags and people cheering and kissing them[642] in big cities. As I had mentioned earlier, a Vietnam Veteran told me that "some" WW2 GIs in "some" American Legion in Minnesota (Western Suburbs) had looked down at him and did not want him there for "you did not fight in a real war" (to be discussed later – LRRP Tom)? Had most of the Americans, even WW2 veterans, been brain-washed by the television news media at home?

At this point (2008), I thought about interviewing some Vietnam Vets and tried to figure out the Why's. This injustice had indeed troubled me a great deal as a school administrator who believed everyone should have her or his rights to social justice. It was so troubling I decided to write the <u>Vietnam: Peace or Freedom</u> book series (2009). How could they, the Heroes, be treated worse than the refugees themselves? It was unbelievable! It was so sad for the most powerful country in the world, a country that had been historically defending the World Freedom and Democracy!

Major James Connelly (name was changed), MACV

Maj. Connelly had come home a few years before the last U.S. troops were withdrawn in 1973. Mission accomplished! He was living in his own house on base with his small family of two. He was a career officer. As usual, as if he was still in combat in Việt-nam, he always kept a weapon under his pillow, either a dagger, or, in most cases, his side arm, a Colt 45 pistol.

When they were still in Việt-nam, their life was routine: whenever he took off for missions, his wife, a military wife just like any others (there were not many, as most GIs were single, or married but whose spouse was left in the U.S.), would take

---

[642] The iconic picture of the WW2 sailor kissing a woman who was welcoming him home.

care of social support for military personnel in the base. He had married this Vietnamese wife in the late 1960s. Loan enjoyed organizing events for the servicemen and women. It was nice as their unit was really tight and the base was not too far from her hometown. Life was beautiful, as long as she could stay away from the Vietnamese civilian crowd whose cheap lives were threatened all the time by terrorism. Attacks could happen anytime, whether on public transportation, or at entertainment events (clubs, movie theaters, market, parks…).

At the end of his second tour, they both moved back to the states. He continued to serve in the U.S. Army with the hope they would get decent earnings and benefits when he retired. As any other good Vietnamese Confucian[643] woman who served her husband, Loan continued to be a housewife instead of taking on a job. When their two boys came to life, it became more interesting and much more meaningful, as she would be able to fulfill her three duties: she had served her father back home, and now her husband, and, finally, her sons.

The only thing that bothered her was that James would oftentimes wake up in the middle of the night, panting and sweaty… mumbling something in his mouth as if he was speaking a different language. It was a nightmare that had terrified James, but he would not talk about it, nor would he look for counseling support. He was certainly not becoming crazy, was he? He was still a competent and dedicated commander, loving and attentive husband and father. Later on, he turned quieter and quieter, and became silent for a long time, sometimes for days at a time. Was he experiencing something new in his thoughts or dreams, and did not feel like sharing it with Loan? He started going to the Veteran Affairs Hospital. What was that term the Hospital kept referring to? Some kind of syndrome! "Post traumatic Syndrome Disorder, PTSD" family friends said. Many of the veterans coming back from the war had experienced one form or another of this disorder… some sort of mental disorder? Was he becoming crazy? Insane?

---

643 According to Confucius, Men and Women had their own virtues and responsibilities in the society. Man's five virtues consisted of "Nhân" Humaneness, "Nghĩa" Loyalty, "Lễ" Respect, "Trí" Wisdom, and "Tín" Trustworthiness. He worshipped three levels of hierarchy: Top was "Quân" the King or Emperor, next was "Sư" the teacher, then "Phụ" the father. Woman had four virtues she had to adhere to as good lady: First would be "Công" household skills (cooking, sewing,…), then "Dung" maintaining beauty of self, "Ngôn" observing respectful speech, and "Hạnh" maintaining good behavior. Besides, she had to serve "Phụ" her father when she was still living with him, then "Phu" her husband when she left her parents' to live with the former, and finally, "Tử" her eldest son, after her husband passed away.

Did the WW2 veterans get it, too? Or just Vietnam Vets?

The war had ended and many refugees from South East Asia had moved into their neighborhood... "How come refugees or the ARVN do not have the syndrome?" Loan was wondering. Eventually, she and James divorced: it was too much for her as he became more and more violent. It was understandable, though. How could a spouse handle a husband waking up in the middle of the night, pulling his 45 caliber pistol out from under his pillow, cocking it and pointing it at the forehead? His lack of communication did not help but precipitated their relationship that was gradually collapsing. What was probably angering them even more was that neither one could discuss their repressed resentment because they were unable to talk it out or even take the time to be there and hear each other speak about their problem. The trauma and despair for seeing a loved one mentally suffer had eventually silently destroyed their marriage.

Tom (unknown), Long Range Reconnaissance Platoon[644]

It was midnight. As usual, I would run my errands or go to my dance practice until later in the evening, then sometimes go home late, and run more errands. What could you say, lifestyle of a retiree... That night, I needed to pick up my medications, so, there I was at the neighborhood 24/7 Walgreen at midnight. After I got my meds, I saw an older Caucasian man checking out the supplements aisle. I had never seen such a petite Caucasian man before. This one was at least one foot shorter than I. When I was working as an assistant principal, even though I was 5'7, most of the time, I would have to look up at students' face when talking with them as they seemed to be as tall as a tower, towering at least a foot above my head... In this case, this man was unusually short, much shorter than I, who was already a short Vietnamese guy. Short and skinny he was. He could have been a "tunnel rat[645]," I told myself.

---

[644] Long-range reconnaissance patrols, or LRRPs (pronounced "Lurps"), are small, heavily armed long-range reconnaissance teams that patrol deep in enemy-held territory. During the Tết Offensive of 1968, Quang Tri was one of the hardest hit city/provinces in the RVN as it was flooded with VC and NVA troops attacking, thus violating the Vietnamese New Year of the Monkey Tết Mậu Thân. The U.S. 1st Cavalry had relied much on recon operations launched by their LRRP to locate enemy units and their movements.

[645] The requirement for U.S. soldiers to become "tunnel rats" (who explored

Tonight, in order to see his face, I had to look down... Tom attracted my attention as soon as I saw him because he was wearing a worn out Việt Nam era Tiger-striped field shirt and a matching boney jungle hat with a newer "LRRP" insignia. Tom and I got into a heavy conversation right away, and we forgot it was getting late. Long Range Recon Platoon was the small unit he was part of: it was severely decimated, reformed, decimated again, and reborn again... in just within a few days in the middle of the Tết Mậu Thân Offensive of 1968, Year of the Monkey. He did not know how he stayed alive while his team members were taking turns getting killed or injured when snooping around the dismantled Quang Tri city walls and among the decomposing cadavers. He often times remembered his team was just on one side of some walls while the enemy was whispering to each other on the other side, or were they not enemy at all, but civilians? He surely wished he had learned some Vietnamese! The hardest thing about Vietnam War, Tom thought, was not being able to make the difference between the enemies and friends. I told him it was the same even for us ARVN: we could not tell who the enemy or friend was from the population. The VCs[646] were usually not wearing a uniform, or if they did, they might wear the ARVN (American-styled) ones so they could mingle with us and fool our allies.

What hurt Tom the most was not how he was treated there in Nam by the enemies, the weather, or the harsh battles... or seeing friends die, seeing civilians killed... What irked him the most was that his own American people were looking down at him and treated him as if he were a criminal when he came back. He would never forget being refused services and "kicked out" of an American Legion Post in the Western suburbs of Minneapolis because "you guys (Vietnam Vets) did not really fight a war... World War II is a real war..." He was discriminated even by his own kind, the veterans of WW2! Unbelievable!

Tom strongly believed that the American public had been poisoned by what they saw on television during the war. Only

---

and "handle" VC in their tactical tunnel), was to be petite. They had to be short and skinny enough to be able to slide through the complicated maze of narrow tunnels.
646 During the war, NVA absolutely denied NVA soldiers were sent to the South and fight. As a result, we rarely saw communist soldiers wearing NVA uniforms or the PDR of VN red flag . VC outfits were usually black or dark brown peasant clothes and sandals. The famous "Bình Trị Thiên" sandals (names of the three "hot" provinces Quảng Bình, Quảng Trị and Thừa Thiên), made with used car or truck tires and ropes, could sometimes be seen worn.

the trusted news media could brainwash his country men and women against their own, for the sake of making their news interesting and becoming the most popular network. Tom remembered very well seeing and hearing what Walter Cronkite had to say, and how Hà-nội Jane had stabbed him and his fellow soldiers right in the back.

I was very puzzled about what he had just said regarding what American Legion Post had done to him. How could WW2 veterans be so narrow-minded about their counterparts returning from Việt-nam? I bet there was no Hà-nội Jane then, and probably the news media were not allowed to tag along with the U.S. Army, show horrendous clips of dismembered corpses and even evoke the public's disdain sentiment about the war (e.g. Cronkite's). War had always been inhumane and non-sense, whether it was WW1, WW2 or even the Civil War. There was always tragic death in war.

The way the television news media had covered war news and their algorithm had convinced the American people that Việt Nam GIs were children, women and elders' killers. Tom would not forget being called by his own neighbors "children's killers." I tried to explain to him that, in most cases, war casualties were inevitable. Children might have been killed in this war, but there was usually a reason behind it that the news media purposefully failed to present or that they were culturally or strategically ignorant. In some cases, children, particularly in Việt-nam where everyone wanted to be patriots, might be led to think they would be patriotic if they participated in what the older comrades wanted them to do, simply because they as children could do it more easily than grown up comrades could. I myself had witnessed children, or women themselves collaborating with the enemies, or being the VC themselves. They were gathering in a group in some secret meetings in Xuân Lộc near my durian and hair fruit orchard. Everyone was pale yellow as they had obviously contracted malaria. Only combatants contracted malaria, especially those who were "hanging out" in the jungle. They usually teamed up with the local VC, these being probably related to them. They usually busied themselves to gathering military intelligence from their own RVN relatives, or from the American GIs at their bases, while selling them cigarettes or pot. These were the easiest tasks they could perform as they would not attract counter-terrorism forces, because they were "innocent and inoffensive." In many cases, they might be among the children who hung out at the base camps waiting for treats from the children-loving soldiers, which most of the GIs were. "All efforts were made to outwit and defeat the invaders."

We had to admit it was very difficult to be civilian during the war because the latter always ended up caught in between the two sides: either they cooperated with the nationalists and allies and received a harsh retaliation from the comrades at night, or they had to side with the communists at night and received the "humane" consequences from the nationalists the next day. Which choice would be the best? Of course, siding with the communists would be the safest! In many cases, they had to side with both to maintain peace in their village and to keep their families safe. Refusing to cooperate with either side always put them in harms' way. If caught by the ARVN in an allegation, they would be suspected as rebels and interrogated or even tortured, but the Geneva Convention would keep them alive and well. If suspected by the VC for cooperating with the ARVN, their family might be massacred for treason or be themselves executed as example for others. As the consequences for cooperating with the RVN were usually deadlier if sanctioned by the VC, it would be to their benefits to side with the communists: at least everyone would stay alive. Such was the unfair side of Vietnam War! Could anyone ever win fighting a humane war against an inhumane enemy?

One thing that profoundly bothered both Tom and me was that the U.S. news media had reported news unfairly toward the nationalist side. The other side, the VC and NVA, had massacred hundred or thousand-fold more civilians, much more than the U.S. or ARVN troops, but was ignored. communist massacres, such as the Tết Offensive ones in Huế that wasted nearly 8,000 civilians, or the 32 lives of all ages taken by VC Capt. Bảy Lốp (Lém), this precipitating Col. Loan to execute him in public. What Capt. Bảy Lốp had done was the old multi-generational sentence used by the Chinese during their domination of An Nam: "Chu Di Tam Tộc" (execute three generations). This barbaric cruelty was designed to prevent retaliation when the victims' family avenged loved ones (e.g. Lady Triệu [647] 225-248AD). By the way, both massacres committed by the VC I just listed were ignored by the press;

---

647   https://en.wikipedia.org/wiki/Lady_Tri%E1%BB%87u Lady Triệu of the 3rd Century AD raised an army to revolt against China Domination after her husband was executed, also for rebellion.

meanwhile, Col. Nguyễn Ngọc Loan's photo[648] became an iconic symbol of the ARVN cruelty that ruined the rest of the colonel's life.

Another discrepancy made Tom and I question the validity of Social Justice. It seemed to be so contradictory that, on one hand, GIs were wanted back home, but on the other hand, they were looked down and despised when they returned. Tom kept wondering why his wife was so afraid of him, and of his nightmares. It should be normal to have nightmares, shouldn't it? As she decided to stop communicating with him which she had always fondly done when they were still in love, Tom started keeping things in, and, soon, he did not feel like talking any longer. Communication was supposed to always be two-way to make it meaningful. He described himself to me as a hermit: he gradually retreated into his niche at home, and felt not like doing anything, as things turned more and more sour at home, until his wife took off with their little daughter. Nights and days meant nothing any more, and, profoundly alone, he started wandering around, with no goals, no thoughts, no feelings, no friends, like a lost man who wished he had dementia… It was just like that for him the night we met, almost 40 years after he returned home.

Tom kept talking and talking insatiably with me as if he had never had a chance to spill his soul and guts out before. I believed he was a talker, a very insightful talker who could conquer an audience easily. He was definitely not a bad guy at all, as he portrayed himself to be. He did not deserve to be abandoned. No one should! Not at all! It was the rest of the world that was oppressing him by making judgment at first sight and labeling him a children murderer, a loser, or a "crazy dude," even before making any attempt to hear him out. He had been silenced by his own people, a voiceless hero, and a hero who was disowned by his own kind, and by his own beloved one! That night, Tom seemed to come back alive. He was able to open up because somebody was there to communicate with him. Minnows could talk with each other, indeed!

He believed he still had friends, though; this belief might have kept him sane and alive. These friends were the imaginary Vietnamese kids he used to give treats to, and play with around villages in Việt-nam, most of whom, I was sure had passed

---

[648] https://en.wikipedia.org/wiki/Execution_of_Nguyễn_văn_Lém Photo of the execution was taken by Eddie Adams (Associate Press photographer) and filmed by Võ Sửu (NBC cameraman).

away. Those imaginary kids actually became snapshots that had maintained him alive. His home seemed to be those stretches of bamboo trees that lined the rice paddies in country-side Central Vietnam. Here in the U.S. he had no one, and his home was now still somewhere in Việt-nam. **PTS could kill you, but it can also soothe your soul, if and only if you know how it works!**

I was sure this conversation was very therapeutic for him, as well as for me. We bid farewell at 3 in the morning. I went home, and was wondering where Tom would go next. To think about it now, I might be the wanderer too, hanging out around the aisles at Walgreen's at this time of the night. The only difference was that I had a home, a loving wife and many loving friends. Tom felt he was free when thinking he was still in Việt-nam, but he felt despised and abandoned here in his homeland, a misfit, and feeling betrayed. Meanwhile, I felt free here in my new country, but definitely repressed if I were still in my homeland. I had sacrificed in this war, but I believed he had sacrificed even more, much more. I was surprised he never talked about a drinking or doping habit. His reasoning was plausible and feelings genuine. Definitely, he was not a drinker or drug user. To me, with my entire luggage of applied psychology I had cumulated during the near 5 decades of career in education, I believed that he was an introvert silent thinker who could so clearly philosophize about how insanely human beings could behave against each other. I surely wish we had exchanged contact information so we could continue this meaningful conversation.

Gene Giunta, a baby-killer[649]

"Spit at, called 'baby killer,' identified as scumbag for having served our country… it happened to me and to others as we came home from Việt-nam. This was another huge emotional milestone hung around the neck of each veteran. For many Americans at home then, behaving this way could be popular and trendy, and the thing to do to fit in with the peers. To vilify the veterans for the perceived transgressions of our nation, to make them the sacrificial lamb, it was in vogue. This lasted for a while, and then faded when we abandoned the

---

[649] Excerpt from Memoirs of the Unsung Hero, 2016, in his own word, via Janette Maring's Facebook Blogspot

people of South Vietnam.

The collateral emotional damage was huge and a nagging burden to those of us who suffered the experience of the war. It was over in 1975 for those who taunted us, we had given into the Jane Fonda academic mentality and the world began to pass us by. I know I tried to keep us up, it wasn't easy. School, raising a family, a job, success, it all became a part of the chemistry of who I needed to be.

Deep inside, in a place no one else gains access, lies an ember odd unabated anger, mistrust and uncertainty. It calls to me at time and now that I am older with more time to reflect, I can hear it more clearly.

As I think back, I remember after I left Việt-nam, focusing on the societal injustices doled out every day and in countless ways. I aim my anger and hostilities towards it. I know now that my anger had deeper roots. The culture of politics, that of the antiwar movement, the fuzziness of our purpose to be in Việt-nam, the absurd rules which applied as we fought the enemy, the carnage, all those things factored into my being.

I longed for the life I knew before I went off to war, but it was only a memory now, and I struggled with that, still do at times. It stole my innocence; it made me face things I never knew I might have to face. It challenged my beliefs and my faith, it confused me to my core.

There is no happy ending; it will always be the way I see it. I know this to be true. I find bits of solace in what I do these days and tighten my bonds with others of like kind.

In general, life has been good to me and I will accept what it offers…but I will always wonder if complete peace will ever come to my spirit."

July 8, 2016 – 5:29AM

Tony Lazzarini and the Forever Brotherhood[650]

Over 40 years ago, groups of people (returning GIs) started arriving in the United States. It seemed the majority of Americans wanted nothing to do with them. They were ridiculed, rejected and despised. The many sacrifices they made for their country would yield no rewards, heal any wounds or fade the visions of war. They sought solitude through anonymity and stealth. They were not Vietnamese refugees, they were Vietnam Veterans. Sadly, casualties of this conflict would mount long after the war's end.

Decades would pass before awareness to this great travesty would break Lady Liberty's heart.

Then the Brotherhood began, the men who saw the dragon. *Those who shed their blood with me today shall be my brothers.* Their numbers grew. The words honor, loyalty and friendship would seal the bond. *Never forget* became their motto. Please honor them today, these men, these warriors, my brothers, these Vietnam Veterans.

July 7, 2016 – 9:34PM

Heart-felt stories and long-scarred memories like the ones above had been in the mind of most returning Vietnam Veterans. Rory Kennedy's awarded documentary film <u>The Last Days in Vietnam</u> as well as many newer documentaries had also expressed Việt-nam Veterans feeling of guilt for having abandoned their allies. Only soldiers who were there until the last minutes would know how it felt leaving comrades behind. Not a few, but over a million, followed by millions more to linger a slow mental torture and death! This had never happened before in humankind history. Over forty years later, those causing this major mass murder, the villains, still smiled and conducted normal life as if they had nothing to do with it.

---

[650] Excerpt from Memoirs of the Unsung Hero, 2016, in his own word, via Janette Maring's Blogspot on Facebook

PTS

Nightmares? Guilt? Flashbacks? Avoidance? Isolation? Loss of motivation? Rage? Denial? Low self-esteem? Depression? Distrust? Shut down? Living in the past?... All of these are triggered by permanently imprinted images and thoughts from the initial Snap Shots perceived that one can never erase... unless we are aware of how we are hit by them and if we can process them as caused by our Reptilian Brain. They are like injuries that can be healed if we use the right antibiotic, suture and pain killer.

I gradually realize that, we, ARVN personnel and now refugees and immigrants and American citizens, or even the civilians who might have experienced traumatic situations during the war, are not exempted from Post Traumatic Syndrome. The truth of the matter is that, when we refuse to think we are mentally ill (which we are not), we have inadvertently allowed the condition to linger. That condition is an "injury" just like any other physical injury a soldier suffers from the battlefield. It has a cause and effect, and it is "healable" (as opposed to "curable"), as proven by a great many cases. Every day, we conduct a normal life and oftentimes have an adventure in our nightmares at night. If we are cognizant enough (by processing) and come up with the realization it is just a dream, we can eventually consider it as if we are watching movies. We will be healed if we wake up and want to return to that nightmare, anxious to know how the ending will be. I cannot really tell what the other ARVN veterans feel about their PTS and nightmares, but I always enjoyed my nightly "movies."

Probably because we ARVN grow up with the war, we have possibly got used to war events that they have become part of our daily routine. Mines and booby traps exploding at the fair ground full of adult civilians and children, or SAM missiles landing in residential areas in the middle of the night, or grenades thrown into city buses or at theaters... occur so frequently we always expect something to happen anytime, anywhere, whenever we have to go out of our house. Not even in our house can we feel safe. My close friend Chí has his eyes seriously damaged when a whole VC rocket lands on his house while he is sleeping at night... We always have in mind when we go out that we might not come back home alive. After all, flashbacks are nothing but extensions of our daily life and a reminiscence of those war days. For me, there is nothing to sweat about when I have a traumatic dream, as I am always aware in my nightmare that I will be unharmed and still alive when I wake up. After a while, being a war movie lover, I just consider flashbacks as a good war movie, in which I am a participating actor, and that is how I start loving them. Goleman-wise, we are so at home with the sense of belonging to daily hell that we have evolved into hell self-actualization.

For those not growing up in the war, it will naturally be the opposite. What a terror to find yourself being dumped from a peaceful environment, e.g. sheltered American high schools or colleges, or even a "hot" gang-controlled ghetto... into a mysterious creature-infested jungle full of undetectable booby traps and invisible snipers, or onto the top of a hill that can be surrounded anytime by an ocean of armed to the teeth, invisible, and eager-to-kill-you ghost VC who appear and disappear at will like zombies! I bet no WW2 soldier has to fight such an enemy! Not even those fighting in Korean War where North Korean and Chinese enemy are easily identifiable (wearing uniforms) and who do not melt into the civilian population! Like fish from a familiar pH water of their home in the U.S., the new army recruits receive the shock of their life being dunked into the lethal pH of Việt Nam jungle... fresh from boot camp. On top of that, think about the changes in climate, language, culture, setting...they have to face overnight! It is indeed a total Maslow transplant! It is definitely not like taking a trip to a foreign country... And the Eternal Resistance strategies used by Hồ Chí Minh who can so precisely predict it will fail the people of the mighty U.S.! Everything the new grunts are used to at home as civilians has in a blink of an eye disappeared out of their life and are replaced by a totally different world.

After one of two "tours of duties," after having witnessed major insane destruction, non-sense death of people of all ages, gender, civilian or military, natives or non-natives, enemies or buddies or allies' life being extinguished or forever incapacitated,...they are sent back home, if still alive, oftentimes with no debriefing or transition... The tragedy does not stop here, but becomes much worse, as if we have been shot by a sniper and, as soon as we come back to base camp to be exploded by our own buddies' grenade! While being still haunted by the atrocity of war from Nam, instead of being welcome back home with honor and respect as WW2

GIs have been, these returning heroes have to face ferociously shaming compatriots, among these can even be their former comrades of arms (now as protesting medal-trashing veterans) or family members, and to find out their beloved spouse and children have moved on... It is as if they are coming back as Rip Van Winkle who has just woken up from a different world to be insulted by people he has thought are his own! Whether these highly shocking changes have affected them directly, or indirectly, their effects have already been permanently imprinted in their mind. That is why they will not open themselves and talk about Nam.

Wherever Tom, Gene, James, or Tony and a great many more... are now, I wish them well, as they deserve much better, in the name of Social Justice. Not until 2009[651] that they had officially received their heroes' Welcome Home event. It was long due! They are truly, not just our national Heroes, but among the world honorable defenders of Freedom and Democracy just like those fighting in WW1, WW2, Middle Eastern Wars, and others. This country should not wait until they all die out to honor them and I do believe the new television press media and the politicians ought to make more efforts to recognize them, because they deserve much better.

I surely hope our men and women who are now serving or have served in the Middle East war (or any others) will be treated with dignity and honor when they come back, whether they win or not, and those with consequential injuries will receive proper treatment and compensation, as they have done their best to serve their country and to preserve the Freedom and Democracy all of us are enjoying and benefitting from. Furthermore, the villains ought to receive the proper consequence or at least be recognized for their wrong doing. It's all about Social Justice.

During my visit to Branson, Missouri, in September 2019 to participate in the U.S. 12th Cavalry Regiment Reunion, I was amazed to see how welcoming and supportive the people of Branson were towards veterans. The participants in the Reunion came from many states. They did not include just Vietnam War Veterans; their spouses, and even one Gold Star family member were also there to celebrate them. They

---

[651] Minnesota: On June 12, 2009 Vietnam Veterans were officially Welcome Home by Minnesota Veterans Affairs in front of the State Capitol. This was the first official nationwide ceremony I have known of. The Minnesota Former ARVN League had also participated on the same side with Vietnam Veterans.

joined in laughter and tight comradery as if they were still in Nam, enjoying each other while telling each other stories. Obviously, many of their snapshots had been replaced by good ones, and they had more or less recovered from their past misery. The performers in their fantastic 1960s musical show at Clay Cooper Theater we attended even had a tribute to these Vietnam Veterans.

In general, I could see that Branson were honoring veterans from all wars, whether they were winners or not. I had visited many cities and many states during my 44 years here, but had seen nothing like that at all until Branson. Our veterans deserved better. They had faithfully done their best and had survived their mission, and that was what should count. Those who had sacrificed deserved the same, if not more. We definitely needed this spirit to echo and reverberate everywhere in the U.S. Should some organizations need an example, they would just spend some time in Branson and watch and learn how to help undo the veterans' PTS.

EPILOGUE

Wars have always been atrocious; no matter how peace-loving one can be, if an enemy is going to attack you, you will have no choice but defend yourself and your loved ones. Either defend yourself or die! When you defend yourselves, you have actually participated in that war. If you wait for the UN to intervene, many have already died or be permanently disabled. Compensation for the dead will not bring them back. Will everyone agree on living in Peace and avoid war? There will be many who will agree to live in Peace, but there will also be some who will not. Power hunger always exists and that is the cause of war.

War, if conventional, will be better- defined and straight forward, like WW1, WW2, or even the Civil War: the combatants attack in their own uniforms, and we defend in ours; they overrun us, allies will come to our rescue. The truth of the matter is that it is not true in all cases! As you have seen when reading through The War, an unconventional war like the one in Việt-nam can be confusing, and it will be even more confusing when it is stirred and torn up by politics and cultural differences, and much more chaotic with misleading information that causes major internal polarization. Such is the Vietnam War. Such are the Middle East Wars.

Concluding an analytic research of such a confusing war is nearly impossible. I have made my account purposefully redundant, not for the sake of boring readers to death, but rather to strengthen the reasoning points that are oftentimes culturally unusual. As a socio-psychological analyst-participant, I have tried to draw your attention on plausible and obvious happenings that might not be obvious to those who are not multicultural and multi-generational insiders. Once the cultural obstacles have been overcome, we can hence suggest a resolution that compromises diverse viewpoints. Nevertheless, this is going to be a long and complicated resolution!

As fiction author Tim O'Brien[652], a 1969-70 Vietnam Veteran and 1968 Macalester College political science Magna Cum Laude graduate (also Post Graduate Harvard recipient[653]),

---

[652] Featured in Ken Burns and Lynn Novick's PBS 18 hours series The Vietnam War (2018)

[653] https://www.goodreads.com/author/show/2330.Tim_O_Brien

has said, during his 2017 interview <u>Writing in Peace and War</u> by Prof. Marlon James[654], that the main and only wish he as well as many of his companions in arms have while fighting in the war is to remain alive. One year earlier, the Tết Offensive has ravaged the South and the news brought back by the television media, particularly Walter Cronkite's viewpoint, might have been devastating and demoralizing for Americans at home and incoming grunts, in spite of great efforts of the war machine to warn people of the potential Communist Domino atrocities.

Probably many Americans have been wondering why the U.S. has managed to get in this mess, and whether it could have been avoided. Seriously, how many times have we seen that this war could have been avoided when reading this book? Many leaders have tried to prevent it from escalating or happening, such as Hồ Chí Minh in the 1940s (Pres. Truman's letters), Pres. Ngô Đình Diệm in the 1962-63 (his "Giải-pháp Bắc Nam" North-South Solution), the U.S. and U.S.S.R Détente in 1962 (Premier Nikita Khrushchev and Pres. J. F. Kennedy's communication on Cuba, nuclear testing, disarmament…), Pres. J. F. Kennedy in 1963 (his Memorandum NSAM 263 ordering the withdrawal of 1,000 troops from South Vietnam), Se. Robert Kennedy (JFK back channel)…, but these leaders are either ignored, eliminated or replaced. War is inevitable when communication is clogged up!

As I had described earlier, this fishing boat (analogy), and its "researcher-angler," while trolling along the lake shore during this "qualitative research" fishing experience, might catch the same fish over and over (redundant materials), but it might at the end catch some new and even more preferred species (more new and relevant materials)… The new species here was Secretary McNamara's role in Pres. Kennedy's Memorandum ordering troop withdrawal. McNamara was actually the person, as he admitted in-person in the autobiography film Fog of War[655], who had successfully convinced Pres. Kennedy (October 1962) to completely withdraw the

---

654 https://www.macalester.edu/news/2017/10/tim-obrien-68-and-marlon-james-in-conversation/ broadcasted by PBS TPT,
655 https://vietnam-war-history.com/review/documentary/mcnamara-the-fog-of-war <u>Fog of War: Eleven Lessons from the Life of Robert Strange McNamara. McNamara talks about his experience when he was the Secretary of Defense under FJK and LBJ, covering from the truth about the two Tonkin Gulf incidents, the influence of Japan firebombing war in WW2 on the Vietnam War, to the Cuban crisis, Pres. Diệm's coup, and JFK's change of course and preparation for complete 16,000 advisor withdrawal from Vietnam.</u>

16,000 American military advisors from Vietnam within two years, 1963-65. Unfortunately, Diệm was shot down in Vietnam, and a few weeks later, Kennedy was also assassinated as soon as the first withdrawal Memorandum was issued (at this, I had presented my conspiracy theories).

He also stated that his relationship with Pres. Johnson was not productive as the latter was pushing for a war escalation while his data-driven analysis, that gave him the nickname "the Man Who Knows Everything McNamara," could prove the U.S. involvement in the war would not be a winning one. LBJ had strongly opposed to McNamara's plan of troop withdrawal, but he just "sat silent" while JFK and McNamara were discussing this plan in October 1962." McNamara jokingly said he was *"fired"* by LBJ. LBJ's passive behavior, in contrast with his assertiveness about escalating troops, and killing and expanding war efforts in Vietnam, more or less confirmed another of my conspiracy theory about the Joint Chiefs of Staff's pressure on both presidents.

The question is still what to do when the repressive Communism (or autocratic governments) continues to expand and propagate menacing your friends while killing innocent people! A sinister preview of communist oppression (the Bolshevik/ Soviet and Chinese Revolutions) is the reason why the U.S. has jumped into the War. Has this expansion stopped yet? The invasion might not be military any longer, but might happen in a different format. Think about it… For many centuries, China is a generational foe that has been defeated by An Nam or Đại Việt many times, so Chair Mao will not likely become a big brother for the State or Republic of Vietnam. The Soviet is a China away, so Premier Stalin cannot intervene into Vietnam without having to go through Uncle Mao. The U.S. has been an ally of Tonkin during the Vietnamese war against Japan (Hồ Chí Minh's Việt Minh being trained by OSS); therefore, it is logically the party that can bring about a peaceful solution for France and the Việt Minh. Who has hidden the letters Hồ has sent to Pres. Truman begging for intervention with the French? A simple mediation, as I have discussed earlier, could have solved the dilemma and prevented a decade of bloodshed in Vietnam and economic disaster to many countries. The U.S. could have intervened as a great ally and helped unify Vietnam since the 1940s without the need for a Geneva Conference. Democrat Truman has been President (Chief of Staff John Steelman also Democrat) for two terms from 1945 to 1953 during the transitional time in Vietnam when France is relinquishing her grip on Indochina. 1947 is also the year when U.S. Congress swings right (see Chronology) for a short period; however, the big swing happens in 1953 when Republican Pres. Eisenhower came to power with a Chief of Staff Sherman Adams who is also Republican. His military Joint Chiefs of Staff are without any doubt the victorious generals from WW2 (this set up continues into J.F. Kennedy's presidency). Am I insinuating something? It seems like I am guiding the discussion into a partisan funnel, but I am actually triangulating pieces of information that lead us to a possible conclusion: that the OSS or CIA must have hidden the Truman letters (as some sources have suggested) and that the sentiment of WW2 victory turns out to be the light guiding the U.S. into the path of war. Hồ has no choice but turn to the communist block for help against the French.

The next question follows: Is the purpose the U.S. has intervened in Việt-nam in the 1960s to rescue RVN from communist aggression or is the RVN getting into the War to help the U.S. fight it? We can see South Vietnam has resented fighting a war, and has tried to plan alternatives. Diệm has refused U.S. direct intervention and just wants to receive aids to rebuild and strengthen the South. He has planned to communicate with the DRV, not to have a general referendum for unification as prescribed by the Geneva Conference (1954), but to negotiate a different plan called "Giải-pháp Bắc Nam" (North South Solution) that might be similar to the one between North and South Korea that has lasted until today. He has even assigned his brother Advisor Nhu the task of preparing to meet with Hồ Chí Minh's team in Hà-nội, but JFK has opposed this plan. Not long afterwards, both are assassinated. Again, War is inevitable!

What happened next is questionable: on two occasions, the U.S.S Maddox destroyer is attacked by NVA Navy torpedo boats (1964). These incidents give the U.S. the reason to wage war against the DRV. Nevertheless, …

…The SIGINT also shows, according to Hanyok, that a second attack, on August 4, 1964, by North Vietnamese torpedo boats on U.S. ships, did not occur despite claims to the contrary by the Johnson administration… According to National Security Archive research fellow John Prados, "the American people have long deserved to know the full truth

about the Gulf of Tonkin incident. The National Security Agency is to be commended for releasing this piece of the puzzle. The parallels between the faulty intelligence on Tonkin Gulf and the manipulated intelligence used to justify the Iraq War make it all the more worthwhile to re-examine the events of August 1964 in light of new evidence."

… A declassified internal Security Agency study (it has been kept secret since 2001) reveals in 2005, as confirmed by McNamara in his Fog of War interview, that the Tonkin Incident intelligence is skewed[656]: it has allowed the Johnson's Presidency a reason to escalate "skirmishes" into a major nine-year war against North Vietnam. An evitable War has become inevitable!

What does that mean? It means that the United States has intentionally violated of both the Geneva Convention[657] finalized in 1949 and the Geneva Conference[658] in 1954 (ending the first Indochina war): based on a big lie, the U.S. has attacked a foreign sovereignty and interfering with the Peace and neutrality of countries in post French Indochina. It also insinuates that Hồ Chí Minh is right about the U.S. invading Việt-nam without reason. It also means the U.S. secret intervention in Laos and Kampuchea are also violations under the same clauses. It might mean that the current Việt-nam, Laos and Kampuchea governments and people as well as the RVN, Lao, Hmong, Khmer survivors can be recipients for some kind of reparations[659]!

We have looked into the evitability of the War. The undeniable piece of Truth is that Peace-minded people are always haunted by losing it; therefore, if one loves Peace, one has to prepare for War. "Si Vis Pacem, Para Bellum," as the Romans have once said. Now, we need to look into the technicality of it.

In earlier chapters, I have written about the various Hồ Chí Minhs or doppelgangers. The story does not stop there. Who is "Uncle Hồ," really? He's been a multifaceted character with multiple appearances and disappearances; nevertheless,

---

656   https://nsarchive2.gwu.edu//NSAEBB/NSAEBB132/press20051201.htm the 2005 NSA Report, http://hnn.us/roundup/entries/17620.html Robert Hanyok research on Tonkin incidents,
657   https://en.wikipedia.org/wiki/Geneva_Conventions#Conventions
658   https://en.wikipedia.org/wiki/1954_Geneva_Conference
659   Aid for rebuilding the country, healthcare assistance et. al.

he is definitely a relentless communist pioneer and preacher. Is he an inhumane blood-thirsty murderer? He might not have directly killed but his "people," his team has. Also, we have to remember "Uncle Hồ" is not the only one in his gang who has master-minded many of the most horrendous genocides of the century. There are others in his team who are the behind-the-scene instigators and planners who has caused atrocities in the south. They are Lê Duẫn (probably the main marionette handler behind the scene), Trường Chinh, Văn Tiến Dũng, Võ Nguyên Giáp, and maybe some others. When we talk about atrocities committed by Uncle Hồ, we are inclusive of the others in this list, of course.

Nevertheless, people pay more attention to Hồ because he has always been the façade of the Party; this being true even nowadays half a century after his death. Is he the Chinese Maj. Hồ Quang[660] disguising as Hồ Chí Minh as a newer Vietnamese nationalist conspiracy theory claims? If only we can solve the conflict of timeline information during Hồ Chí Minh's supposedly death in 1932 or 1933 and the supposedly apparition of Hồ Chí Minh's by the Chinese Hồ Quang. The latter's efforts have been supposedly geared towards a Chinese silent invasion of Việt-nam. However, if this is true, the imposter Hồ Quang has then been a Vietnamese patriot fighting French colonizer in the 1950s? Can there be a Chinese politician who is a Vietnamese patriot and who is completely devoted to fighting against an incoming American invasion until he dies in 1969? This is impossible because Vietnamese and Chinese have been eternal enemy for many centuries in the past, the present and, possibly, in the future. Then a newer name comes up, Hồ Tập Chương[661], believed to be the last Hồ Chí Minh version because the Nguyễn Ái Quốc or Nguyễn Tất Thành who has been to Harlem is not highly fluent in Chinese. If so, who is Maj. Hồ Quang? An imposter who is the actual Hồ Chí Minh? Something still does not click here with this Hồ Quang and Hồ Tập Chương conspiracy theory! The more one uses rational reasoning to understand, the more it takes the truth far away, into the cloud. This might be a mystery forever! In the meantime, Vietnam still worships him with an altar in all public buildings and dedicated "disciples[662]."

---

660 http://vietinfo.eu/tin-the-gioi/trung-quoc-cong bo ho-chi-minh-la-thieu-ta-ho-quang-thuoc.html in Vietnamese,

661 https://www.vietthuc.org/that-bai-dau-tien-cua-cong-san-vn-ho-chi-minh-la-thieu-ta-ho-quang-cua-quan-doi-trung-cong/ in Vietnamese

662 The "network."

Policy-wise, the War since the beginning has been a total American lie from all directions: the White House, the Joint Chiefs of Staff and Pentagon, the CIA assigned to Việt-nam projects, the U.S. Military High command, the news media (dirty competition, "secret agenda"), the repressive Paris Treaty orchestrated by Hà-nội and the White House, the Việt Communist Party (lying to peasants, Buddhists), the RVN (corruption), the instigation by communist instigators infiltrated in Antiwar movement using Americans' Humanity and Conscience... The victims end in all cases up to always be the Việt civilians caught in between, and the "little soldiers" on all sides (incl. all ARVN, U.S. and Freedom fighting allies as well as the youth coming from the North to "liberate" the south). They are all naive Minnows placed in a pond to be used and eaten by politician sharks gathered under the star spangled Old Glory[663] and the "Blood Flag[664]."

As far as strategies are concerned, one has to realize the Vietnam War is fought in the U.S. terms: fight the war the White House and the Pentagon prescribe if the RVN wants to receive aids. There are conditions the RVN have to abide to in order to receive the so-needed aids that millions of lives have to depend on. As I have pointed out earlier, ARVN generals start rebelling against the American conditions for receiving aid. Is this rebellion another reason why the White House has decided to pull out of South Vietnam when aid is a life and death matter?

The Asian Sun Tzu's strategies that have played a vital role for Vietnamese combatants throughout all wars in the history of Việt-nam have been ignored by the Pentagon, thus, as I have explained, causing discontentment among the ARVN high commands[665], and trickling down to confused unit operations. Furthermore, sadly to say, the RVN, which is supposed to be one of the main combating parties in the war between the North and the South, has usually been treated as a non-existent pawn, as verified by its tokenistic role in the Paris negotiations and by its abandonment by the U.S. in 1973. By the same token, the fate of the "unknown allies" Royal Lao, Thai, Miao and Khmer falls in the same category of "non-

---

663   United States Flag.

664   Democratic Republic of Vietnam Communist Red Flag with yellow star in the center "Cờ Máu."

665   This might be true only at the high command level, not battalion or below.

existing" combatants. It turns out to be strictly a two-hand card game between the U.S. and the DRV! No wonder why Hà-nội refuses to talk with Sài-gòn! The RVN has no control over the war at all!

The Vietnam War has gone through many phases with four American presidencies (Kennedy, Johnson, Nixon, and Ford) and the multiple facets of both Democrat and Republican environments have put it through pivoting events. That is a lot of Maslow adjustments for all the pawns to endure during the ten years of war. Furthermore, drastic socio-political changes have morphed the United States through a total Civil Right Revolution, the rebellion of individual and group of Freedom fighters, and the devastating economic turmoil, all of which being stirred up by the fury cauldron of the misleading media... There is no way for the nationalists, American, allies and South Vietnam, to win the Vietnam War militarily! Why? Because of the collusion of the two groups of "liberators" who see they have **a common cause: the Americans at home realizing the South Vietnamese ought to be liberated and the communist Vietnamese who are claiming their liberation from U.S. imperialism. Both forces have used well the Conscience of America hand of cards.**

At the home front, the American people have definitely won their liberation war for civil rights and acquired their precious Freedom; meanwhile, the communists have also "won their war of liberation from invaders." However, there is a big difference: on the one hand, some are truly getting their Democracy to work for them, but, on the other hand, some others were just being moved from one jar of "liberation from the invaders" into a jar where liberation seems to be very vague and murky. The freed people in South Vietnam have reached their Peace, but they will soon realize the Democracy they have been promised is not really the one they and their parents have heard of and dreamt about. As time goes by, it actually turns out that **Democracy is really just for a privileged minority** which they, as the common people, and even a great many liberators themselves, are now not a part of.

Therefore, we end up with the final question: Who has won the War? Who has lost it? On many occasions, the card game could have been overturned, but the grand finale is obvious the DRV has militarily won it in 1975, because of a cold-turkey betrayal. Nevertheless, that final question is not really the final one as the world keeps turning. As I have demonstrated towards the book final chapters, the Freedom Fighters (RVN, U.S., Allies…) who have lost the military war in 1975 have actually, rationally speaking, won it, indeed sociopolitically, as the majority of the civilians in South and Central Vietnam, many of whom I am sure used to be sympathizers, if not collaborators, or NVA VC themselves, have slowly and subtly made a total switch of position and allegiance against the current political policy of their government. If you travel there and are lucky enough to connect with them, they will tell you all about it. Don't be fooled by Việt-nam's current economic splendor! Just be open and think about what the "little people" have to say, loudly in open public…

    If we talk with the "little people" in Việt-nam nowadays, we will not know how to feel and what to do to help this lowest social echelon that represents at least 75% of this "renewed and revamped nation" population, the un-connected and un-networked ones." During my last visit in 2019, we will be shocked by what people openly tell us to exhaust the pain they have been storing deep in their chest. No one in the country will listen to them because everyone is suffering the same repression by the authorities and the same extortion by the rich. The poor majority struggles and still sinks poorer, while the rich elevates much wealthier along with their loan-shark and corrupt network. Some more knowledgeable younger generation has openly stated that, if given weapons and leadership to fight back, probably no one will rise as everyone is stuck in the own status quo, either of contentment if they make enough, or of self-appeasement as they blame their misfortune on their destiny or fate if they are gradually sinking in the mud of disaster. By subjugating the population to a total dependence on cumulated wealth or debts, the system has successfully brought everyone under shackle-like control. The Vietnamese people are stuck in the basic needs bottom stage of Maslow Hierarchy of Needs while trying to develop a sense of belonging to something (?) and trying to understand what is going on at the pyramid top of self-actualization and self-determination. Nevertheless, the basic needs of making a difficult living are occupying them the most, mind and body.

    During my last trip to Việt-nam in 2019, almost everyone I talked with, taxi drivers, software specialist, singer, pharmacist, coffee farmers, restaurateurs, former Party civil servants, current Party members, former NVA officers, tour guide, commercial ad designer, restaurant servers… whether intellect or not, old school or new trendy, younger or older, men or women, in big city like Sài-gòn or small like Đà Lạt, Bến Tre or Vĩnh Long … All are poor and working their "buns off" 10 to 12 hours a day, 7 days a week, to barely make enough for food. Loans at .02% monthly interest (24% per year) had to be taken to fulfill the other needs. From birth to 16 years of age, most children never knew what a movie theater was, needless to say amusement park or trip taking. Debts lead mothers to giving up their own child (usually girls) to adoption or labor job abroad, and risking sex trafficking, deadly organ harvesting, slavery, and concubinary. An older Party member kept wondering why his family was getting so indebted. Their Party membership helps only when they grow old (maybe, if they can live that long with meager nutrition and health care). In the meantime, his son, a Party member, had to work manually in road and public construction, while his two beautiful daughters, also Party members, had no choice but become sex workers. Sex workers were available everywhere, in massage parlors, men's' hair salon, men's ears and nail spas, on the street, online… Young mothers would not hesitate to solicit tourists for their daughters' adoption (I was solicited there) or to send them abroad for marriage (I was asked to help). Meanwhile, more organizations from China had been corroborating with some officials (bribes) and lurking around to find organ harvest victims or sex slaves. A substantial work force from Việt-nam had been working abroad to send money home, many never came back. Others, like the case of the 39 found dead in a refrigerated container truck in the U.K.[666] (November 2019), were still trying to find their way out.

It is becoming clearer that those who have lost the war militarily 45 years ago are actually the winners as there is evidence proving that they have fought it for a generous and good cause. However, it is sad to see that there is very little, or perhaps none, that can be done to save the victims who now realize they are trapped in a deadly quick sand swamp, unless they, together hands in hands, wake up and save themselves, as the Chinese people are doing in the spirit of Tiananmen Square[667] in 1989 and Hong Kong[668] in 2019. .

---

666   https://www.independent.co.uk/news/uk/home-news/essex-lorry-deaths-migrants-vietnamese-arrests-human-trafficking-a9181911.html
667   https://en.wikipedia.org/wiki/1989_Tiananmen_Square_protests
668   https://en.wikipedia.org/wiki/2019%E2%80%9320_Hong_Kong_protests

While the "little people" in Việt-nam seem to have given up all hopes to their fate, those first generation "Việt-kiều" (immigrants) in Minnesota (and perhaps in other states, as my friends there complain "not much is going on at the legislative level!"), in spite of their financial well-being, seem to have withdrawn into their sheltered mono-cultural mono-lingual cocoon and are content with their status-quo no matter how little it is . Meanwhile, the younger generation "minds its own business" and live their own socio-political life : they are more or less assimilated and might not care much about their RVN root. The RVN legacy has perhaps come to its end, and both groups (in Việt-nam and in the U.S.) live their own Democracy which they were given to by either government. They are still struggling trying to understand it, in deprivation (in Vietnam) or in ignorance (in the U.S.); nevertheless, the more educated ones seem to grasp it better as they can think rationally (Daniel Goleman's Emotional Intelligence).

In my case, I am also confused and baffled by Democracy. I have risked my life fleeing Peace that has arrived in my Homeland and searching for Freedom and Democracy. I have achieved my goal finding them in the U.S. Now, in retrospect, I realize it is that same Democracy that has caused America betray Việt-nam. The same Democracy is currently polarizing America into two "gangs" that are trying to outwit and outgun each other at the expense of its children's mental and physical wellbeing. My Facebook Vietnam Veteran friends might be experiencing the same dilemma. I can clearly distinguish those who can rationally think and analyze versus those who eagerly and blindly respond in a reactive way (Daniel Goleman's "Reptilian Brain" and Emotional Intelligence). The socio-political situation is making people question their belonging status while confusing their self-actualization and self-determination process (Maslow).

Are we going overboard with Democracy now? Which Democracy is nurturing enough to elevate everyone? Only the future can tell, but definitely, a true Democracy is the one that can make everyone, or at least a majority of the people, happy and kind to each other, like the one in Scandinavia, whether it's in Việt-nam, Laos, Kampuchea, the U.S. or in Europe, Asia, Australia… I wonder if that Utopia will ever exist? I wonder whether there will be a day when Democracy will truly and genuinely value human lives more than out beating the opposing party. Is partisanship making people phase away from being caring and kind to each other. One deceptive find is that, it is not just the Democrats that have betrayed South Vietnam with the antiwar movement (late 1960s to early 1970s), and both the Frank-Church total cut in military aids and the War Power Act that paralyze the agonizing SE Asian allies (early 1970s). The Republicans are also deceptive when they have "discreetly " led Vietnam into the proxy war right from the beginning (even though executed by Democrats presidencies), the "cold turkey ally dumping" plan colluding with the enemy around the negotiations table of Paris , the "discreet" switching from a current ally to Middle East to rescue the other "ally …" No matter how humanitarian U.S. intervention in Third World countries is, partisanship will always, as shown in History, intervene to spoil the pot, stirring up the course of events and abandoning at the cost of a great many human lives. That is 2013 Iraq. That is 2019 Syrian Kurds. What else in the Middle East 2020? Whenever there is a change of guard in the White House, policy might change. Does policy depend on the ruling party or the opposing one? Who knows! It does not seem to reflect a stable sovereignty as it does in other countries. Such is the reason why America cannot sustain a long war: politics changes too fast. For some reason, I believe Hồ Chí Minh has sensed this when he led his eternal resistance war: keep fighting until the enemy collapses!

    One more time, I am demonstrating that Democracy, by being divisive, can become a detriment to Victory. We have won the WW1 and 2 because we fought them fast, and the people do not know much what is going on there. They are hard and extremely costly wars, but they happen fast. We lost Vietnam because it is too long and American spirits cannot withstand it to be that long. What about the Middle East? Can we win it with our superiority in weaponry and human power? It has been going on for more than two decades. What if ISIS turns guerilla and fight right in our own backyard? Can drone war out power asymmetric war? Can we afford more debts spent on insatiable status quo foreign aids and on bottomless defense budget? One thing we cannot afford to ignore is that, among the friends and the foes we have made, which one is our true ally and which one might destroy us? Do we "know our foes, and know ourselves" well enough so that, "out of one hundred battles, we will win one hundred victories?"

On a final note, Việt-nam is an extremely precious lesson that all other allied countries ought to learn, so that mistakes can be avoided in the future. The United States is truly a beautiful democratic country that has been somewhat chaotic since the 1969 Woodstock, Stonewall and Pentagon Paper events. Nevertheless, from the mud (literally, it was actually raining at Woodstock) of the 1969 events, people have enjoyed more and more the fruits of true Democracy. We cannot deny that, also since 1969, the more democratic the country grows, the more polarized and chaotic it becomes when everyone wants her/his voice heard and needs met. As people's wish that their own point of view is honored, they will also begin to feel and realize that the more politically diverse the nation becomes, the more unstable policies will be, when their individual or group aspirations start clashing each other. Meanwhile, politicians are like jugglers trying to handle in each of their tosses many political items (demands to be addressed), most of which are oftentimes mutually contradicting and antagonizing, while trying to caress their constituency for their votes. The primary elections which used to have a low profile, are now major events. Presidential campaigns become national tumultuous commotions as people become more and more polarized. As a result, policies will no longer be the same when politics and presidencies change at the end of terms, in a never-ending tug war. One thing for sure is that someone is benefitting from all of this turmoil.

What have people learned so far? Not much as we look back at the 1960s and 2010s and compare! The younger generation nowadays does not know much about the good from the bad, as they have been polarized and see things differently based on different value systems they have been exposed to. Nevertheless, if they do, it might be just, as we say it in Vietnamese, "nửa nạc, nửa mỡ," "half fat, half meat" (well-marbled meat), meaning "some truth mixed with a lot of confusing misleading information." Have we really made it a priority to teach and help people learn the factual truth from past history? Not much if the truth continues to be masked! Look around yourself every day: aren't you scared seeing that people are much more divided nowadays and become aggressive with each other over differences that used to be simple to resolve? Ignorant and divided we are! What if we come to the brink of mutually mutilating, mentally and physically? Meanwhile, American war machine keeps on running and we minnows keep being sacrificed!

"…The House that is divided within cannot stand…"

Lincoln (1886)

"Một Cây làm chẳng nên Non, Ba Cây chụm lại nên hòn Núi cao" (One tree can't make a mound, three trees together will become a tall mountain)

Vietnamese proverb

Glossary

ARVN	Army of the Republic of Vietnam
Aussie	Australian forces. New Zealand also had their armed forces fighting along side
CIA	Central Intelligence Agency of the United States of America. During World War 2, it was Office of Strategic Services OSS
DAO	The U.S. Defense Attaché Office was the liaison between the US military operation in Saigon and the US White House and Pentagon.
DMZ	De-Militarized Zone at the 17th Parallel separating North and South Vietnam
DRV	Democratic Republic of Vietnam, or North Vietnam
HCM	Hồ Chí Minh (e.g. HCM trail)
JCS	Joint Chiefs of Staff (U.S.)
JGS	Joint General Staff (RVN) "Tổng Tham Mưu"
Khmer	Cambodia under the French, Kampuchea as of Pres. Lon Nol Republic
MACV	U.S. Military Advisory Command was the first advisory units consisting of military as well as civilian personnel that were sent to serve in Vietnam during the war. It was also the last one to leave when the war ended with the Paris Accords.
MACV SOG	the Studies and Observations Group of MACV was a highly classified, multi-service United States special operations unit which conducted covert unconventional warfare operations prior to and during the Vietnam War. It had created many Vietnamese

	commando units, including the Strategic Technical Directorate later operated by the RVN General Staff.
Meo, Miao	Pre 1970s appellation referring to the Hmong tribe. Hmong being new as of the early 1970s meaning "free people"
MIKE	Mobile Strike Force Command commandos were recruited among minority ethnic groups in the RVN, mainly in the IV Corps among Khmer-Krom villages.
NLF	National Liberation Front, referring to the VC group that rises against the RVN
NVA	North Vietnam Army
PAVN	People's Army of Vietnam is the other appellation for the Army of North Vietnam to avoid using the term North that insinuate that Vietnam is not unified
RM#	Région Militaire # in Laos - Military Region number…
PAVN	People's Army of Vietnam
PLA	Pathet Lao Army, Communist insurgence army in the Kingdom of Laos
PLO	Palestinian Liberation Organization
POW	Prisoner of war
PRG	Provisional Revolutionary Government (Communist, VC, NLF National Liberation Front
RLA	Royal Lao Army
RVN	Republic of Vietnam (South Vietnam)
SE Asia	South East Asia referring to Laos, Cambodia and Vietnam (and Some other countries not

	involved in the war). In this book, it refers to the first three countries.
SGU	Special Guerilla Unit formed and trained by the CIA mainly in Laos Regions Militaires 2, 3 and 4 as counter-insurgency measure against the invasion of the NVA. Members of these irregular units consist of Lao Theung (mainly in RM#3 and RM#4), Miao (mainly in RM#2), and other ethnics and Montagnards.
SDRV	Socialist Democratic Republic of Vietnam or North Vietnam during the war, inferring that it is a democracy
SRV	This appellation is also used for DRV, in contrast with RVN that is not socialist. The DRV is also known as the People's Socialist Republic of Vietnam PSRV.
USSR	Union of Soviet Socialist Republics
VC	Việt-cộng, Communist insurgent in South Vietnam, members of the National Liberation Front NLF
VP	Vice President
WW1 or 2	World War 1 or 2

Some Samples from my Bibliography Collection

BIBLIOGRAPHY

BOOKS and ARTICLES

Berman, Larry (2001). "*No Peace, No Honor.*" New York, New York: The Free Press, A Division of Simon and Schuster, Inc. **Recommended for expanded research.**

Bluhm, Col. Raymond K. (2012). "*The Vietnam War: A Chronology of War.*" New York, New York: Rizzoli International Publications, Inc. Reference. Timeline with photographs from the military bases and battlefields of Vietnam.

Briggs, Thomas (2009). "*Cash on Delivery: CIA Special Operations During the Secret War in Laos.*" Rosebank Press. Many people interested in Vietnam War researches have heard more about the Secret War in Laos fought by Gen. Vang Pao's Special Guerilla Units to help the CIA protect the Radar Station in the mountain and to rescue many American lives. Many have been told via Hmong Oral History these brave soldiers have been fighting along the Hồ Chí Minh Trail to control the NVA troops invading South Vietnam, that without them, the U.S. could have lost much more lives. The truth is that the Miao SGUs were in RM#2 while the Trail was in Région Militaire RM#3 and RM#4. The Truth is that the indigenous Lowland Lao airborne commandos (Royal Lao Army) were the real heroes spying on and directing bombardment of the Trail. The Truth is that Région Militaire #3 and #4 also had CIA case Officers assisting there, not just in Vang Pao's #2. The Truth is that these groups are also called Special Guerilla Units. This book unveils those truths that have not been talked about, and Thomas Briggs was a CIA Case Officer in Pakse RM#3. He is now a researcher on CIA in Laos during the War against Communist Aggression 1959-1975. **Recommended for expanded research**. One of the three reliable documents on Gen.

Vang Pao and his Miao SGU battalions operating in RM#2 Laos. The other two are RM#2 war photographer J. Hamilton-Meritt's book "*Tragic Mountains: The Hmong, the Americans, and the Secret Wars for Laos, 1942-1992*" and RM#2 CIA case officer J. Parker's book "*Battle of Skyline Ridge: The CIA Secret War in Laos.*"

Brigham, Robert K. (2007). "*ARVN, Life and Death in the South Vietnamese Army.*" Lawrence, Kansas: University Press of Kansas. **Recommended for expanded research.**

Cao, Viên văn, Gen. (2005). "*The Final Collapse – Part of Indochina Monographs.*" Honolulu, Hawaii: University Press of the Pacific. Gen. Viên, one of the two four-star generals (the other one was Gen. Trần Thiện Khiêm), was the Chief of the RVN Joint General Staff (1965-75) and Defense Minister (1960s). This account of the last years of the RVN presents the viewpoint of one of the most involved officers in the History of South Vietnam from her birth in the mid-1950s until her last breath in April 1975. The drastic effects of the Paris Accords resulting in the hasty and neglected re-deployment of the armed forces from I and II Corps was well explained. It is available free on-line at https://history.army.mil/html/books/090/90-26/CMH_Pub_90-26.pdf . **Recommended for expanded research.**

Cao, Viên văn, Gen., Ngô, Trương Quang, Gen., Đồng, Quyên văn, Gen., and others (1977, 2011). "*The U.S. Adviser – Part of Indochina Monographs*" Washington, DC:U.S. Government Printing Office/U.S. Army Center for Military History as part of the Indochina Monographs. This collection of viewpoints about the role of the U.S. Advisers in the Vietnam War by at least 7 high ranking Republic of Vietnam commanders. "To analyze and evaluate the United States advisory experience in its entirety is not an easy task. It cannot be accomplished thoroughly and effectively by a single author since there were several types of advisers representing different areas of specialty but all dedicated to a common goal. Therefore, each member of the Control Group for the Indochina Refugee Authored Monograph Program has made a significant contribution as we presented the Vietnamese point of view." **Recommended for expanded research.**

Conboy, Kenneth and Andrade, Dale (2000). "*Spies and Commandos: How America Lost the Secret War in North Vietnam (Modern War Studies)*". Lawrence, Kansas: University Press of Kansas. **Recommended for expanded research.** During the Vietnam War, the United States sought to undermine Hà-nội's subversion of the Sài-gòn regime by sending Vietnamese operatives behind enemy lines. A secret to most Americans, this covert operation was far from secret in Hà-nội: all of the commandos were killed or captured, and many were turned by the communists to report false information. These RVN commandos were recruited and trained by MACV-SOG then later transferred to the RVN government to operate under the Strategic Technical Directorate (RVN General Staff). The book vividly describes scores of dangerous missions-including raids against North Vietnamese coastal installations and the air-dropping of dozens of agents into enemy territory— as well as psychological warfare designed to make Hà-nội believe the "resistance movement" was larger than it actually was.

Daugherty, Leo (2002). "*The Vietnam War – Day by Day.*" New York, New York: Chartwell Books, Inc. 2011. Reference.

Defourneaux, René J. (2000). "*The Winking Fox: Twenty-Two Years in Military Intelligence.*" Indiana: Indiana Creative Arts. The Winking Fox is a 22 year self-account of a French-born who joined the American OSS

(Office of Strategic Services, original phase of the CIA) for espionage and resistance missions in France during the WW2. After France recovered its independence from Germany, he was sent to French Indochina as the second in command of the American team of 6 to assist Hồ Chí Minh and Gen. Võ Nguyên Giáp in training the Việt Minh for guerilla insurgence against the Japanese. In 1956, he was reassigned to the U.S. Army Pacific Command in Laos, and Japan as an intelligence officer. **Recommended for expanded research.**

Defourneaux, René J. (2002). *"The Tracks of the Fox with "Between Two Crosse"* uncovering the secrets of wartime covert operations. Indiana: Indiana Creative Arts. This book reveals what the author has not covered in the previous one "The Winking Fox" regarding post war France-Vietnam activities. What he has learned from his team in France during the resistance against the Nazi about the communist agenda taking over France at the end of WW2 can be now applied to Vietnam. Communist France has something in common with communist Vietnam, and the link might be the patriotic anti-French Hồ Chí Minh. As a CIA officer, he and his team is to cooperate with Hồ against the occupying Japanese. **Recommended for expanded research.**

Đinh, Thanh Lâm (2009). *"Tôn-giáo Trong Vai-trò Tranh-đấu Tự Do, Dân-chủ và Nhân-quyền cho Việt Nam* "(The Role of Religion in the Fight for Freedom, Democracy and Humn Rights for Vietnam). Article in Vietnamese. Garden Grove, California: Chiến-Sĩ Cộng Hoà Magazine.

Đỗ, Uyển Ngọc (2009). *"Từ Cải Tạo: Tội Ác Chống Nhân Loại Của Cộng Sản Việt Nam"* (Re-Education Camp: The Crime Against Humanity of Communist Vietnam). Article in Vietnamese. Garden Grove, California: Chiến-Sĩ Cộng Hoà Magazine.

Foisie, Jack (1972). Article: "*The Strength and Weaknesses of the ARVN.*" Los Angeles, California: Los Angeles Times. A study of the factors that affect the success and failure of the Army of the Republic of Vietnam.

Freire, Paulo (2000). "*Pedagogy of the Oppressed.*" New York, New York: Bloomsbury Academic Publishing. Freire examines the effects of subjugation among people who are oppressed. **Recommended.**

Giles, Lionel (translation and commentary), Honig, Jan Willem (introduction) and Kaihko, Ilmari (supplementary material) (2012). "*The Art of War – Sun Tzu*" New York, New York: Barnes and Noble. The teaching of Sun Tzu in the arts of living with others and fighting wars. Other similar books are also available. **Recommended for expanded research.**

Goleman, Daniel (1997). "*Emotional Intelligence: Why It Can Matter More Than IQ*". New York, New York: Bantam Books – Random House. Goleman explains how our brain processes stimuli from the initial stage reptilian brain to the rational thinking one, and how the signals can affect our emotions. **Recommended for expanded research.**

Hamilton-Meritt, Jane (1993,1999). "*Tragic Mountains: The Hmong, the Americans, and the Secret Wars for Laos, 1942-1992.*" Bloomington and Indianapolis, Indiana. The author was a journalist and photographer in SE Asia since the mid-1960s during the war. This is a well-documented about the war in Laos from its early birth involving the Miao and other tribes irregular Special Guerilla Units, the Lao Theung and Royal Lao Army, and the Americans, both open (diplomatic) and covert (CIA), while interacting with the Việt Minh invasion in favor of the Pathet Lao. The secret involvement of Thailand is also reported. Various American-operated radar sites overlook into North Vietnam to aid Rolling Thunder B52 bombing in North Vietnam. A large portion of this book is devoted to Gen, Vang Pao interaction with the Kingdom of Laos, for which he serves as a full fledge general, the Royal Lao Army, but most significantly, with the CIA in between Na Khang (forward base) and Long Cheng of Laos Région Militaire #2. This might be the most accurate account of the Miao tribes and their leader Gen. Vang Pao vigorous involvement in the war in Laos RM#2, especially in the Plaine des Jarres (SW Corner of RM#2). **Recommended for expanded research**. One of the three reliable documents about Gen. Vang Pao and his SGU battalions operating in RM#2. The other two are RM#4 Pakse CIA case officer T. Briggs's book "*Cash on Delivery: CIA Special Operations During the Secret War in Laos*" and RM#2 CIA case officer J. Parker's book "*Battle of Skyline Ridge: The CIA Secret War in Laos.*"

Hoàng, Thành Đình PhD (2009). "*Công và Tôi của Chủ-tich Hồ Chí Minh và Đảng Công-sản Việt-nam 1945-2006 - Lịch-sử Hiện-đại Việt-nam.*" San Jose, California: Nghĩa Phú Publisher. This is historian Dr. Hoàng report on his research about what right and wrong Hồ Chí Minh and the Vietnamese Communist Party have done between 1945 and 2006. Text is in Vietnamese, a companion of a DVD. **Recommended for expanded research**.

Hứa, Du (2012). "*The Escapes and My Journey to Freedom*" Bloomington, Indiana: Authorhouse Publishing. It is about Hứa's personal experience as a four year old boy trying to escape many times the totalitarian regime of communist Vietnam. He finally succeeded on his 11th attempt and was rescued by a German ship. Recommended.

Isaacs, Jeremy and Downing, Taylor (1998). "*Cold War – An Illustrated History 1945-1991.*" Jeremy Isaacs Productions and Turner Original Productions Limited, Canada: Little, Brown and Company. A companion illustration of CNN series on the Cold War.

Katcher, Philip (1980). "*Armies of the Vietnam War 1962-75.*" London, Great Britain: Osprey Publishing Limited. An illustration of various armed forces involved in the Vietnam War.

Kissinger, Henry PhD (2003). "*Ending the Vietnam War: A History of America's Involvement in and Extrication from the Vietnam War.*" New York, New York: Simon and Schuster. This book is based on Dr. Kissinger's memoirs, with new and updated materials. Recommended, along with a video-research "*The Trials of Henry Kissinger.*" by Alex Gibney and Eugene Jarecki. Jigsaw Productions (2002) and Sundance Channel, LLC (2003).

Kolko, Gabriel. (1994). "*Anatomy of a War: Vietnam, the United States, and the Modern Historical Experience*" – Excerpts. New York, New York: New Press. Originally: Pantheon Books, Toronto, Canada (1985)

Kolko, Gabriel (2013). "*The 1973 Paris Peace Agreement Reconsidered*" (Article). Internet Blog: Counter Punch - Tells the Facts and Names the Names.

Lartéguy, Jean and Coutard, Raoul (1975-76). "*L'Adieu à Saigon.*" Paris, France: Presses Pocket (1993 Pocket).

Lê, Điều Tất (2009). "*Cổ Lai Chinh Chiến*" (From Battlefields, Since Ancient Past, Few Warriors Expect to Ever

Return). Excerpt in English The Last Seven Words, from one of the Letters to James Keeran Bloomington, Illinois. Garden Grove, California: Chiến-Sĩ Cộng Hoà Magazine.

Lê, Nhân Trung (2009). "*Vua Quang Trung Đại Phá Quân Thanh - Cuộc Quật Khởi của Toàn Dân Việt*" (King Quang Trung Destroyed Tang Dynasty Army – the Rise of All the People of Vietnam). Article in Vietnamese. Garden Grove, California: Chiến-Sĩ Cộng Hoà Magazine.

Li, Xiaobing (2010). "*Voices for the Vietnam War*" – Stories from American, Asian, and Russian Veterans. Lexington, Kentucky: The University Press of Kentucky. Personal accounts from American, Soviet and North Vietnam Army veterans.

McNamara, Robert S., with VanDeMark, Brian (1995) "*In Retrospect: The Tragedy and Lessons of Vietnam*." New York, New York: Random House. In Retrospect tells the story of one man's journey throughout the trials and tribulations of what seems to be the United States utmost fatality: the Vietnam War. It discusses the signs, symptoms and facts that have led to the failure and damages. **Recommended for expanded research..**

Maslow, Abraham H. (1943, 2011). "*Hierarchy of Needs: A Theory of Human Motivation.*" This is Maslow's newer theory expanded from his original 1943 version of Hierarchy of Needs that has been widely applied by educators and psychologists when they work with individuals who are exposed to a new environment. Seattle, Washington: Kindle Publisher. **Recommended for expanded research.** https://www.goodreads.com/book/show/13518096-hierarchy-of-needs Saul McLeod had also updated the 1943 version and published it in https://simplypsychology.org/maslow.html

Nguyễn, Hưng Tiến, Ph.D. and Jerrold Schecter (1986). *"The Palace File"* Dr. Tiến Nguyễn, as a special assistant to South Việt-nam Pres. Nguyễn văn Thiệu, has access to 31 letters from both U.S. Pres. Richard Nixon and Gerald Ford during the Second Republic of Vietnam. Many promises have been made, but never kept. New York City, New York: HarperCollins Publishers. **Recommended for expanded research (the official RVN point of view)**

Nguyễn, Hưng Tiến, Ph.D. (2005). *"Khi Đồng Minh Tháo Chạy"* (When Our Allies All Fled). Written for a Vietnamese audience San Jose, California: Hứa Chấn Minh Publishing. This book was written expressively in Vietnamese exposing the failure of the U.S. government in its commitment to the Republic of Vietnam, as he had observed via correspondence and interaction between the White House / Pentagon and Thiệu's presidency. Nontraditional publisher (?). **Recommended for expanded research**. https://vi.wikipedia.org/wiki/Nguyễn_Tiến_Hưng

Nguyễn, Hưng Tiến, Ph.D. (2014). *"Tâm Tư Tổng Thống Thiệu"* (Pres. Thiệu's Concerns). This 700 page long book was written for a Vietnamese audience. It exposed new materials, 150 pages of them were declassified into English for the younger generation to read. Many of Pres. Thiệu had also revealed many of his inner thoughts just before he passed away in 2001. Policies from Pres. L.B. Johnson through Pres. Nixon had also been explained, including how Việt-nam had been used to bridge the U.S. to China, with a conclusion that the Paris Accords were nothing but a façade. Nontraditional publisher (?).**Recommended for expanded research**.

Nguyễn, Hưng Tiến, Ph.D. (2016). *"Khi Đồng Minh Nhảy Vào."* (When Our Allies Jumped In) in Vietnamese. Discussion of the years when the RVN allies were so eager to join into the Vietnam War. Hứa C. Minh Publishing Company. http://www.worldcat.org/title/khi-ong-minh-nhay-vao/oclc/953289539?loc=

Nguyễn, Kỳ Cao (1976). "*How We Lost the Vietnam War.*" New York, New York: Cooper Square Press. Vice President Kỳ, as a participant in part of the second Republic of Việt-nam had witnessed the conflicts between the U.S. and the RVN policies. His concerns were about the Press inaccurately reporting the truth about the war. He is determined that the Vietnam War is a proxy war between the communist block and the free world. **Recommended for expanded research..**

Nguyễn, Phong Kỳ (2006). "*Vũng Lầy của Bạch Ốc - Người Mỹ và Chiến Tranh Việt Nam*" (The White House in Quick Sand – The American and the Vietnam War) in Vietnamese. Falls Church, Virginia: Tiếng Quê Hương Book Club Publisher. An overview view point about the Americans' involvement in the Vietnam War and how they are stuck in it, as seen by a Vietnamese author as he is now an immigrant in the U.S.

Page, Tim and Pimlott, John (1988). "*Nam – The Vietnam Experience 1965-1975*" New York, New York: Mallard Press. Collection of photographs and events during the Vietnam War between 1965 and 1975.

Parker, James E. Jr. (2019). "*Battle of Skyline Ridge: The CIA Secret War in Laos.*" Haverton, Pennsylvania: Casemate Publishers. .Parker was a CIA case officer in Région Militaire RM#2 where Gen. Vang Pao and his Miao Secret Guerrilla Units operated. Parker's well documented materials described that the Miao SGUs were not the only SGUs created and supported by the CIA, but that, oftentimes, they had to rely on the support of the Royal Lao SGUs from Régions Militaires RM#1, 3 and 4 in order to be able to counter balance to NVA divisions that were there to assist the Pathet Lao and to cause havoc. **Recommended for expanded research**. One of the three reliable documents on Gen. Vang Pao and the Miao SGU. The other two are Pakse RM#4 CIA case officer T. Briggs's "*Cash of Delivery: CIA Special Operations During the Secret War in Laos*" and RM#2 war photographer J. Hamilton-Meritt's "*Tragic Mountains: The Hmong, the Americans, and the Secret Wars for Laos, 1942-1992.*"

Pozdol, Thomas (2009). "*Tam Kỳ, The battle of Núi Yon Hill*". Bloomington, Indiana: iUniverse Publishing. One of the bloodiest battles happened near the Provincial Headquarters Tam Kỳ and Núi Yon in Quảng Tín Province, I Corps, in May 1969. It involved many large American and RVN units of Infantry, Airborne, Regional Forces, Artillery, Cavalry, Air force…at regiment and division level. The Hill was overrun by NVA

Snepp, Frank (1999). "*Irreparable Harm – A Firsthand Account of How One Agent Took On the CIA in an Epic Battle Over Secrecy and Free Speech.*" New York, New York: Random House.

Sorley, Lewis (1999). "*A Better War – The Unexamined Victories and Final Tragedy of America's Last Year in Vietnam.*" Orlando, Florida: Hartcourt Books.

Thinh, Quang (2009). "*Tổ Tông Dân-tộc Việt Không Phải Chỉ Con Rồng Cháu Tiên, Mà Còn Vượt Cả Thời-gian và Không-gian Nữa*" (Vietnamese Ancestors Were Not Just Children of the Dragons and Angels, but They Also Extended Beyond Time and Space). Article in Vietnamese. Garden Grove, California: Chiến-Sĩ Cộng Hoà Magazine.

Tourison, Sedgwick (1995-96). "*Secret Army, Secret War: Washington's Tragic Spy Operation in North Vietnam (Naval Institute Special Warfare Series)*"., Annapolis, Maryland: Naval Institute Press. In a disastrous effort to undermine the North Vietnamese, the CIA in 1961 began to parachute small teams of Vietnamese covert agents into North Vietnam. By 1964 the Pentagon was certain those men had been killed, captured, or "turned" to work for the North and began sending new agents north. Dubbed Operation Plan 34-Alpha, this renewed effort was part of a shadowy covert action force known as the Studies and Observations Group (SOG). **Recommended for expanded research** (companion book of Conboy's "*Spies and Commandos: How America Lost the Secret War in North Vietnam*").

Trần, Lý (2009). "*Trận Suối Lòng 1967 (TĐ 52 BĐQ) và Cái Chết của Tướng BV Nguễn Chí Thanh*" (The 1967 Battle ò Suối Lòng by ARVN Ranger Battalion 52, and the Death of North Vietnam Army Gen. Nguyễn Chí Thanh). Article in Vietnamese. Garden Grove, California: Chiến-Sĩ Cộng Hoà Magazine

Trần, Phụng Gia (2009). "*Giải-pháp Bảo Đại*" (the Bảo Đại Solution). Article in Vietnamese. Garden Grove, California: Chiến-Sĩ Cộng Hoà Magazine. **Recommended for expanded research** on pre-Vietnam War.

Trần, Trung Chính (2009). "*Chiến-tranh, Bản-chất và Mục-đích - Phần 2: Kinh-nghiệm Lịch-sử*" (War, Characteristics and Objectives – Part 2: History Experience). Article in Vietnamese. Garden Grove, California: Chiến-Sĩ Cộng Hoà Magazine.

Tường, Hà H. (2019). "V*ietnam: Peace or Freedom - Two Minnows.*" Minneapolis, Minnesota: Vietnam: Peace or Freedom Publisher. The author grew up during the last years of the French Protectoral ruling of Indochina, through the first and second Republics of Vietnam. He fought in the war from 1973 (Paris Accords) through the last day of the Republic of Vietnam and escaped to the United States as a Boat People refugee at the end of the war. This book records his memoir of an ARVN officer through many stages of his life, from early years attending a French Lycée, through a Vietnamese university, boot camp and services in the Army of the RVN… to his escape to Malaysia and successful resettlement in Minnesota, from an empty-handed immigrant to a successful principal of Minneapolis Public Schools. Two Minnows and The War started as one book, but kept growing as more discoveries came up, until it had to be split into two separate books. Two Minnows is the Memoir part, and The War the research-based analysis of the war. Recommended as pre-reading for this book The War (answers to why and how-questions), as the latter unveils the answers to many questions that appear in Two Minnows (what had happened? Why? How?).

"Tướng Ngô Quang Trưởng Qua Đời" (Gen. Ngo Quang Truong Has Passed Away) Article excerpt from BBC Vietnamese.com Jan 22, 2007.

Võ, Nghĩa Minh (2003). *"The Bamboo Gulag: Political Imprisonment in Communist Vietnam.*" This 254 page long book is a comprehensive review of the concentration camp system instituted by the Socialist Republic of Việt-nam under the name of "Re-Education Camps." It had incarcerated nearly one million military as well as civil service personnel of the Republic of Việt-nam between April 1975 and 1997. The author was one of the inmates. Jefferson, North Carolina: McFarland & Company. **Recommended for expanded research.** https://www.goodreads.com/book/show/691664.The_Bamboo_Gulag

Võ, Nghĩa Minh (2005). *"The Vietnamese Boat People, 1954*

and 1975-1992." This is a comprehensive account of the fate of more than a million of Boat People who had chosen to face the dangerous ocean for Freedom in a new country, rather than staying behind with the incoming communists. It also refreshed our memory about the million refugees from North Vietnam Tonkin who had done the same thing in 1954 during the few months of Passage to Freedom to the South. Jefferson, North Carolina: McFarland & Company. **Recommended for expanded research.** .https://www.goodreads.com/book/show/1672501.TheVietnamese_Boat_People_1954 and_1975_1992

Vũ, D Hiếu (2011). "*Republic of Vietnam Commandos*" is a collection of stories by the MACV-SOG operated ARVN SEAL commandos (Airborne and Frogmen) who had been inserted in clandestine operations in North Vietnam, Laos and Cambodia/ Kampuchea between late 1950s and 1975. The organization was transferred from MACV-SOG to the ARVN General Staff Bộ Tổng Tham Mưu - Quân Lực Việt Nam Cộng Hoà. Note that the U.S. SEAL was created by Pres. J.F. Kennedy in 1962 with the original Teams 1 and 2 in Vietnam, also giving birth to the ARVN SEAL and, later on, Thai SEAL.CreateSpace Independent Publishing Platform. **Recommended for expanded research.**

Vương, Anh Hồng (2009). "*Trung Tướng Đỗ Cao Trí*" (Lt. Gen. Đỗ Cao Trí)." Article in Vietnamese. Garden Grove, California: Chiến-Sĩ Cộng Hoà Magazine

Wiest, Andrew (2008). "*Vietnam's Forgotten Army.*" New York, New York: New York University Press. **Recommended for expanded research.**

AUDIO AND VIDEO DOCUMENTARY MEDIA

1968 Fall (2018). Documentary by CNN. Protests led to the end of war, and Nixon's Victory.

1969 Generation Woodstock (2019). Documentary by KSTP. Youth movements embraced Freedom at Woodstock and spurred Stonewall Inn riots. A Civil Rights Revolution stemmed out and expanded at lightning speed, uprooting the traditional norms and causing major socio-political instability. This was a good occasions for enemy insertion and manipulation, as both sides, the U.S. protesters and the Vietnamese communist insurgence have found a common cause, fight for Liberation.

American Sniper: The Real Story (2016). By REELSHD. American hero Chris Kyle of the U.S. Navy SEAL is an example of a mental health corroded sniper of a platoon in Iraq. In order to continue his legend legacy in his back-to-civilian life, he has fabricated many stories, such as the contested case of punching MN Governor Jessie Ventura in a bar, the proven untrue case of killing two car hijackers, or the claim of hundreds of looters in Katrina hurricane. This illustrates how the psychology of Maslow can be applied to explain the reaction of individuals when they cannot fit into the new environment (self-confidence and sense of belonging phases). They attempt to create a new Comfort Zone by fabricating legacy stories or creating external "heroic" images (such as wear uniforms with unearned medals) that will facilitate the feeling of being respected.

Apocalypse Now Movie: The Shocking Truth (2018). By REELSHD. Col. Kurtz and the CIA's plot to assassinate him in the movie actually reflects the real CIA policy implemented by the U.S. Green Beret Col. Robert Roe executing a Viet team member who is suspected for being a double agent also working for the VC. The is the precedence for the Special Forces to start a campaign killing suspected collaborators ("kill them all, and let God sort them out").

The Assassination of President Kennedy. The Sixties on CNN Special by Herzog and Company (2014). Recommended

Bobby Kennedy: After JFK (2017). By National Geographic "between the two assassinations." Account on Human Rights movement, his support for Israel, and "get out of Vietnam.

The Class that went to war / ABC News. CRM Films (distributor), 1977. An estimated forty percent of Vietnam War veterans have had problems adjusting to civilian life. Close to a quarter million are unemployed, and thousands have become the forgotten wounded, the ones nobody wants to talk about. Through the microcosm of one New Jersey high school class, focuses on the war and its legacy.

Cold War: Vietnam: 1954-1968, part 11 volume 4 / Richard Melman; Neil Cameron. Warner Home Video (distributor), 1998 Vietnam has been divided since the end of French colonial rule. The North is run by communists, the South by anti-Communists. Ignoring warnings against involvement in a national struggle, the United States commits its armed forces. American protests against the war mount. The United States realizes this is not a war it can win. Narrated by Kenneth Branagh. (Also on this tape: "Cuba: 1959-1962" and "Mad: 1960-1972")

Cold War: Make love not war: The Sixties, part 13 volume

5 / James Barker; Cate Haste; Neil Cameron. Warner Home Video (distributor), 1998. Series: Cold War Western economies grow and prosper, fueled partly by armaments production. Rejecting their parents' affluence and the Cold War, many of the young protest and rebel. There is racial violence in U.S. inner cities. Rock music expresses the mood of a disenchanted generation. Narrated by Kenneth Branagh. (Also on this tape: "Red Spring: The Sixties" and "China: 1949-1972").

Cold War: Detente: 1969-1975, part 16 volume 6 / James Barker; Tessa Coombs; Brian Moser.. Warner Home Video (distributor), 1998.North Vietnam launches a new offensive against the South. The U.S. steps up its bombing campaign but seeks peace through diplomacy. Nixon and Brezhnev sign the strategic Arms Limitation Treaty (SALT). The U.S. finally withdraws from Vietnam. Detente culminates in the Helsinki Declaration of 1975. Narrated by Kenneth Branagh. (Also on this tape: "Good Guys, Bad Guys: 1967-1978" and "Backyard: 1954-1990").

Declassified Archival Conversation 793-006: President Nixon and Henry Kissinger on South Vietnamese President Nguyễn văn Thiệu. (Oct. 6, 1972) WhiteHouseTapes.org. Charlottesville, VA: Miller Center. **Highly recommended for further factual study**

Dick Cavett's Vietnam (2015). By PBS. Interviews on Dick Cavett's Show, network news broadcasts and materials from National Archives show how the Vietnam War had impacted America. Recommended for expanded research.

The Fog of War: Eleven Lessons from the Life of Robert Strange McNamara (2004). By Sony. Defense Sec.

Robert McNamara discusses military issues during the Vietnam War with film maker Eric Morris. He covers from WW2 fire-bombing in Japan to how it has influenced the one in Vietnam (2 to 3 times more tons than bombs dropped in WW2). Details about the U.S. on the brink of nuclear war because of Cuba (192 nuclear missiles), his plan with Pres. Kennedy to completely withdraw all 16,000 military advisors from Vietnam starting 1963, and the two Tonkin Gulf incidents that gave Pres. Johnson the reason to escalate the war are well explained. This film is a visual documentary companion of his book In Retrospect: The Tragedy and Lessons of Vietnam he wrote in 1995. **Highly recommended for expanded research.**

The Ghosts of My Lai (2009). By NewsMax. My Lai massacre is revisited with interviews of survivors. These peasants still vividly remember what they perceive that has happened over forty years ago. It is interesting to find out that G.I.s who have tried to help the victims are misunderstood they are there to massacre.

Inside the Vietnam War (2008). By National Geographic NGC HD. Actual accounts by Vietnam Veterans, archival footage and never-seen-before photos show covert operations and military strategies during the conflict.

Israel Indivisible (2016). Aired on Jewish JBS Channel 1948-current. Synapses on Israel rights to the land.

JFK: The Lost Bullet (2017). Documentary on JFK's assassination by National Geographic.

John F. Kennedy: A President Betrayed. A documentary based on Something to Talk About, by Darin Nellis. Fine Stripes Productions and Agora Productions – Brain Storm Media and DirectTV special (2013). 2 hours. . **Highly recommended for further factual study.**

The Last Day of JFK Jr. (2019). Documentary by KSTP. Stressful life of JFK Junior (the son of Pres. Kennedy) trying to meet up with people's expectations from Kennedy legacy leading to his death in plane crash with wife Carolyn Bessette and her sister Lauren.

Life by the Sword (5/3/18). Documentary on the Israel struggle through History for her national identity. Aired on Jewish Life television Series 1, Episode 2 from War to War.

Navy SEALs: American's Secret Warriors by History Channel HD 2017. The making of the U.S. Navy SEALS from WW2 predecessors (Navy commandos on Europe D-Day, Japan and Korea Wars, through the test period in Vietnam War of the early 1960s (developing the first ARVN Frogmen) and creation of Team 1 and 2 in 1962 in Việt-nam by Pres Kennedy… until the current issues (e.g. SEAL Team 6) Panama, Granada, Bin Laden...

"Người Lính Việt Nam Cộng Hoà" (the Army of the Republic of Vietnam Soldiers) / Thuý Anh Productions; technical producers, Nam Trần, J.J. Wehrly. (SRLF). Westminster, Calif.: Thuy Anh Productions, [199?]; Documentary in Vietnamese of the Army of the RVN efforts since 1954. It did not have to wait until U.S. Vietnamization to fight the war. **Recommended.**

Nixon by Nixon (2014). Aired on HBSeHD. Pres. Nixon's secretly recorded his private conversation in the White House between 1971 and 1973. **Recommended for expanded research.**

Our Nixon (2013). By CNN. Nixon's home movies by his closest aides looking into his home environment as a President. **Recommended for expanded research.**

Secrets of War: Vietnam, A War Unwanted, (1 DVD, 5 episodes, 4 hours). Narration by Charlton Heston. MM the DocuMedia Group, Mill Creek Entertainment (2008). .

The Seventies Series by CNN (2015).
The State of the Union is not good (S1 E5)
Peace with Honor (S1 E3)
US vs. Nixon (S1 E2).
Recommended.

The Sixties Series by CNN (2014).
Sex, Drugs and Rock N' Roll (S1 E10)
The Time, They are Changing (S1 E9)
1968 (S1 E8)
The Assassination of Pres. Kennedy (S1 E3)
The World on a Brink (S1 E2).
Recommended for expanded research.

The Shining stars in battles, "Những Vị Sao Thời Lửa Đạn", a presentation by the Vietnamese Broadcasting Company; a documentary by Le Binh; screen story by Vương Hồng Anh. Westminster, Calif.: VBC International Media (200?), in Vietnamese, with screen credits in English Documentary about the famous Army of the Republic of Vietnam battles during the Vietnam War and the heroes who fought to the last minute of the war.

A Television history, Legacies: Vietnam (1975-thereafter):, episode 13 / Elizabeth Deane; Richard Ellison.
WGBH Boston Video (distributor), 1983 Vietnam is in the Soviet orbit, poorer than ever, at war on two fronts; America's legacy includes more than 500,000 refugees, 2.5 million Vietnam Veterans and some questions that won't go away. Volume 7, the final volume in this 7 volume series on the Vietnam War. A detailed visual and oral account of the war that changed a generation and continues to color American thinking on many military and foreign policy issues. The series offers an analysis of the costs and consequences of this controversial but intriguing war.

The Trials of Henry Kissinger, by Alex Gibney and Eugene Jarecki. Jigsaw Productions (2002) and Sundance Channel, LLC (2003). 80min. . **Highly recommended for further factual study.**

Tricky Dick (2019). Documentary by CNN Parts 1 and 2. From losing for California Governors' seat to events with JFK, then retreating from politics, a determination to shed away his loser's image... and finally presidency in 1968.

United Shades of America (2019). Report by Kamau Bell on the Hmong Community and the Secret War. Meetings with Hmong political leaders in the Twin Cities, Minnesota. Multi-generational stories of loss, perseverance and hope.
Caution: most archival documentary reference videos used as reference are of the Royal Lao Army as commented by Pakse CIA and RLA commander. Besides, many oral history based stories are unverified.

Vietnam – America's Conflict. 4 DVD collection on Roots and Evolution of the conflict (23 hours). Mill Creek Entertainment (2008)

Vietnam: A Television history Series / a co-production of WGBH Boston, Central Independent Television/ UK and Antenne-2, France, in association with LRE Productions. [Boston, Mass.] : WGBH Boston Video, 2004 Version. Disc 1: Roots of a war, 1945-1953; America's Mandarin, 1954-1963 -- Disc 2: LBJ goes to war, 1964-1965; America takes charge, 1965-1967; America's enemy, 1954-1967 -- Disc 3: Tet 1968; Vietnamization of the war, 1968-1973; Cambodia/ Kampuchea and Laos -- Disc 4: "Peace is at hand," 1968-1973; Home front U.S.A; The End of the tunnel, 1973-1975.

Vietnam: A Television history (1945-54). Roots of a war, episode 1 / Judith Vecchione; Richard Ellison. WGBH Boston Video (distributor), 1983. Despite cordial relations between American intelligence officers and communist leader Ho Chi Minh in the turbulent closing months of World War II, French and British hostility to the Vietnamese revolution laid the groundwork for a new war. Volume 1 in this 7 volume series on the Vietnam War. A detailed visual and oral account of the war that changed a generation and continues to color American thinking on many military and foreign policy issues. The series offers an analysis of the costs and consequences of this controversial but intriguing war. (Also on this tape: The First Vietnam War (1945-1954).

Vietnam: A Television history (1954-67). Andrew Pearson; Richard Ellison. WGBH Boston Video (distributor), 1983. In two years, the Johnson Administration's troop built-up by dispatching 1.5 million Americans to Việt-nam to fight a war they found baffling, tedious, exciting, deadly, and unforgettable. Volume 3 in this 7 volume series on the Vietnam War. A detailed visual and oral account of the war that changed a generation and continues to color American thinking on many military and foreign policy issues. The series offers an analysis of the costs and consequences of this controversial but intriguing war. (Also on this tape: America's Enemy (1954-1967).

Vietnam: A Television history, The First Vietnam War (1945-1954), episode 2 / Judith Vecchione; Richard Ellison. WGBH Boston Video (distributor), 1983. The French general expected to defeat Ho's rag-tag Vietminh guerrillas easily, but after eight years of fighting and $2.5 billion in U.S. aid, the French lost a crucial battle at Dien Bien Phu — and with it their Asian Empire. Volume 1 in this 7 volume series on the Vietnam War. A detailed visual and oral account of the war that changed a generation and continues to color American thinking on many military and foreign policy issues. The series offers an analysis of the costs and consequences of this controversial but intriguing war. (Also on this tape: The Roots of a War).

Vietnam: A Television history, Tet 1968, episode 7 / Austin Hoyt; Richard Ellison. WGBH Boston Video (distributor), 1983. The massive enemy offensive at the lunar new year decimated the Vietcong and failed to topple the Sài-gòn government but led to the beginning of America's military withdrawal from Vietnam. Volume 4 in this 7 volume series on the Vietnam War. A detailed visual and oral account of the war that changed a generation and continues to color American thinking on many military and foreign policy issues. The series offers an analysis of the costs and consequences of this controversial but intriguing war. (Also on this tape: Vietnamizing the War (1968-1973).

Vietnam: A Television history, Peace is at hand (1968-1973), episode 10 / Martin Smith; Richard Ellison. WGBH Boston Video (distributor), 1983. Series: The American Experience. While American and Vietnamese soldiers continued to clash in battle, diplomats in Paris argued

Vietnam – America's Conflict. 4 DVD collection on Roots and Evolution of the conflict (23 hours). Mill Creek Entertainment (2008)

Vietnam: A Television history Series / a co-production of WGBH Boston, Central Independent Television/ UK and Antenne-2, France, in association with LRE Productions. [Boston, Mass.] : WGBH Boston Video, 2004 Version. Disc 1: Roots of a war, 1945-1953; America's Mandarin, 1954-1963 -- Disc 2: LBJ goes to war, 1964-1965; America takes charge, 1965-1967; America's enemy, 1954-1967 -- Disc 3: Tet 1968; Vietnamization of the war, 1968-1973; Cambodia/ Kampuchea and Laos -- Disc 4: "Peace is at hand," 1968-1973; Home front U.S.A; The End of the tunnel, 1973-1975.

Vietnam: A Television history (1945-54). Roots of a war, episode 1 / Judith Vecchione; Richard Ellison. WGBH Boston Video (distributor), 1983. Despite cordial relations between American intelligence officers and communist leader Ho Chi Minh in the turbulent closing months of World War II, French and British hostility to the Vietnamese revolution laid the groundwork for a new war. Volume 1 in this 7 volume series on the Vietnam War. A detailed visual and oral account of the war that changed a generation and continues to color American thinking on many military and foreign policy issues. The series offers an analysis of the costs and consequences of this controversial but intriguing war. (Also on this tape: The First Vietnam War (1945-1954).

Vietnam: A Television history (1954-67). Andrew Pearson; Richard Ellison. WGBH Boston Video (distributor), 1983. In two years, the Johnson Administration's troop built-up by dispatching 1.5 million Americans to Việt-nam to fight a war they found baffling, tedious, exciting, deadly, and unforgettable. Volume 3 in this 7 volume series on the Vietnam War. A detailed visual and oral account of the war that changed a generation and continues to color American thinking on many military and foreign policy issues. The series offers an analysis of the costs and consequences of this controversial but intriguing war. (Also on this tape: America's Enemy (1954-1967).

Vietnam: A Television history, The First Vietnam War (1945-1954), episode 2 / Judith Vecchione; Richard Ellison. WGBH Boston Video (distributor), 1983. The French general expected to defeat Ho's rag-tag Vietminh guerrillas easily, but after eight years of fighting and $2.5 billion in U.S. aid, the French lost a crucial battle at Dien Bien Phu — and with it their Asian Empire. Volume 1 in this 7 volume series on the Vietnam War. A detailed visual and oral account of the war that changed a generation and continues to color American thinking on many military and foreign policy issues. The series offers an analysis of the costs and consequences of this controversial but intriguing war. (Also on this tape: The Roots of a War).

Vietnam: A Television history, Tet 1968, episode 7 / Austin Hoyt; Richard Ellison. WGBH Boston Video (distributor), 1983. The massive enemy offensive at the lunar new year decimated the Vietcong and failed to topple the Sài-gòn government but led to the beginning of America's military withdrawal from Vietnam. Volume 4 in this 7 volume series on the Vietnam War. A detailed visual and oral account of the war that changed a generation and continues to color American thinking on many military and foreign policy issues. The series offers an analysis of the costs and consequences of this controversial but intriguing war. (Also on this tape: Vietnamizing the War (1968-1973).

Vietnam: A Television history, Peace is at hand (1968-1973), episode 10 / Martin Smith; Richard Ellison. WGBH Boston Video (distributor), 1983. Series: The American Experience. While American and Vietnamese soldiers continued to clash in battle, diplomats in Paris argued about making peace. After more than four years, they reached an accord that proved to be a preface to further bloodshed. Volume 5 in this 7 volume series on the Vietnam War. A detailed visual and oral account of the war that changed a generation and continues to color American thinking on many military and foreign policy issues. The series offers an analysis of the costs and consequences of this controversial but intriguing war. (Also on this tape: Kampuchea and Laos).

Vietnam: A Television history, Homefront U.S.A, episode 11 / Elizabeth Deane; Richard Ellison. WGBH Boston Video (distributor), 1983.Through troubled years of controversy and violence, U.S. casualties mounted, victory remained elusive and American opinion moved from general approval of, to general dissatisfaction with the Vietnam War. Volume 6 in this 7 volume series on the Vietnam War. A detailed visual and oral account of the war that has changed a generation and continues to color American thinking on many military and foreign policy issues. The series offers an analysis of the costs and consequences of this controversial but intriguing war. (Also on this tape: The End of the Tunnel (1973-1975).

Vietnam: A Television history, The End of the tunnel (1973-1975), episode 12 / Elizabeth Deane; Richard Ellison. WGBH Boston Video (distributor), 1983 South Vietnamese leaders believed that America would never let them go down to defeat — a belief that died as North Vietnamese tanks smashed into Sài-gòn on April 30, 1975, and the long war ended with South Vietnam's surrender. Volume 6 in this 7 volume series on the Vietnam War. A detailed visual and oral account of the war that has changed a generation and continues to color American thinking on many military and foreign policy issues. The series offers an analysis of the costs and consequences of this controversial but intriguing war. (Also on this tape: Homefront U.S.A).

Vietnam in HD Series (2011-2016). The newest series by History Channel Serves Up A Fitting And Personal Six Part Miniseries Tribute To Those Who Served During Vietnam as seen on History Channel by K. Harris HALL OF FAMETOP 50 REVIEWER on November 14, 2011 Episodes: (1) The Beginning 1964-1965, (2) Search & Destroy 1966-1967, (3) The Tet Offensive 1968, (4) An Endless War 1968-1969, (5) A Changing War 1969-1970, and (6) Peace With Honor 1971-1975.
- Very Benedetto patrolling the Mekong Delta
- James Anderson lead a battalion into Cambodia
- Bon Clewell helicopter under fire
- Joe Galloway returning to Vietnam

.Highly recommended for further factual study

Vietnamizing the war (1968-1973): Vietnam: A Television history, episode 8 / Martin Smith; Richard Ellison. WGBH Boston Video (distributor), 1983Richard Nixon's program of troop pull-outs, stepped-up bombing and huge arms shipments to Sài-gòn changed the war and left GIs wondering which of them would be the last to die in Vietnam. Volume 4 in this 7 volume series on the Vietnam War. A detailed visual and oral account of the war that changed a generation and continues to color American thinking on many military and foreign policy issues. The series offers an analysis of the costs and consequences of this controversial but intriguing war. (Also on this tape: Tet 1968).

The Vietnam War (2015). Documentary by NEWSMX.
Jungle of Death: perspectives from the U.S. and the DRV.
Defeat and Departure by NEWSX
The invisible enemy NEWSX

The Vietnam War (2018), Documentaries by PBS – Ken Burns. 12 episodes totaling 18 hours. Old documentary films from the past produced by various media networks are illustrated by new in-person interviews of Vietnam Veterans, former combatants of the North Vietnamese

Army and the Việt-cộng (National Liberation Front), and the Army of the RVN.
Use this series with caution, preferably with guidance by a Vietnam Veteran and an ARVN Veteran. As U.S. and ARVN veterans believe it is pro-communist, portrays U.S. Veterans as victims and ARVN veterans as coward and incompetent, while glorifying the Communist Party and armed forces. Three co-producers from the DRV were involved in this 10-year long production.

A Year of War, Turmoil and Beyond: 1968 (2018). Documentary by AXSTV. Tet Offensive convinces many Americans that the Vietnam War won't end soon. Champions for changes. Assassination of Rev. Dr. Martin Luther King and Sen. Robert F Kennedy.

INTERNET – BLOGS

Beck, Sanderson (On Going). BECK index. Book: South Asia 1800-1950. Ngô, Linh Thế (On Going). *(Ongoing Tribute to ARVN Heroes)*. Pictures, bibliographies, unit historical documentaries, maps, memorials. Most ARVN commando and frogman units are cited. http://ngothelinh.tripod.com/

Nguyễn, Chồi Bá (2014). *Letter from a Former Soldier of the National Liberation Front for South Vietnam.* Danlambao Blog.

Phạm, Hoa (2000-curent). *http://historynkt.blogspot.com/* Nha Kỹ Thuật NKT or Strategic Technical Directorate STD Blogspots created and maintained by Hoa Pham is one of the major sources for Two Minnows updates on the Republic of Việt-nam strategic repositioning abandoning I and II Corps in 1975. **Recommended for expanded research..**

Staff, Sgt Grit (unknown). Gen. Bui Tin Describes North Vietnam's Victory. NVA Gen. Bùi Tín's interview by Stephen Young for Wall Street Journal and posted in SGT GRIT by Sgt Grit blog staff. **Recommended for expanded research** http://www.grunt.com/corps/scuttlebutt/marine-corps-stories/gen-bui- tin-describes-north-vietnams-victory/

The Communist War of Aggression (2018-2025). This website is sponsored by the CAVWV Coalition of Allied Vietnam War Veterans. It links all CIA/ MACV/ General Staff-verifiable information and documentaries of military and political events with stories and interviews collected from living participants and witnesses who had survived the war in the Republic of Cambodia, the Kingdom of Laos and the Republic of Vietnam. **Recommended for research projects and references.** http://www.thecommunistwar.org/

A multitude of other website resources were also used as listed in text foot notes.

MOTION PICTURES

Apocalypse now / directed by Francis Ford Coppola. Paramount Home Video (distributor), 1979. Francis Ford Coppola's epic vision of the Vietnam Conflict, inspired by Joseph Conrad's novel, "Heart of Darkness." When Lieutenant Willard receives orders to seek out and destroy a renegade military outpost led by the mysterious Colonel Kurtz, he embarks on an epic journey to himself, as he ultimately recognizes an aspect of his own soul in the strange colonel. Cast includes Martin Sheen, Marlon Brando, and Robert Duvall. Also see the documentary: **Apocalypse Now Movie: The Shocking Truth (2018).**

Chúng Tôi Muốn Sống (We want to live) / film story & director, Vinh Noan. Westminster, Calif.: Vinh Noan, [unknown year of production]. This film portrays the story of how Ho Chi Minh's communists had used the People Court to prosecute and execute the rural land owners in North Vietnam between 1952 and 1954. All the actors and extras in the film are refugees who fled from his reign of terror in the North to the Republic of South Vietnam (1954 Passage to Freedom). **Recommended for expanded research on pre-Vietnam War.**

Điện Biên Phủ. Produced by Jacques Kirshner, Patrick Delauneux, Patrick Miller and Yves Dutier. Distributed by France Pathé AMLF in 1992. Dubbed in Vietnamese by Secofilm. Commentary in French by Directeur Pierre Schoendoerffer. This black and white movie is one of the closest account by British-born American Howard Simpson who was present in Vietnam during the Battle of Điện Biên Phủ as a writer-reporter for San Francisco Chronicle daily newspaper. In spite of being seriously outnumbered by the Việt Minh, the French allied army had done their best to defend the undefendable base. It also showed the Vietnamese airborne units being transitioned into the first army of the State of Vietnam (the 1st Airborne Battalion), that eventually becomes the Republic of Vietnam. The French surrendered after

55 days sieged by the Việt Minh forces. **Recommended for expanded research on pre-Vietnam War.**

Full metal jacket / directed by Stanley Kubrick. Warner Home Video. 1986. Stanley Kubrick's brilliant saga about the Vietnam War and the dehumanizing process that turns people into trained killers. Young recruits are followed from boot camp hell to the battlefields of Vietnam. Based on the novel "The Short-Timers" by Gustav Hasford. Cast includes Matthew Modine, Adam Baldwin, Vincent D'Onfrio, Lee Ermey, Dorian Harewood, Arliss Howard, Kevyn Major Howard, Ed O'Ross. Screenplay by Stanley Kubrick, Michael Herr, Gustav Hasford.

Go Tell the Spartans / Allan F. Bodoh; directed by Ted Post. Vestron Home Video (distributor), 1977. In Vietnam, 1964, a hard-boiled major is ordered to establish a garrison with a platoon of burned-out Americans and Vietnamese mercenaries. Based on the novel, "Incident at Muc Wa," by Daniel Ford. Cast includes Burt Lancaster, Craig Wasson, David Clennon, and Marc Singer.

Hamburger Hill (1987) / is a realistic interpretation of the bloody war that has made no meaning to many men and women in the armed forces: sacrifice many lives to take over a hill, then be ordered to abandon it upon successful mission. The film carries some anti-war message according to RottenTomatoes commentators. https://www.rottentomatoes.com/m/hamburger_hill

Heaven and earth (1993) / Oliver Stone; Arnon Milchan; Robert Kline; A. Kitman Ho. Warner Home Video (distributor). It's about a Vietnamese woman's painful odyssey from a peaceful childhood in a peasant village to a lifetime of upheaval both in Việt-nam and the U.S. She is caught between the forces of North and South in her native country, disc version revises the opening narration and some early scenes as they appeared in the theatrical version of the film. This movie won one Golden Globe Award. https://www.imdb.com/title/tt0107096/

Indochine (1992) / Việt-nam under French Protectorate in the early 1940s. This fiction movie depicts well the life and politics of the Annamites living in the French controlled Indochina. Catherine Deneuve and Linh Dan Pham are the main actresses. Recommended. https://www.imdb.com/title/tt0104507/

Journey of the Fall (2006) / recommended movie about the Boat People and Re-Education camp inmates. This film was produced by independent writer/director Ham Tran and financed entirely by the Vietnamese-American Community. **Recommended.** https://en.wikipedia.org/wiki/Journey_from_the_Fall

Last Day in Việt-nam (2014) / Produced by Rory Kennedy. This documentary is recommended because of it is indeed a factual presentation of what the last American personnel and the South Vietnamese had gone through as the Republic of Việt-nam by-default Pres. Dương văn Minh surrendered on April 30th, 1975. This documentary movie is regarded by Vietnamese-American Vietnam War researchers as the most accurate symbol of American-Vietnamese friendship-until-death: the last Americans, including Ambassador Martin Graham, tried to save as many Vietnamese lives as they can. **Highly recommended for further factual study.** https://www.imdb.com/title/tt3279124/

Passage to India (1984) / India under British control in the 1920s, in comparison with Indochina under the French Protectorate in "Indochine." Interaction between indigenous Indian anti-colonials, collaborators, the pacifists and the ruling British establishment. https://www.imdb.com/title/tt0087892/

Platoon (1986) / written and directed by Oliver Stone. U.S. Army base under VC NVA attack Metro-Goldwyn-Mayer, 1986. Upon arrival in Vietnam, a young, naive American soldier learns the realities of war. Not only will he battle the Viet Cong, but also his own fear, physical exhaustion, and the intense anger growing within himself, if he is to survive the war. The young soldiers not only face enemy fierce attack, but also have to deal with internal turmoil and self-destructing crisis. Written and directed by Oliver Stone. https://www.imdb.com/title/tt0091763/

The Deer hunter / Barry Spikings; Michael Deeley; Michael Cimino; John Peverall. MCA/Universal Home Video (distributor), 1978. A group of steelworker pals from Pennsylvania are followed from the blast furnace of the factory to the battle fields of Vietnam. A searing drama of friendship and courage, and what happens to those qualities under stress. Cast includes Robert De Niro, Christopher Walken, Meryl Streep, John Savage, John Cazale. Directed by Michael Cimino.

The Green berets (1968) / directed by John Wayne and Ray Kellogg. Warner Home Video. Hollywood's first feature film about the Vietnam War and the tone of the film is more in tune with World War II films than the Vietnam War films produced in the 1970s to present day. This patriotic film tells the story of the soldiers of the Special Forces — crack troops skilled in the techniques of unconditional warfare. The film is based on the novel by Robin Moore. Cast includes John Wayne, David Janssen, Jim Hutton, Aldo Ray, Raymond St. Jacques, Bruce Cabot, Jack Soo, George Takei, Patrick Wayne, Luke Askew, Irene Tsu. Screenplay by James Lee Barrett. Music by Miklos Rosza. This DVD version features interactive menus; scene access; widescreen version; production notes; featurette: "Making of the

Green Berets;" seven theatrical trailers. The Green Berets became a classic film about Vietnam War and the first group of U.S. Special Forces in the Việt-nam arena. https://www.imdb.com/title/tt0063035/

The Quiet American (2002 Paperback, 2003 Movie) / A remake movie by Miramax Films and Intermedia Films, a Mirage Enterprises, Saga Pictures, IMF production; screenplay by Christopher Hampton and Robert Schenkkan based on Graham Greene's original 1955 novel; directed by Phillip Noyce.
(YRL and IML) - [United States]: Miramax Home Entertainment; Burbank, Calif.: Buena Vista Home Entertainment (distributor), 2002. A political thriller set in Việt-nam in the early 1950s. A young American falls in love with the beautiful mistress of a British journalist. As war rages around them, these three sink deeper into a world of drugs, passion, and betrayal where nothing is as it seems. **Recommended.** https://www.rottentomatoes.com/m/1118347_quiet_american

Rambo Series: First blood, part I and II / directed by George P. Cosmatos. Avid Home Entertainment (distributor) or Artisan Entertainment (distributor),1985. In search of American POW's, John Rambo is sent on mission in Vietnam... only to discover he's been double-crossed by his superiors.

Ride the Thunder / available on Watch Movie on Amazon or DVD. The realistic film, produced by Scott Jones, is based on the book Ride the Thunder written by Rich Botkin, portraying the well-coordinated efforts between a RVN Marines commander and his American advisor. It is highly endorsed by syndicated talk show host Hugh Hewitt, and a pride of the Vietnamese Community in the United States. **Highly Recommended for expanded research on side-by-side operation** . http://www.ridethethundermovie.com/

We were soldiers / Paramount Pictures and Icon Productions present an Icon/Wheelhouse Entertainment production, a Randall Wallace film; written for the screen and directed by Randall Wallace. Hollywood, Calif.: Paramount, 2002. During the Pleiku Campaign of 1965, Lt. Col. Moore finds his battalion surrounded by North Vietnamese troops in the Ia Drang Valley. In this first major battle of the Vietnam War, the courageous American soldiers refuse to yield, in spite of heavy losses of life.

Thank you for Supporting any Factual History project on the Vietnam War

Thank you for Honoring and Supporting all U.S. and allied veterans of any American War